# Algebraic and Analytic Aspects of Integrable Systems and Painlevé Equations

# CONTEMPORARY MATHEMATICS

**651**

# Algebraic and Analytic Aspects of Integrable Systems and Painlevé Equations

AMS Special Session
Algebraic and Analytic Aspects of Integrable Systems
and Painlevé Equations
January 18, 2014
Baltimore, Maryland

Anton Dzhamay
Kenichi Maruno
Christopher M. Ormerod
Editors

**American Mathematical Society**
Providence, Rhode Island

2010 *Mathematics Subject Classification.* Primary 34M55, 34M56, 37K10, 39A10, 35Q55, 14E07, 17B80, 33C20, 41A05, 81R12.

---

**Library of Congress Cataloging-in-Publication Data**

Algebraic and analytic aspects of integrable systems and Painlevé equations : AMS special session on algebraic and analytic aspects of integrable systems and Painlevé equations : January 18, 2014, Baltimore, MD / Anton Dzhamay, Kenichi Maruno, Christopher M. Ormerod, editors.

pages cm. – (Contemporary mathematics ; volume 651)
Includes bibliographical references and index.
        ISBN 978-1-4704-1654-6 (alk. paper)
1. Painlevé equations–Congresses. 2. Algebra–Congresses. I. Dzhamay, Anton, 1970– editor.
II. Maruno, Kenichi, 1971– editor. III. Ormerod, Christopher M., 1982– editor.
QA372.A37 2015
515′.39–dc23                                                                                              2015011076

Contemporary Mathematics ISSN: 0271-4132 (print); ISSN: 1098-3627 (online)

DOI: http://dx.doi.org/10.1090/conm/651

---

# Contents

# Preface

The theory of integrable systems has been at the forefront of some of the most important developments in mathematical physics in the last 50 years. The techniques to study such systems have solid foundations in algebraic geometry, differential geometry, and group representation theory. The analytic tools developed to study integrable systems have countless applications in random matrix theory, statistical mechanics and quantum gravity. One of the most exciting developments has been the emergence of good and interesting discrete analogues of the classical integrable differential equations, such as the Painlevé equations and soliton equations. Many algebraic and analytic ideas developed for continuous integrable systems generalize in a beautifully natural manner to discrete integrable systems.

The theory of integrable systems has been enriched by some very powerful tools of a more geometric nature. In particular, these tools have elucidated the central role played by elliptic curves in constructing interesting generalizations of some important classes of integrable systems. Systems whose geometry is described by the most nondegenerate elliptic curves, such as the elliptic Painlevé equation, the discrete and continuous Krichever-Novikov equation, the elliptic solutions to the Yang-Baxter equation, and elliptic hypergeometric functions all play the role of the master class of systems from which many other cases may be derived as degenerations. Understanding and developing these ideas has led to many new current research directions and improved our understanding of known integrable models.

Another active area of research concerns the well-established theme of integrable systems admitting nontrivial symmetries. The symmetries of integrable systems often define examples of representations and realizations of interesting algebraic structures. Generalizations of these structures to quantum and $q$-deformed versions allow for the construction of new integrable systems. These ideas have been responsible for the continuing fruitful interaction between representation theory and integrable systems.

A related important field of research is the theory of Painlevé equations and special functions. The motivation behind the original definition of Painlevé equations is to obtain new genuinely nonlinear special functions as their general solutions. Such solutions, called the *Painlevé transcendents*, are playing an increasingly important role in a wide range of applications in mathematics and physics. In the opposite direction, Painlevé equations often have important exact solutions expressed in terms of special functions, such as the hypergeometric functions. Currently a lot of research activity has been directed on constructing a discrete analogue of Painlevé equations and their higher dimensional generalization. Yet again, algebraic geometry has been playing a very important role in these developments.

Many powerful techniques for studying nonlinear dynamical systems have been obtained by connecting them, usually in a highly nontrivial way, to some auxiliary linear problems. The paradigm for such techniques is the Lax pair formalism and at present a considerable effort has been devoted to understanding the connections between this linear framework and a geometric framework for integrable systems. But there are also many other interesting applications of the classical theory of linear systems for integrable systems. For example, the Riemann-Hilbert and the inverse scattering techniques continue to be powerful tools in analyzing important problems in statistical mechanics, random matrix theory, and the physical sciences.

This volume arose as a result of the AMS Special Session on *Algebraic and Analytic Aspects of Integrable Systems and Painlevé Equations* at the 2014 Joint Mathematics Meetings in Baltimore. In organizing this special session we aimed at presenting a wide range of new research ideas and directions in the theory of integrable equations, Painlevé equations, and applications. In a similar spirit, the present volume contains a collection of expository and research articles that represent a good cross-section of ideas in these active research areas.

We start our volume with *Padé Interpolation and Hypergeometric Series* by Masatoshi Noumi. This article is representative of the origins of integrable systems through its links with special functions. Of great importance in the theory of many integrable systems is the idea of special function solutions expressible in terms of hypergeometric and basic hypergeometric series and their orthogonal polynomial degenerations. This article delves into the relatively new and very active area of elliptic hypergeometric functions.

Our second article, *A q-analogue of the Drinfeld-Sokolov Hierarchy of Type A and q-Painlevé System* by Takao Suzuki, considers higher order analogues of the q-analogue of the sixth Painlevé equation arising as similarity reductions of the q-analogue of the Drinfeld-Sokolov hierarchy. In this way, this article ties together two interesting discrete analogues of classical integrable differential equations. The article details the construction of a Lax formalism for these systems and presents special solutions in terms of the basic hypergeometric functions of type $_n\phi_{n-1}$.

We may also generalize classical integrable systems by assuming the systems lie in some specified noncommutative space such as skew fields over $\mathbb{C}$. Results of this nature are presented in the third article, *Fractional Calculus of Quantum Painlevé Systems of Type $A_l^{(1)}$* by Hajime Nagoya. The author realizes a representation of the affine Weyl group of type $A_l^{(1)}$ on the skew field of Schrödinger operators, which defines a quantum Painlevé system.

The fourth article, *Spectral Curves and Discrete Painlevé Equations* by Christopher Ormerod, is an exploratory piece that considers the spectral curve for the associated linear problems of the discrete Painlevé equations as a way of examining the geometry of Lax pairs and how they are related to the geometry of the discrete Painlevé equations.

The fifth article, *Geometric Analysis of Reductions from Schlesinger Transformations to Difference Painlevé Equations* by Anton Dzhamay and Tomoyuki Takenawa, continues the theme of Lax pairs by presenting two discrete Painlevé equations $\left(d\text{-P}\left(A_2^{(1)*}\right)\right.$ and $\left.d\text{-P}\left(A_1^{(1)*}\right)\right)$ as Schlesinger transformations of Fuchsian differential equations. They thoroughly describe and connect the geometry of both the Lax pairs and the resulting nonlinear difference equations.

The sixth article, *Beta Ensembles, Quantum Painlevé Equations and Isomonodromy Systems* by Igor Rumanov, is a review article of some of the recent developments in random matrix theory and their relations with conformal field theory. The main point is that the integrable structures arising in connection with the quantum Painlevé equations for $\beta$-ensembles where $\beta = 2$ may be extended to more arbitrary $\beta$ values.

The last article, *Inverse Scattering Transform for the Focusing Nonlinear Schrödinger Equation with a One-Sided Non-Zero Boundary* by Barbara Prinari and Federica Vitale, presents results in studies of integrable systems that are of a more classical nature. The authors use inverse scattering to solve the focusing nonlinear Schrödinger equation with a one-sided non-zero boundary condition. The focusing NLS with these boundary conditions is not only interesting from the point of view of integrability; it has tangible applications in the investigation of rogue waves and perturbed soliton solutions in physical media.

We would like to thank Christine Thivierge, AMS Associate Editor for Proceedings, for her valuable help and constant support during the preparation of this volume. We hope that you enjoy the articles presented here.

THE EDITORS

# List of Participants

Gino Biondini
State University of New York
at Buffalo, Buffalo, NY, USA

Anton Dzhamay
University of Northern Colorado,
Greeley, CO, USA

Nalini Joshi
The University of Sydney, Sydney,
NSW, Australia

Sarah Lobb
The University of Sydney, Sydney,
NSW, Australia

Kenichi Maruno
Waseda University, Tokyo, Japan

Hajime Nagoya
Rikkyo University, Tokyo, Japan

Masatoshi Noumi
Kobe University, Kobe, Japan

Christopher M. Ormerod
California Institute of Technology,
Pasadena, CA, USA

Virgil Pierce
University of Texas–Pan American,
Edinburg, TX, USA

Barbara Prinari
University of Colorado at Colorado
Springs, Colorado Springs, CO, USA

Takao Suzuki
Kinki University, Osaka, Japan

Tomoyouki Takenawa
Tokyo University of Marine Science
and Technology, Tokyo, Japan

Ralph Willox
The University of Tokyo, Tokyo, Japan

Contemporary Mathematics
Volume **651**, 2015
http://dx.doi.org/10.1090/conm/651/13034

# Padé Interpolation and Hypergeometric Series

## Masatoshi Noumi

ABSTRACT. We propose a class of Padé interpolation problems whose general solution is expressible in terms of determinants of hypergeometric series.

## 1. Introduction

In this paper we investigate a class of Padé interpolation problems to which the solutions are expressible in terms of determinants of hypergeometric series. Padé interpolation problems have been discussed by Spiridonov–Zhedanov [**14**] from the viewpoint of biorthogonal rational functions. They are also sources of the Lax pairs for discrete Painlevé equations constructed by Yamada [**16**], [**17**], and by Noumi–Tsujimoto–Yamada [**11**]. The goal of this paper is to clarify how hypergeometric series arise in Padé interpolation problems, by analyzing the determinantal expression of the general solution.

In Section 2, we formulate a general Padé interpolation problem and a universal determinant formula for the general solution (Theorem 2.1). We also show that the determinants expressing the general solution can be condensed to smaller determinants by a variation of the Dodgson condensation (Theorem 2.2). After these preliminaries, we investigate in Section 3 a class of Padé interpolation problems relevant to generalized hypergeometric series $_{r+1}F_r$. We propose there two types of formulas expressing the solutions in terms of determinants of generalized hypergeometric series. The first one (Theorem 3.1), derived through Theorem 2.2, is based on the condensation of determinants and Krattenthaler's determinant formula, while the second (Theorem 3.2) is constructed by means of the Saalschütz summation formula for terminating $_3F_2$ series. We remark that Padé approximations to generalized hypergeometric functions have been discussed by Luke [**7**], [**8**]. It would be an important question to clarify the relationship between interpolations and approximations in the context of generalized hypergeometric functions. Section 4 is devoted to the extension of these results to three types of very well-poised hypergeometric series including basic (trigonometric) and elliptic hypergeometric series. The two determinant formulas of Theorem 4.1 and Theorem 4.2 are obtained by Warnaar's elliptic extension of the Krattenthaler determinant and by the Frenkel–Turaev summation formula for terminating $_{10}V_9$ series.

---

2010 *Mathematics Subject Classification.* Primary 41A05, 33C20, 33E20.
*Key words and phrases.* Padé interpolation, hypergeometric series, Dodgson condensation, Krattenthaler determinant.

Two fundamental tools of our approach are the condensation of determinants *along a moving core* (an identity of Sylvester type), and variations of Krattenthaler's determinant formula. For the sake of convenience, these subjects are discussed separately in Appendix A and Appendix B respectively. Generalization of Sylvester's identity on determinants has been developed extensively by Mühlbach–Gasca [9] (see also [1]). The version we use in this paper (Lemma A.2), based on the *Neville elimination strategy*, is originally due to Gasca–López-Carmona–Ramirez [2]. We also remark that Sylvester's identity and its extensions play important roles in recent studies of integrable systems (see Spicer–Nijhoff–van der Kamp [13] for example). As to Appendix B, basic references are the works of Krattenthaler [5], [6] and Warnaar [15] (see also Normand [10] for recent works on the evaluation of determinants involving shifted factorials). Although the contents of these appendices are basically found in the literature, we include them as self-contained expositions which might be helpful to the reader.

Throughout this paper we use the following notation of submatrices and minor determinants. For an $m \times n$ matrix $X = (x_{ij})_{1 \leq i \leq m, 1 \leq j \leq n}$ (with entries in a commutative ring), we denote by

$$(1.1) \qquad X^{i_1,\ldots,i_r}_{j_1,\ldots,j_s} = \begin{bmatrix} x_{i_1 j_1} & \cdots & x_{i_1 j_s} \\ \vdots & \ddots & \vdots \\ x_{i_r j_1} & \cdots & x_{i_r j_s} \end{bmatrix} = (x_{i_a j_b})_{1 \leq a \leq r, 1 \leq b \leq s}$$

the $r \times s$ submatrix of $X$ with row indices $i_1, \ldots, i_r \in \{1, \ldots, m\}$ and column indices $j_1, \ldots, j_r \in \{1, \ldots, n\}$. When $r = s$, we denote by $\det X^{i_1,\ldots,i_r}_{j_1,\ldots,j_r}$ the corresponding minor determinant.

## 2. Padé interpolation problems and their determinant solutions

In this section we formulate general Padé interpolation problems and propose some universal determinant formulas for the solutions.

Let $f_0(x)$, $f_1(x)$, ..., $f_m(x)$ and $g_0(x)$, $g_1(x)$, ..., $g_n(x)$ be two sequences of linearly independent meromorphic functions in $x \in \mathbb{C}$ and set $N = m + n$. We consider a pair $(P_m(x), Q_n(x))$ of two functions

$$P_m(x) = p_{m0} f_0(x) + p_{m1} f_1(x) + \cdots + p_{mm} f_m(x),$$

$$(2.1) \qquad Q_n(x) = q_{n0} g_0(x) + q_{n1} g_1(x) + \cdots + q_{nn} g_n(x),$$

which are expressed as $\mathbb{C}$-linear combinations of $f_j(x)$ and $g_j(x)$ respectively. Noting that the ratio $P_m(x)/Q_n(x)$ contains $N + 1 = m + n + 1$ arbitrary constants, we investigate the interpolation problem

$$(2.2) \qquad \frac{P_m(u_0)}{Q_n(u_0)} = v_0, \quad \frac{P_m(u_1)}{Q_n(u_1)} = v_1, \quad \ldots, \quad \frac{P_m(u_N)}{Q_n(u_N)} = v_N$$

for a set of $N + 1$ generic reference points $x = u_0, u_1, \ldots, u_N$ and a set of $N + 1$ prescribed values $v_0, v_1, \ldots, v_N$. This problem is equivalently rewritten as

$$(2.3) \qquad P_m(u_k) : Q_n(u_k) = \lambda_k : \mu_k \qquad (k = 0, 1, \ldots, N)$$

for $v_k = \lambda_k / \mu_k$ $(k = 0, 1, \ldots, N)$. We remark that the *Padé interpolation problem* defined as above contains the *Lagrange interpolation problem* as a special case where $n = 0$ and $g_0(x) = 1$.

A general solution of this Padé interpolation problem is given as follows in terms of $(N+2) \times (N+2)$ determinants:

(2.4)
$$P_m(x) = \det \begin{bmatrix} f_0(x) & \cdots & f_m(x) & 0 & \cdots & 0 \\ \mu_0 f_0(u_0) & \cdots & \mu_0 f_m(u_0) & \lambda_0 g_0(u_0) & \cdots & \lambda_0 g_n(u_0) \\ \vdots & & \vdots & \vdots & & \vdots \\ \mu_N f_0(u_N) & \cdots & \mu_N f_m(u_N) & \lambda_N g_0(u_N) & \cdots & \lambda_N g_n(u_N) \end{bmatrix},$$

(2.5)
$$Q_n(x) = -\det \begin{bmatrix} 0 & \cdots & 0 & g_0(x) & \cdots & g_n(x) \\ \mu_0 f_0(u_0) & \cdots & \mu_0 f_m(u_0) & \lambda_0 g_0(u_0) & \cdots & \lambda_0 g_n(u_0) \\ \vdots & & \vdots & \vdots & & \vdots \\ \mu_N f_0(u_N) & \cdots & \mu_N f_m(u_N) & \lambda_N g_0(u_N) & \cdots & \lambda_N g_n(u_N) \end{bmatrix}.$$

THEOREM 2.1. *The pair of functions $(P_m(x), Q_n(x))$ defined by (2.4), (2.5) solves the Padé interpolation problem (2.3) if $(P_m(u_k), Q_n(u_k)) \neq (0,0)$ for $k = 0, 1, \ldots, N$.*

In order to prove that this pair $(P_m(x), Q_n(x))$ gives a solution of the interpolation problem, we introduce two parameters $\lambda, \mu$ and consider the $(N+2) \times (N+2)$ determinant
(2.6)
$$R_{m,n}(x; \lambda, \mu) = \det \begin{bmatrix} \mu f_0(x) & \cdots & \mu f_m(x) & \lambda g_0(x) & \cdots & \lambda g_n(x) \\ \mu_0 f_0(u_0) & \cdots & \mu_0 f_m(u_0) & \lambda_0 g_0(u_0) & \cdots & \lambda_0 g_n(u_0) \\ \vdots & & \vdots & \vdots & & \vdots \\ \mu_N f_0(u_N) & \cdots & \mu_N f_m(u_N) & \lambda_N g_0(u_N) & \cdots & \lambda_N g_n(u_N) \end{bmatrix}.$$
By decomposing the top row as
(2.7) $\qquad \mu\left(f_0(x), \cdots, f_m(x),\ 0, \cdots, 0\right) + \lambda(0, \cdots, 0,\ g_0(x), \cdots, g_n(x)),$
we have
(2.8) $$R_{m,n}(x; \lambda, \mu) = \mu\, P_m(x) - \lambda\, Q_n(x).$$
On the other hand, the determinantal expression of $R_{m,n}(x; \lambda, \mu)$ implies
(2.9) $\quad R_{m,n}(u_k; \lambda_k, \mu_k) = \mu_k P_m(u_k) - \lambda_k Q_n(u_k) = 0 \qquad (k = 0, 1, 2 \ldots, N),$
and hence
(2.10) $$P_m(u_k) : Q_n(u_k) = \lambda_k : \mu_k \qquad (k = 0, 1, \ldots, N)$$
as desired.

The $(N+2) \times (N+2)$ determinants (2.4), (2.5) representing $P_m(x)$ and $Q_n(x)$ can be *condensed* into an $(m+1) \times (m+1)$ and $(n+1) \times (n+1)$ determinants respectively, by means of a variation of the Dodgson condensation (see Appendix A).

We denote by
(2.11) $\qquad F = \bigl(f_j(u_i)\bigr)_{0 \le i \le N,\ 0 \le j \le m}, \qquad G = \bigl(g_j(u_i)\bigr)_{0 \le i \le N,\ 0 \le j \le n}$

the matrices defined by the values of the functions $f_j(x)$ $(0 \le j \le m)$ and $g_j(x)$ $(0 \le j \le n)$, respectively, at the reference points $u_i$ $(0 \le i \le N)$. We assume that the configuration of reference points $u_k$ $(k = 0, 1, \ldots, N)$ is *generic* in the sense that the minor determinants

$$(2.12) \qquad \det F_{0,\ldots,n}^{i,\ldots,i+n} \quad (0 \le i \le m), \quad \det G_{0,\ldots,m}^{i,\ldots,i+m} \quad (0 \le i \le n)$$

of maximal size with consecutive rows are all nonzero.

Assuming that $\lambda_k \ne 0$, $\mu_k \ne 0$ for $k = 0, 1, \ldots, N$, we set
$$(2.13)$$
$$U_{i,j} = \frac{\lambda_i}{\mu_i} \det \begin{bmatrix} \frac{\mu_i}{\lambda_i} f_j(u_i) & g_0(u_i) & \cdots & g_n(u_i) \\ \frac{\mu_{i+1}}{\lambda_{i+1}} f_j(u_{i+1}) & g_0(u_{i+1}) & \cdots & g_n(u_{i+1}) \\ \vdots & \vdots & & \vdots \\ \frac{\mu_{i+n+1}}{\lambda_{i+n+1}} f_j(u_{i+n+1}) & g_0(u_{i+n+1}) & \cdots & g_n(u_{i+n+1}) \end{bmatrix} \left( \det G_{0,1,\ldots,n}^{i+1,\ldots,i+n+1} \right)^{-1}$$

for $0 \le i < m$, $0 \le j \le m$ and
$$(2.14)$$
$$V_{i,j} = \frac{\mu_i}{\lambda_i} \det \begin{bmatrix} \frac{\lambda_i}{\mu_i} g_j(u_i) & f_0(u_i) & \cdots & f_m(u_i) \\ \frac{\lambda_{i+1}}{\mu_{i+1}} g_j(u_{i+1}) & f_0(u_{i+1}) & \cdots & f_m(u_{i+1}) \\ \vdots & \vdots & & \vdots \\ \frac{\lambda_{i+m+1}}{\mu_{i+m+1}} g_j(u_{i+m+1}) & f_0(u_{i+m+1}) & \cdots & f_m(u_{i+m+1}) \end{bmatrix} \left( \det F_{0,1,\ldots,m}^{i+1,\ldots,i+m+1} \right)^{-1}$$

for $0 \le i < n$, $0 \le j \le n$. Then by Lemma A.2 (of *condensation along a moving core*), the $(N+2) \times (N+2)$ determinants (2.4), (2.5) are condensed as follows into $(m+1) \times (m+1)$ and $(n+1) \times (n+1)$ determinants respectively (see also (A.18), (A.19)).

THEOREM 2.2. *The two functions $P_m(x)$, $Q_n(x)$ defined in Theorem 2.1 are expressed as follows in terms of $(m+1) \times (m+1)$ and $(n+1) \times (n+1)$ determinants respectively:*

$$(2.15) \quad P_m(x) = \prod_{i=0}^{m-1} \mu_i \prod_{i=0}^{n} \lambda_{m+i} \det G_{0,1,\ldots,n}^{m,\ldots,m+n} \det \begin{bmatrix} f_0(x) & \cdots & f_m(x) \\ U_{0,0} & \cdots & U_{0,m} \\ \vdots & & \vdots \\ U_{m-1,0} & \cdots & U_{m-1,m} \end{bmatrix},$$

$$(2.16) \quad Q_n(x) = \epsilon_{m,n} \prod_{i=0}^{n-1} \lambda_i \prod_{i=0}^{m} \mu_{n+i} \det F_{0,1,\ldots,m}^{n,\ldots,n+m} \det \begin{bmatrix} g_0(x) & \cdots & g_n(x) \\ V_{0,0} & \cdots & V_{0,n} \\ \vdots & & \vdots \\ V_{n-1,0} & \cdots & U_{n-1,n} \end{bmatrix},$$

*where $\epsilon_{m,n} = (-1)^{mn+m+n}$.*

By expanding the determinants $U_{ij}$ and $V_{ij}$ along the first column we further obtain the series expansions

(2.17)
$$U_{ij} = \sum_{k=0}^{n+1} (-1)^k \frac{\mu_{i+k}\,\lambda_i}{\lambda_{i+k}\,\mu_i} f_j(u_{i+k}) \frac{\det G_{0,1,\ldots,n}^{i,\ldots,\widehat{i+k},\ldots,i+n+1}}{\det G_{0,1,\ldots,n}^{i+1,\ldots,i+n+1}} \qquad (0 \le i < m,\ 0 \le j \le m),$$

(2.18)
$$V_{ij} = \sum_{k=0}^{m+1} (-1)^k \frac{\lambda_{i+k}\,\mu_i}{\mu_{i+k}\,\lambda_i} g_j(u_{i+k}) \frac{\det F_{0,1,\ldots,m}^{i,\ldots,\widehat{i+k},\ldots,i+m+1}}{\det F_{0,1,\ldots,m}^{i+1,\ldots,i+m+1}} \qquad (0 \le i < n,\ 0 \le j \le n).$$

Hence the problem to determine $P_m(x)$ and $Q_n(x)$ is reduced to the computation of minor determinants of the matrices $F = (f_j(u_i))_{i,j}$ and $G = (g_j(u_i))_{i,j}$. We remark that these formulas for $P_m(x)$ and $Q_n(x)$ hold *universally* for any choice of the functions $f_j(x)$ and $g_j(x)$.

In Sections 3 and 4, we show that these expansion formulas (2.17), (2.18) in fact give rise to *hypergeometric series* of various types for appropriate choices of the functions $f_j(x)$, $g_j(x)$, the reference points $u_k$ and the prescribed values $v_k = \lambda_k/\mu_k$.

## 3. Hypergeometric series arising from determinants

We explain below how the series expansions (2.17), (2.18) can be used for generating hypergeometric series. In this section we use the notation of shifted factorials

(3.1)     $(a)_n = a(a+1)\cdots(a+n-1) = \Gamma(a+n)/\Gamma(a) \qquad (n = 0,1,2,\ldots).$

As a typical example, we consider the two sequences of rational functions

(3.2)     $f_j(x) = \dfrac{(a+x)_j}{(b+x)_j}, \qquad g_j(x) = \dfrac{(c+x)_j}{(d+x)_j} \qquad (j = 0,1,2,\ldots)$

with four complex parameters $a$, $b$, $c$, $d$, to form a pair $(P_m(x), Q_n(x))$ of rational functions

(3.3)     $P_m(x) = \displaystyle\sum_{j=0}^{m} p_{m,j} \frac{(a+x)_j}{(b+x)_j}, \qquad Q_n(x) = \sum_{j=0}^{n} q_{n,j} \frac{(c+x)_j}{(d+x)_j}.$

Taking an arithmetic progression $u_k = u + k$ $(k = 0,1,2,\ldots,N;\ N = m+n)$ of points in $\mathbb{C}$, we consider the Padé interpolation problem

(3.4)     $\dfrac{P_m(u+k)}{Q_n(u+k)} = \dfrac{\lambda_k}{\mu_k} \qquad (k = 0,1,\ldots,N).$

THEOREM 3.1. *Consider the Padé interpolation problem* (3.3), (3.4) *for the rational functions* $f_j(x)$, $g_j(z)$ *in* (3.2) *and the reference points* $u_k = u + k$ ($k = 0,1,\ldots,N; N = m+n$). *Then the solution* $(P_m(x), Q_n(x))$ *of Theorem 2.1 is explicitly given by*

$$(3.5) \quad P_m(x) = \frac{\prod_{i=1}^n i! \, (d-c)_i}{\prod_{i=0}^n (d+u_{m+i})_n} \prod_{i=0}^{m-1} \mu_i \prod_{i=0}^n \lambda_{m+i} \det \begin{bmatrix} f_0(x) & \cdots & f_m(x) \\ U_{0,0} & \cdots & U_{0,m} \\ \vdots & & \vdots \\ U_{m-1,0} & \cdots & U_{m-1,m} \end{bmatrix},$$

$$(3.6) \quad Q_n(x) = \epsilon_{m,n} \frac{\prod_{i=1}^m i! \, (b-a)_i}{\prod_{i=0}^m (b+u_{n+i})_m} \prod_{i=0}^{n-1} \lambda_i \prod_{i=0}^m \mu_{n+i} \det \begin{bmatrix} g_0(x) & \cdots & g_n(x) \\ V_{0,0} & \cdots & V_{0,n} \\ \vdots & & \vdots \\ V_{n-1,0} & \cdots & U_{n-1,n} \end{bmatrix},$$

*where*

$$(3.7) \quad U_{ij} = \frac{(a+u_i)_j}{(b+u_i)_j} \sum_{k=0}^{n+1} \frac{(-n-1)_k (d+u_{i+n})_k}{k! \, (d+u_i)_k} \frac{(a+u_{i+j})_k (b+u_i)_k}{(a+u_i)_k (b+u_{i+j})_k} \frac{\mu_{i+k} \lambda_i}{\lambda_{i+k} \mu_i},$$

$$(3.8) \quad V_{ij} = \frac{(c+u_i)_j}{(d+u_i)_j} \sum_{k=0}^{m+1} \frac{(-m-1)_k (b+u_{i+m})_k}{k! \, (b+u_i)_k} \frac{(c+u_{i+j})_k (d+u_i)_k}{(c+u_i)_k (d+u_{i+j})_k} \frac{\lambda_{i+k} \mu_i}{\mu_{i+k} \lambda_i}.$$

As we remarked in the previous section, the functions $U_{ij}$ in Theorem 2.2 are expressed as

$$(3.9) \quad U_{ij} = \sum_{k=0}^{n+1} (-1)^k \frac{\mu_{i+k} \lambda_i}{\lambda_{i+k} \mu_i} f_j(u_{i+k}) \frac{\det G_{0,1,\ldots,n}^{i,\ldots,\widehat{i+k},\ldots,i+n+1}}{\det G_{0,1,\ldots,n}^{i+1,\ldots,i+n+1}}.$$

Since $u_k = u + k$ $(k = 0, 1, \ldots, N)$, we have

$$(3.10) \quad f_j(u_{i+k}) = \frac{(a+u_{i+k})_j}{(b+u_{i+k})_j} = \frac{(a+u_i)_j}{(b+u_i)_j} \frac{(a+u_{i+j})_k (b+u_i)_k}{(b+u_{i+j})_k (a+u_i)_k}.$$

The $(n+1) \times (n+1)$ minor determinants of the matrix

$$(3.11) \quad G = \left( \frac{(c+u_i)_j}{(d+u_i)_j} \right)_{0 \le i \le N, \, 0 \le j \le n}$$

can be computed by means of a special case of Krattenthaler's determinant formula [5] (see Appendix B). In fact by (B.3), we have

$$\det G_{0,1,\ldots,n}^{i,i+1,\ldots,i+n} = \frac{\prod_{l=1}^n l! \, (d-c)_l}{\prod_{l=0}^n (d+u_{i+l})_n},$$

$$(3.12) \quad \det G_{0,1,\ldots,n}^{i,\ldots,\widehat{i+k},\ldots,i+n+1} = \frac{\prod_{l=1}^n l! \, (d-c)_l}{\prod_{l=1}^{n+1} (d+u_{i+l})_n} \frac{(-1)^k (-n-1)_k (d+u_{i+n})_k}{k! \, (d+u_i)_k}.$$

Hence $U_{ij}$ is computed as

$$(3.13) \quad U_{ij} = \frac{(a+u_i)_j}{(b+u_i)_j} \sum_{k=0}^{n+1} \frac{(-n-1)_k (d+u_{i+n})_k}{k! \, (d+u_i)_k} \frac{(a+u_{i+j})_k (b+u_i)_k}{(a+u_i)_k (b+u_{i+j})_k} \frac{\mu_{i+k} \lambda_i}{\lambda_{i+k} \mu_i}.$$

The corresponding formula for $V_{ij}$ is obtained by exchanging the roles of $(m, n)$, $(a, c)$ and $(b, d)$.

If we choose the prescribed values appropriately, the series $U_{ij}$ and $V_{ij}$ in Theorem 3.1 give rise to *generalized hypergeometric series*

$$(3.14) \qquad {}_{r+1}F_r \left[ \begin{matrix} \alpha_0, \alpha_1, \ldots, \alpha_r \\ \beta_1, \ldots, \beta_r \end{matrix} ; z \right] = \sum_{k=0}^{\infty} \frac{(\alpha_0)_k (\alpha_1)_k \cdots (\alpha_r)_k}{(1)_k (\beta_1)_k \cdots (\beta_r)_k} z^k.$$

Consider the case where

$$(3.15) \qquad v_k = \frac{\lambda_k}{\mu_k} = \frac{(s_1)_k \cdots (s_r)_k}{(t_1)_k \cdots (t_r)_k} \left( \frac{z}{w} \right)^k \qquad (k = 0, 1, 2, \ldots, N)$$

with complex parameters $s_1, \ldots, s_r$ and $t_1, \ldots, t_r$. Since

$$(3.16) \qquad \frac{\mu_{i+k} \lambda_i}{\lambda_{i+k} \mu_i} = \frac{(t_1 + i)_k \cdots (t_r + i)_k}{(s_1 + i)_k \cdots (s_r + i)_k} \left( \frac{w}{z} \right)^k,$$

$U_{ij}$ and $V_{ij}$ are determined as

$$U_{ij} = \frac{(a + u_i)_j}{(b + u_i)_j} {}_{r+4}F_{r+3} \left[ \begin{matrix} -n-1, d+u_{i+n}, a+u_{i+j}, b+u_i, t_1+i, \ldots, t_r+i \\ d+u_i, a+u_i, b+u_{i+j}, s_1+i, \ldots, s_r+i \end{matrix} ; \frac{w}{z} \right],$$

$(3.17)$

$$V_{ij} = \frac{(c + u_i)_j}{(d + u_i)_j} {}_{r+4}F_{r+3} \left[ \begin{matrix} -m-1, b+u_{i+n}, c+u_{i+j}, d+u_i, s_1+i, \ldots, s_r+i \\ b+u_i, c+u_i, d+u_{i+j}, t_1+i, \ldots, t_r+i \end{matrix} ; \frac{z}{w} \right].$$

If we choose the prescribed values

$$(3.18) \quad v_k = \frac{\lambda_k}{\mu_k} = \frac{(b+u)_k (c+u)_k}{(a+u)_k (d+u)_k} \frac{(s_1)_k \cdots (s_r)_k}{(t_1)_k \cdots (t_r)_k} \left( \frac{z}{w} \right)^k \qquad (k = 0, 1, 2, \ldots, N),$$

then $U_{ij}$ and $V_{ij}$ are slightly simplified as

$$U_{ij} = \frac{(a + u_i)_j}{(b + u_i)_j} {}_{r+3}F_{r+2} \left[ \begin{matrix} -n-1, d+u_{i+n}, a+u_{i+j}, t_1+i, \ldots, t_r+i \\ c+u_i, , b+u_{i+j}, s_1+i, \ldots, s_r+i \end{matrix} ; \frac{w}{z} \right],$$

$(3.19)$

$$V_{ij} = \frac{(c + u_i)_j}{(d + u_i)_j} {}_{r+3}F_{r+2} \left[ \begin{matrix} -m-1, b+u_{i+n}, c+u_{i+j}, s_1+i, \ldots, s_r+i \\ a+u_i, d+u_{i+j}, t_1+i, \ldots, t_r+i \end{matrix} ; \frac{z}{w} \right].$$

As for the Padé interpolation problem for the rational functions $f_j(x)$ and $g_j(x)$ as in (3.2), one can construct another type of determinant formula for $P_m(x)$ and $Q_n(x)$ involving hypergeometric series.

THEOREM 3.2. *Consider the Padé interpolation problem* (3.3), (3.4) *for the rational functions* $f_j(x)$, $g_j(z)$ *in* (3.2) *and the reference points* $u_k = u + k$ ($k = 0, 1, \ldots, N$; $N = m + n$). *Then the solution* $(P_m(x), Q_n(x))$ *of Theorem 2.1 is expressed as*

$$(3.20) \qquad P_m(x) = K_{m,n}(c, d) \prod_{i=0}^{N} \lambda_i \det \begin{bmatrix} f_0(x) & \cdots & f_m(x) \\ \Phi_{0,0} & \cdots & \Phi_{0,m} \\ \vdots & & \vdots \\ \Phi_{m-1,0} & \cdots & \Phi_{m-1,m} \end{bmatrix},$$

$$(3.21) \qquad Q_n(x) = \epsilon_{m,n} K_{n,m}(a, b) \prod_{i=0}^{N} \mu_i \det \begin{bmatrix} g_0(x) & \cdots & g_n(x) \\ \Psi_{0,0} & \cdots & \Psi_{0,n} \\ \vdots & & \vdots \\ \Psi_{n-1,0} & \cdots & \Psi_{n-1,n} \end{bmatrix},$$

8                                    MASATOSHI NOUMI

*where*
(3.22)
$$\Phi_{ij} = \frac{(a+u)_j}{(b+u)_j} \sum_{k=0}^{N} \frac{(-N)_k(d+u+N-1-i)_k}{(1)_k(c+u-i)_k} \frac{(c+u)_k}{(d+u)_k} \frac{(a+u_j)_k(b+u)_k}{(b+u_j)_k(a+u)_k} \frac{\mu_k}{\lambda_k},$$

(3.23)
$$\Psi_{ij} = \frac{(c+u)_j}{(d+u)_j} \sum_{k=0}^{N} \frac{(-N)_k(b+u+N-1-i)_k}{(1)_k(a+u-i)_k} \frac{(a+u)_k}{(b+u)_k} \frac{(c+u_j)_k(d+u)_k}{(d+u_j)_k(c+u)_k} \frac{\lambda_k}{\mu_k}.$$

We remark that if the prescribed values are given by
(3.24)
$$v_k = \frac{\lambda_k}{\mu_k} = \frac{(s_1)_k \cdots (s_r)_k}{(t_1)_k \cdots (t_r)_k} \left(\frac{z}{w}\right)^k \qquad (k=0,1,2,\dots,N),$$

then $\Phi_{ij}$ and $\Psi_{ij}$ give rise to generalized hypergeometric series

$$\Phi_{ij} = \frac{(a+u)_j}{(b+u)_j} \,_{r+5}F_{r+4}\left[\begin{matrix}-N,d+u+N-1-i,c+u,a+u+j,b+u,t_1,\dots,t_r\\ c+u-i,d+u,b+u+j,a+u,s_1,\dots,s_r\end{matrix}; \frac{w}{z}\right],$$

(3.25)
$$\Psi_{ij} = \frac{(c+u)_j}{(d+u)_j} \,_{r+5}F_{r+4}\left[\begin{matrix}-N,b+u+N-1-i,a+u,c+u+j,d+u,s_1,\dots,s_r\\ a+u-i,b+u,d+u+j,c+u,t_1,\dots,t_r\end{matrix}; \frac{z}{w}\right].$$

If we choose
(3.26) $$v_k = \frac{\lambda_k}{\mu_k} = \frac{(b+u)_k(c+u)_k}{(a+u)_k(d+u)_k} \frac{(s_1)_k \cdots (s_r)_k}{(t_1)_k \cdots (t_r)_k} \left(\frac{z}{w}\right)^k \qquad (k=0,1,2,\dots,N),$$

then $\Phi_{ij}$ and $\Psi_{ij}$ are simplified as

$$\Phi_{ij} = \frac{(a+u)_j}{(b+u)_j} \,_{r+3}F_{r+2}\left[\begin{matrix}-N,d+u+N-1-i,a+u+j,t_1,\dots,t_r\\ c+u-i,b+u+j,s_1,\dots,s_r\end{matrix}; \frac{w}{z}\right],$$

(3.27) $$\Psi_{ij} = \frac{(c+u)_j}{(d+u)_j} \,_{r+3}F_{r+2}\left[\begin{matrix}-N,b+u+N-1-i,c+u+j,s_1,\dots,s_r\\ a+u-i,d+u+j,t_1,\dots,t_r\end{matrix}; \frac{z}{w}\right].$$

In order to obtain the expression of Theorem 3.2, we first rewrite (2.4) as

$$P_m(x) = \prod_{i=0}^{N}\lambda_i \, \det\begin{bmatrix} f_0(x) & \cdots & f_m(x) & 0 & \cdots & 0 \\ \frac{\mu_0}{\lambda_0}f_0(u_0) & \cdots & \frac{\mu_0}{\lambda_0}f_m(u_0) & g_0(u_0) & \cdots & g_n(u_0) \\ \vdots & & \vdots & \vdots & & \vdots \\ \frac{\mu_N}{\lambda_N}f_0(u_N) & \cdots & \frac{\mu_N}{\lambda_N}f_m(u_N) & g_0(u_N) & \cdots & g_n(u_N) \end{bmatrix}$$

(3.28) $$= \prod_{i=0}^{N}\lambda_i \, \det\begin{bmatrix} f(x) & 0 \\ \widetilde{F} & G \end{bmatrix}.$$

We construct an $(N+1)\times(N+1)$ invertible matrix $L = (L_{ij})_{i,j=0}^{N}$ such that $(LG)_{ij} = 0$ for $i+j < N$, and define $M = (M_{ij})_{i,j=0}^{n}$ by $M_{ij} = (LG)_{m+i,j}$. If we set $\Phi = L\widetilde{F}$, we have

(3.29) $$\begin{bmatrix}1 & \\ & L\end{bmatrix}\begin{bmatrix}f(x) & 0 \\ \widetilde{F} & G\end{bmatrix} = \begin{bmatrix}f(x) & 0 \\ \Phi & LG\end{bmatrix} = \begin{bmatrix}f(x) & 0 \\ \Phi' & 0 \\ \Phi'' & M\end{bmatrix}.$$

Hence, by taking the determinants of the both sides we obtain

$$(3.30) \qquad P_m(x) = \lambda_0 \cdots \lambda_N \det \begin{bmatrix} f_0(x) & \cdots & f_m(x) \\ \Phi_{00} & \cdots & \Phi_{0m} \\ \vdots & \ddots & \vdots \\ \Phi_{m-1,0} & \cdots & \Phi_{m-1,m} \end{bmatrix} \frac{\det M}{\det L},$$

which will give formula (3.20) with $K_{m,n}(c,d) = \det M / \det L$. In view of

$$(3.31) \qquad g_j(u_k) = \frac{(c+u_k)_j}{(d+u_k)_j} = \frac{(d+u)_k}{(c+u)_k} \frac{(c+u_j)_k}{(d+u_j)_k} \frac{(c+u)_j}{(d+u)_j},$$

we recall the Saalschütz sum

$$(3.32) \quad {}_3F_2 \left[ \begin{matrix} -N, d+u+N-1-i, c+u+j \\ c+u-i, d+u+j \end{matrix} ; 1 \right] = \frac{(d-c)_N(-i-j)_N}{(c+u-i)_N(d+u+j)_N},$$

namely,

$$(3.33) \qquad \sum_{k=0}^{N} \frac{(-N)_k(d+u+N-1-i)_k(c+u_j)_k}{(1)_k(c+u-i)_k(d+u_j)_k} = \frac{(d-c)_N(-i-j)_N}{(c+u-i)_N(d+u+j)_N}.$$

With this observation, we define the matrix $L = \left( L_{ij} \right)_{i,j=0}^{N}$ by

$$(3.34) \qquad L_{ij} = \frac{(-N)_j(d+u+N-1-i)_j}{(1)_j(c+u-i)_j} \frac{(c+u)_j}{(d+u)_j} \qquad (0 \le i,j \le N).$$

Then we have

$$(LG)_{ij} = \sum_{k=0}^{N} \frac{(-N)_k(d+u+N-1-i)_k}{(1)_j(c+u-i)_k} \frac{(c+u_j)_k}{(d+u_j)_k} \frac{(c+u)_j}{(d+u)_j}$$

$$(3.35) \qquad = \frac{(d-c)_N(-i-j)_N}{(c+u-i)_N(d+u+j)_N} \frac{(c+u)_j}{(d+u)_j}$$

by the Saalschütz sum. In particular $(LG)_{ij} = 0$ $(i+j < N)$. The determinant of the matrix $M$ is computed as

$$(3.36)$$
$$\det M = (-1)^{\binom{n+1}{2}} \prod_{j=0}^{n} (LG)_{N-j,j} = (-1)^{\binom{n+1}{2}} \prod_{j=0}^{n} \frac{(d-c)_N(-N)_N}{(c+u-N+j)_N(d+u+j)_N} \frac{(c+u)_j}{(d+u)_j}.$$

Also, the entires of $\Phi = L\widetilde{F}$ are expressed as

$$\Phi_{ij} = \sum_{k=0}^{N} L_{ik} \frac{\mu_k}{\lambda_k} f_j(u_k)$$

$$= \sum_{k=0}^{N} \frac{(-N)_k(d+u+N-1-i)_k}{(1)_k(c+u-i)_k} \frac{(c+u)_k}{(d+u)_k} \frac{\mu_k}{\lambda_k} \frac{(a+u_k)_j}{(b+u_k)_j}$$

$$(3.37) \qquad = \frac{(a+u)_j}{(b+u)_j} \sum_{k=0}^{N} \frac{(-N)_k(d+u+N-1-i)_k}{(1)_k(c+u-i)_k} \frac{(c+u)_k}{(d+u)_k} \frac{(a+u_j)_k(b+u)_k}{(b+u_j)_k(a+u)_k} \frac{\mu_k}{\lambda_k}.$$

The determinant of $L$ can be computed again by Krattenthaler's formula:

$$\det L = \det \left( \frac{(d+u+N-1-i)_j}{(c+u-i)_j} \right)_{i,j=0}^{N} \prod_{j=0}^{N} \frac{(-N)_j}{(1)_j} \frac{(c+u)_j}{(d+u)_j}$$

$$(3.38) \qquad = (-1)^{\binom{N+1}{2}} \prod_{j=0}^{N} \frac{(c-d-N+1)_j}{(c+u-j)_N} \frac{(-N)_j(c+u)_j}{(d+u)_j}.$$

The constant factor in (3.20) is determined as $K_{m,n}(c,d) = \det M/\det L$.

## 4. Three types of very well-poised hypergeometric series

In this section we consider three classes of hypergeometric series

(0) rational        ...  ordinary hypergeometric series
(1) trigonometric   ...  basic (or $q$-)hypergeometric series
(2) elliptic        ...  elliptic hypergeometric series

corresponding to the choice of a "fundamental" function $[x]$:

(0) rational      : $[x] = e^{c_0 x^2 + c_1} x$           $(\Omega = 0)$
(1) trigonometric : $[x] = e^{c_0 x^2 + c_1} \sin(\pi x/\omega)$  $(\Omega = \mathbb{Z}\omega)$
(2) elliptic      : $[x] = e^{c_0 x^2 + c_1} \sigma(x|\Omega)$   $(\Omega = \mathbb{Z}\omega_1 \oplus \mathbb{Z}\omega_2)$

where $\sigma(x|\Omega)$ is the Weierstrass sigma function associated with the period lattice $\Omega = \mathbb{Z}\omega_1 \oplus \mathbb{Z}\omega_2$. It is known that these classes of functions $[x]$ are characterized by the so-called *Riemann relation*: For any $x, \alpha, \beta, \gamma \in \mathbb{C}$,
(4.1)
$$[x+\alpha][x-\alpha][\beta+\gamma][\beta-\gamma]+[x+\beta][x-\beta][\gamma+\alpha][\gamma-\alpha]+[x+\gamma][x-\gamma][\alpha+\beta][\alpha-\beta] = 0.$$

By the notation $[x \pm y] = [x+y][x-y]$ of the product of two factors, this relation is expressed as

$$(4.2) \qquad [x \pm \alpha][\beta \pm \gamma] + [x \pm \beta][\gamma \pm \alpha] + [x \pm \gamma][\alpha \pm \beta] = 0.$$

In what follows, we fix a nonzero entire function $[x]$ satisfying this functional equation.

Fixing a generic constant $\delta$, we define the $\delta$-*shifted factorials* $[x]_k$ by

$$(4.3) \qquad [x]_k = [x]_{\delta,k} = [x][x+\delta]\cdots[x+(k-1)\delta] \qquad (k = 0, 1, 2, \ldots).$$

Then we define the *very well-poised* hypergeometric series $_{r+5}V_{r+4}\big[a_0; a_1 \cdots a_r \big| z\big]$ associated with $[x]$ by
(4.4)
$$_{r+5}V_{r+4}\left[a_0; a_1, \cdots, a_r \Big| z\right] = \sum_{k=0}^{\infty} \frac{[a_0 + 2k\delta]}{[a_0]} \frac{[a_0]_k [a_1]_k \cdots [a_r]_k}{[\delta]_k[\delta + a_0 - a_1]_k \cdots [\delta + a_0 - a_r]_k} z^k.$$

In this paper we use this notation only for terminating series assuming that $a_i$ is of the form $-n\delta$ $(n = 0, 1, 2, \ldots)$ for some $i$. When $z = 1$ we also write
(4.5)
$$_{r+5}V_{r+4}\left[a_0; a_1, \cdots, a_r\right] = \sum_{k=0}^{\infty} \frac{[a_0 + 2k\delta]}{[a_0]} \frac{[a_0]_k [a_1]_k \cdots [a_r]_k}{[\delta]_k[\delta + a_0 - a_1]_k \cdots [\delta + a_0 - a_r]_k}.$$

In this notation, the celebrated Frenkel-Turaev sum is expressed as

$$
(4.6) \quad
\begin{aligned}
&{}_{10}V_9\left[a_0; a_1, a_2, a_3, a_4, a_5\right] \\
&= \frac{[\delta + a_0]_N [\delta + a_0 - a_1 - a_2]_N [\delta + a_0 - a_1 - a_3]_N [\delta + a_0 - a_2 - a_3]_N}{[\delta + a_0 - a_1]_N [\delta + a_0 - a_2]_N [\delta + a_0 - a_3]_N [\delta + a_0 - a_1 - a_2 - a_3]_N},
\end{aligned}
$$

under the balancing condition $a_1 + \cdots + a_5 = 2a_0 + \delta$ and the termination condition $a_5 = -N\delta$ ($N = 0, 1, 2, \ldots$). (See for example [3], [4].)

We remark that, in the rational case where $[x] = x$ and $\delta = 1$, the $_{r+5}V_{r+4}$ series defined above is expressed in terms of a $_{r+2}F_{r+1}$-series:

$$
(4.7) \quad {}_{r+5}V_{r+4}\left[a_0; a_1, \cdots, a_r \,\middle|\, z\right] = {}_{r+2}F_{r+1}\left[\begin{matrix} a_0, \frac{1}{2}a_0 + 1, a_1, \ldots a_r \\ \frac{1}{2}a_0, \; b_1, \; \ldots, \; b_r \end{matrix}; z\right]
$$

where $b_i = 1 + a_0 - a_i$ ($i = 1, \ldots, r$). Also, in the trigonometric case where $[x] = e^{cx/2} - e^{-cx/2}$,

$$
\begin{aligned}
{}_{r+5}V_{r+4}\left[a_0; a_1, \cdots, a_r \,\middle|\, z\right] &= \sum_{k=0}^{\infty} \frac{1 - q^{2k}t_0}{1 - t_0} \frac{(t_0; q)_k \, (t_1; q)_k \, \cdots \, (t_r; q)_k}{(q; q)_k (qt_0/t_1; q)_k \cdots (qt_0/t_r)_k} s^k \\
(4.8) \qquad\qquad &= {}_{r+3}W_{r+2}\left[t_0; t_1, \ldots, t_r; q, s\right]
\end{aligned}
$$

in the notation of very well-poised $q$-hypergeometric series [3], where $q = e^{c\delta}$, $t_i = e^{cx_i}$ ($i = 0, 1, \ldots, r$) and $s = (qt_0)^{\frac{r-1}{2}} z / t_1 \cdots t_r$. We discuss below a class of Padé interpolation problems that can be formulated in an unified manner in the three types of very well-poised hypergeometric series.

Taking the two sequence of meromorphic functions

$$
(4.9) \quad
\begin{aligned}
f_j(x) &= \frac{[a \pm x]_j}{[b \pm x]_j} = \frac{[a + x]_j [a - x]_j}{[b + x]_j [b - x]_j}, \\
g_j(x) &= \frac{[c \pm x]_j}{[d \pm x]_j} = \frac{[c + x]_j [c - x]_j}{[d + x]_j [d - x]_j} \qquad (j = 0, 1, 2, \ldots)
\end{aligned}
$$

and the reference points $u_k = u + k\delta$ ($k = 0, 1, 2, \ldots$), we consider the Padé interpolation problem

$$
(4.10) \quad \frac{P_m(u_k)}{Q_n(u_k)} = v_k = \frac{\lambda_k}{\mu_k} \qquad (k = 0, 1, \ldots, N)
$$

for a pair of functions

$$
(4.11) \quad
\begin{aligned}
P_m(x) &= p_{m,0}\, f_0(x) + p_{m,1}\, f_1(x) + \cdots + p_{m,m}\, f_m(x), \\
Q_n(x) &= q_{n,0}\, g_0(x) + q_{n,1}\, g_1(x) + \cdots + q_{n,n}\, g_n(x)
\end{aligned}
$$

where $N = m + n$. The prescribed values $v_k = \lambda_k / \mu_k$ ($k = 0, 1, 2 \ldots, N$) will be specified later.

THEOREM 4.1. *Consider the Padé interpolation problem* (4.10), (4.11) *for the functions* $f_j(x)$, $g_j(x)$ *in* (4.9) *and the reference points* $u_k = u + k\delta$ ($k = 0, 1, \ldots, N; N = m + n$). *Then the solution* $(P_m(x), Q_n(x))$ *of Theorem 2.1 is explicitly given by*

(4.12)

$$P_m(x) = C_n(c,d) \frac{\prod_{l=1}^{n}[2u_m + l\delta]_l[\delta]_l}{\prod_{l=0}^{n}[d \pm u_{m+l}]_n} \prod_{i=0}^{m-1}\mu_i \prod_{i=0}^{n}\lambda_{m+i} \det \begin{bmatrix} f_0(x) & \cdots & f_m(x) \\ U_{0,0} & \cdots & U_{0,m} \\ \vdots & & \vdots \\ U_{m-1,0} & \cdots & U_{m-1,m} \end{bmatrix},$$

(4.13)

$$Q_n(x) = \epsilon_{m,n} C_m(a,b) \frac{\prod_{l=1}^{m}[2u_n + l\delta]_l[\delta]_l}{\prod_{l=0}^{m}[b \pm u_{n+l}]_m} \prod_{i=0}^{n-1}\lambda_i \prod_{i=0}^{m}\mu_{n+i} \det \begin{bmatrix} g_0(x) & \cdots & g_n(x) \\ V_{0,0} & \cdots & V_{0,n} \\ \vdots & & \vdots \\ V_{n-1,0} & \cdots & U_{n-1,n} \end{bmatrix},$$

*where*

(4.14) $\quad C_n(c,d) = (-1)^{\binom{n+1}{2}} \prod_{k=1}^{n} [d-c]_k [c+d+(k-1)\delta]_k, \quad \epsilon_{m,n} = (-1)^{mn+m+n}$

*and*

$$U_{ij} = \frac{[a \pm u_i]_j}{[b \pm u_i]_j} \sum_{k=0}^{n+1} \frac{[2u_i + 2k\delta]}{[2u_i]} \frac{[2u_i]_k}{[\delta]_k} \frac{[-(n+1)\delta]_k}{[2u_i+(n+2)\delta]_k} \frac{[u_i-d+\delta]_k}{[u_i+d]_k} \frac{[u_i+d+n\delta]_k}{[u_i-d+(1-n)\delta]_k}$$

(4.15)

$$\cdot \frac{[u_i-a+\delta]_k[u_i+a+j\delta]_k[u_i+b]_k[u_i-b+(1-j)\delta]_k}{[u_i+a]_k[[u_i-a+(1-j)\delta]_k[u_i-b+\delta]_k[u_i+b+j\delta]_k} \frac{\mu_{i+k}\lambda_i}{\lambda_{i+k}\mu_i},$$

$$V_{ij} = \frac{[c \pm u_i]_j}{[d \pm u_i]_j} \sum_{k=0}^{m+1} \frac{[2u_i + 2k\delta]}{[2u_i]} \frac{[2u_i]_k}{[\delta]_k} \frac{[-(m+1)\delta]_k}{[2u_i+(m+2)\delta]_k} \frac{[u_i-b+\delta]_k}{[u_i+b]_k} \frac{[u_i+b+m\delta]_k}{[u_i-b+(1-m)\delta]_k}$$

(4.16)

$$\cdot \frac{[u_i-c+\delta]_k[u_i+c+j\delta]_k[u_i+d]_k[u_i-d+(1-j)\delta]_k}{[u_i+c]_k[[u_i-c+(1-j)\delta]_k[u_i-d+\delta]_k[u_i+d+j\delta]_k} \frac{\lambda_{i+k}\mu_i}{\mu_{i+k}\lambda_i}.$$

As before we consider the expansion

(4.17) $$\qquad U_{ij} = \sum_{k=0}^{n+1} (-1)^k \frac{G_{0,1,\ldots,n}^{i,\ldots,\widehat{i+k},\ldots,i+n+1}}{G_{0,1,\ldots,n}^{i+1,\ldots,i+n+1}} f_j(u_{i+k}) \frac{\mu_{i+k}\,\lambda_i}{\lambda_{i+k}\,\mu_i}$$

of the determinant of (2.13). In this case we have

(4.18)

$$f_j(u_{i+k}) = \frac{[a \pm u_{i+k}]_j}{[b \pm u_{i+k}]_j} = \frac{[a \pm u_i]_j}{[b \pm u_i]_j} \frac{[u_i-a+\delta]_k[u_i+a+j\delta]_k[u_i+b]_k[u_i-b+(1-j)\delta]_k}{[u_i+a]_k[[u_i-a+(1-j)\delta]_k[u_i-b+\delta]_k[u_i+b+j\delta]_k}.$$

The $(n+1) \times (n+1)$ minor determinants of the matrix

(4.19) $$\qquad G = \big(g_j(u_i)\big)_{0 \le i \le N,\, 0 \le j \le n} = \left( \frac{[c \pm u_i]_j}{[d \pm u_i]_j} \right)_{0 \le i \le N,\, 0 \le j \le n}$$

can be computed by means of an elliptic extension of Krattenthaler's formula (see Appendix B). In fact, by (B.14) we have

$$\det G_{0,1,\ldots,n}^{i,i+1,\ldots,i+n} = C_n(c,d)\frac{\prod_{l=1}^{n}[2u_i+l\delta]_l[\delta]_l}{\prod_{l=0}^{n}[d\pm u_{i+l}]_n},$$

$$\det G_{0,1,\ldots,n}^{i,\ldots,\widehat{i+k},\ldots,i+n+1} = C_n(c,d)\frac{\prod_{l=1}^{n}[2u_{i+1}+l\delta]_l[\delta]_l}{\prod_{l=0}^{n}[d\pm u_{i+1+l}]_n}$$

$$(4.20) \qquad \cdot(-1)^k\frac{[2u_i+2k\delta]}{[2u_i]}\frac{[2u_i]_k[-(n+1)\delta]_k}{[\delta]_k[2u_i+(n+2)\delta]_k}\frac{[u_i-d+\delta]_k[u_i+d+n\delta]_k}{[u_i+d]_k[u_i-d+(1-n)\delta]_k}.$$

Hence $U_{ij}$ is computed as

$$U_{ij} = \frac{[a\pm u_i]_j}{[b\pm u_i]_j}\sum_{k=0}^{n+1}\frac{[2u_i+2k\delta]}{[2u_i]}\frac{[2u_i]_k}{[\delta]_k}\frac{[-(n+1)\delta]_k}{[2u_i+(n+2)\delta]_k}\frac{[u_i-d+\delta]_k}{[u_i+d]_k}\frac{[u_i+d+n\delta]_k}{[u_i-d+(1-n)\delta]_k}$$

$$(4.21)$$

$$\cdot\frac{[u_i-a+\delta]_k[u_i+a+j\delta]_k[u_i+b]_k[u_i-b+(1-j)\delta]_k}{[u_i+a]_k[[u_i-a+(1-j)\delta]_k[u_i-b+\delta]_k[u_i+b+j\delta]_k}\frac{\mu_{i+k}\lambda_i}{\lambda_{i+k}\mu_i}.$$

The corresponding formula for $V_{ij}$ is obtained by exchanging the roles of $(m,n)$, $(a,c)$ and $(b,d)$.

Consider the case where the prescribed values are specified as

$$(4.22) \qquad v_k = \frac{\lambda_k}{\mu_k} = \left(\frac{z}{w}\right)^k\prod_{s=1}^{r}\frac{[u-e_s+\delta]_k}{[u+e_s]_k} \qquad (k=0,1,\ldots,N).$$

Then we obtain very well-poised series

$$U_{ij} = \frac{[a\pm u_i]_j}{[b\pm u_i]_j}\,{}_{r+12}V_{r+11}\Big[2u_i;-(n+1)\delta,u_i-d+\delta,u_i+d+n\delta,$$

$$(4.23)$$

$$u_i-a+\delta,u_i+a+j\delta,u_i+b,u_i-b+(1-j)\delta,u_i+e_1,\ldots,u_i+e_r\Big|\frac{w}{z}\Big],$$

$$V_{ij} = \frac{[c\pm u_i]_j}{[d\pm u_i]_j}\,{}_{r+12}V_{r+11}\Big[2u_i;-(m+1)\delta,u_i-b+\delta,u_i+b+n\delta,$$

$$(4.24)$$

$$u_i-c+\delta,u_i+c+j\delta,u_i+d,u_i-d+(1-j)\delta,u_i-e_1+\delta,\ldots,u_i-e_r+\delta\Big|\frac{z}{w}\Big].$$

When

$$(4.25) \quad v_k = \frac{\lambda_k}{\mu_k} = \left(\frac{z}{w}\right)^k\frac{[u-a+\delta]_k[u+b]_k}{[u+a]_k[u-b+\delta]_k}\frac{[u+c]_k[u-d+\delta]_k}{[u-c+\delta]_k[u+d]_k}\prod_{s=1}^{r}\frac{[u-e_s+\delta]_k}{[u+e_s]_k}$$

we obtain simpler very well-poised hypergeometric series

$$U_{ij} = \frac{[a\pm u_i]_j}{[b\pm u_i]_j}\,{}_{r+10}V_{r+9}\Big[2u_i;-(n+1)\delta,u_i-c+\delta,u_i+d+n\delta,$$

$$u_i+a+j\delta,u_i-b+(1-j)\delta,u_i+e_1,\ldots,u_i+e_r\Big|\frac{w}{z}\Big],$$

$$V_{ij} = \frac{[c\pm u_i]_j}{[d\pm u_i]_j}\,{}_{r+10}V_{r+9}\Big[2u_i;-(m+1)\delta,u_i-a+\delta,u_i+b+n\delta,$$

$$(4.26)$$

$$u_i+c+j\delta,u_i-d+(1-j)\delta,u_i-e_1+\delta,\ldots,u_i-e_r+\delta\Big|\frac{z}{w}\Big].$$

Another type of determinantal expression for $P_m(x)$ and $Q_n(x)$ is formulated as follows. We remark that this type of determinant formulas has also been discussed in [**11**]. In what follows, we use the notation

$$(4.27) \qquad V^{(k)}\left[a_0; a_1, \cdots, a_r\right] = \frac{[a_0 + 2k\delta]}{[a_0]} \frac{[a_0]_k \, [a_1]_k \, \cdots \, [a_r]_k}{[\delta]_k [\delta + a_0 - a_1]_k \cdots [\delta + a_0 - a_r]_k}.$$

THEOREM 4.2. *Consider the Padé interpolation problem* (4.10), (4.11) *for the functions* $f_j(x)$, $g_j(z)$ *in* (4.9) *and the reference points* $u_k = u + k\delta$ ($k = 0, 1, \ldots, N; N = m + n$). *Then the solution* $(P_m(x), Q_n(x))$ *of Theorem 2.1 is expressed as*

$$(4.28) \qquad P_m(x) = K_{m,n}(c,d) \prod_{i=0}^{N} \lambda_i \, \det \begin{bmatrix} f_0(x) & \cdots & f_m(x) \\ \Phi_{0,0} & \cdots & \Phi_{0,m} \\ \vdots & & \vdots \\ \Phi_{m-1,0} & \cdots & \Phi_{m-1,m} \end{bmatrix},$$

$$(4.29) \qquad Q_n(x) = \epsilon_{m,n} K_{n,m}(a,b) \prod_{i=0}^{N} \mu_i \, \det \begin{bmatrix} g_0(x) & \cdots & g_n(x) \\ \Psi_{0,0} & \cdots & \Psi_{0,n} \\ \vdots & & \vdots \\ \Psi_{n-1,0} & \cdots & \Psi_{n-1,n} \end{bmatrix},$$

*where*

$$\Phi_{ij} = \frac{[a \pm u]_j}{[b \pm u]_j} \sum_{k=0}^{N} V^{(k)}\Big[2u; -N\delta, u - c + \delta + i\delta, u + d + (N-1)\delta - i\delta, u + c, u - d + \delta,$$

$$(4.30)$$

$$u + a + j\delta, u - b + \delta - j\delta, u - a + \delta, u + b\Big] \frac{\mu_k}{\lambda_k},$$

$$\Psi_{ij} = \frac{[c \pm u]_j}{[d \pm u]_j} \sum_{k=0}^{N} V^{(k)}\Big[2u; -N\delta, u - a + \delta + i\delta, u + b + (N-1)\delta - i\delta, u + a, u - b + \delta,$$

$$(4.31)$$

$$u + c + j\delta, u - d + \delta - j\delta, u - c\delta, u + d\Big] \frac{\lambda_k}{\mu_k}.$$

Consider the case where the prescribed values are specified as

$$(4.32) \qquad v_k = \frac{\lambda_k}{\mu_k} = \left(\frac{z}{w}\right)^k \prod_{s=1}^{r} \frac{[u - e_s + \delta]_k}{[u + e_s]_k} \qquad (k = 0, 1, \ldots, N).$$

Then we obtain very well-poised series

$$\Phi_{ij} = \frac{[a \pm u]_j}{[b \pm u]_j} {}_{r+14}V_{r+13}\Big[2u; -N\delta, u - c + \delta + i\delta, u + d + (N-1)\delta - i\delta, u + c, u - d + \delta,$$

(4.33)

$$u + a + j\delta, u - b + \delta - j\delta, u - a + \delta, u + b, u + e_1, \ldots, u + e_r \Big| \frac{w}{z}\Big],$$

$$\Psi_{ij} = \frac{[c \pm u]_j}{[d \pm u]_j} {}_{r+14}V_{r+13}\Big[2u; -N\delta, u - a + \delta + i\delta, u + b + (N-1)\delta - i\delta, u + a, u - b + \delta,$$

(4.34)

$$u+c+j\delta, u-d+\delta-j\delta, u-c+\delta, u+d, u-e_1+\delta, \ldots, u-e_r+\delta \Big| \frac{z}{w}\Big].$$

When

$$(4.35) \quad v_k = \frac{\lambda_k}{\mu_k} = \left(\frac{z}{w}\right)^k \frac{[u-a+\delta]_k[u+b]_k}{[u+a]_k[u-b+\delta]_k} \frac{[u+c]_k[u-d+\delta]_k}{[u-c+\delta]_k[u+d]_k} \prod_{s=1}^r \frac{[u-e_s+\delta]_k}{[u+e_s]_k}$$

we obtain simpler very well-poised hypergeometric series

$$\Phi_{ij} = \frac{[a \pm u]_j}{[b \pm u]_j} {}_{r+10}V_{r+9}\Big[2u; -N\delta, u - c + \delta + i\delta, u + d + (N-1)\delta - i\delta,$$

(4.36)

$$u + a + j\delta, u - b + \delta - j\delta, u + e_1, \ldots, u + e_r \Big| \frac{w}{z}\Big],$$

$$\Psi_{ij} = \frac{[c \pm u]_j}{[d \pm u]_j} {}_{r+10}V_{r+9}\Big[2u; -N\delta, u - a + \delta + i\delta, u + b + (N-1)\delta - i\delta,$$

(4.37)

$$u+c+j\delta, u-d+\delta-j\delta, u-e_1+\delta, \ldots, u-e_r+\delta \Big| \frac{z}{w}\Big].$$

Theorem 4.2 can be proved by a procedure similar to the one we used in the previous section. In this case we define the matrix $L = \big(L_{ij}\big)_{i,j=0}^N$ by

(4.38)

$$L_{ij} = V^{(j)}\Big[2u; -N\delta, u - c + (1+i)\delta, u + d + (N-1-i)\delta, u + c, u - d + \delta\Big]$$

for $0 \le i, j \le N$. Then one can show

(4.39)

$$(LG)_{ij} = \frac{[c + d + (j - i - 1)\delta]_N[-(i+j)\delta]_N[2u+\delta]_N[d-c]_N}{[u+d+j\delta]_N[u-c+(1-j)\delta]_N[u+c-i\delta]_N[-u+d-(1+i)\delta]_N} \frac{[c \pm u]_j}{[d \pm u]_j}$$

by means of the Frenkel-Turaev sum, and hence $(LG)_{ij} = 0$ for $i + j < N$. Then the series $\Phi_{ij}$ are obtained by computing the product $L\widetilde{F}$ as before. We remark that in this case

$$\det M = (-1)^{\binom{n+1}{2}} \prod_{j=0}^n \frac{[c \pm u]_j}{[d \pm u]_j}$$

(4.40)

$$\cdot \prod_{j=0}^n \frac{[c + d - (N + 1 - 2j)\delta]_N[-N\delta]_N[2u+\delta]_N[d-c]_N}{[u+d+j\delta]_N[u-c+(1-j)\delta]_N[u+c-(N-j)\delta]_N[-u+d-(N+1-j)\delta]_N}.$$

The determinant of $L$ can also be computed in a factorized form by the elliptic version (B.14) of Krattenthaler's formula:

$$\det L = \frac{\prod_{j=1}^{N}[\delta]_j[c+d+(N-1-2j)\delta]_j[c-d-(N-1)\delta]_j[2u+j\delta]_j}{\prod_{i=0}^{N}[u+c-i\delta]_N[u-d-(N-2-i)\delta]_N}$$

$$(4.41) \qquad \cdot \prod_{j=0}^{N}\frac{[2u+2j\delta]}{[2u]}\frac{[2u]_j[-N\delta]_j[u+c]_j[u-d+\delta]_j}{[\delta]_j[2u+(N+1)\delta]_j[u-c+\delta]_j[u+d]_j}.$$

The constant in (4.28) is given by $K_{m,n}(c;d) = \det M/\det L$.

## Acknowledgment

The author would like to express his thanks to the anonymous referee for providing various informations on preceding works relevant to the subject of this paper.

## Appendix A. Condensation of determinants

In this Appendix A, we give a review on the variation of Dodgson condensation (Sylvester identity) of determinants due to Gasca–López-Carmona–Ramirez [2], which we call the *condensation along a moving core*. For further generalizations of Sylvester's identity, we refer the reader to Mühlbach–Gasca [9].

We first recall a standard version of the Dodgson condensation (Sylvester's identity) for comparison. For a general $m \times n$ matrix $X = (x_{ij})_{1 \le i \le m, 1 \le j \le n}$ (with entries in a commutative ring), we denote by $X_{j_1,\ldots,j_s}^{i_1,\ldots,i_r} = (x_{i_a j_b})_{1 \le a \le r, 1 \le b \le s}$ the $r \times s$ submatrix of $X$ with row indices $i_1,\ldots,i_r \in \{1,\ldots,m\}$ and column indices $j_1,\ldots,j_r \in \{1,\ldots,n\}$. When $r = s$, we denote by $\det X_{j_1,\ldots,j_r}^{i_1,\ldots,i_r}$ the corresponding minor determinant.

LEMMA A.1 (Dodgson condensation, Sylvester's identity). *Let* $X = (x_{ij})_{i,j=1}^{n}$ *an* $n \times n$ *matrix and set* $n = r+s$ ($r,s \ge 1$). *We define an* $r \times r$ *matrix* $Y = (y_{ij})_{i,j=1}^{r}$ *by using the* $(s+1) \times (s+1)$ *minor determinants* $y_{ij} = \det X_{j,r+1,\ldots,n}^{i,r+1,\ldots,n}$ *of* $X$. *Then the determinant of* $Y$ *is expressed as*

$$(A.1) \quad \det Y = \det X\,(\det X_{r+1,\ldots,n}^{r+1,\ldots,n})^{r-1}; \quad Y = (y_{ij})_{i,j=1}^{r}, \quad y_{ij} = \det X_{j,r+1,\ldots,n}^{i,r+1,\ldots,n}.$$

PROOF. Define an $n \times n$ upper triangular matrix $Z = (z_{ij})_{i,j=1}^{n}$ by setting

$$(A.2) \qquad z_{ij} = \begin{cases} \delta_{i,j}\det X_{r+1,\ldots,n}^{r+1\ldots,n} & (1 \le i,j \le r) \\ (-1)^{j-r}\det X_{r+1,\ldots,n}^{i,r+1,\ldots,\widehat{j},\ldots,n} & (1 \le i \le r;\ r+1 \le j \le n) \\ \delta_{ij} & (\text{otherwise}). \end{cases}$$

Then for $i = 1,\ldots,r$, the $(i,j)$-component of the product $ZX$ is given by

$$(ZX)_{ij} = z_{ii}\,x_{ij} + \sum_{k=r+1}^{n} z_{ik}\,x_{kj}$$

$$= \det X_{r+1,\ldots,n}^{r+1,\ldots,n}\,x_{ij} + \sum_{k=r+1}^{n}(-1)^{k-r}\det X_{r+1,\ldots,n}^{r+1,\ldots,\widehat{k},\ldots,n}\,x_{kj}$$

$$(A.3) \qquad = \det X_{j,r+1,\ldots,n}^{i,r+1,\ldots,n}.$$

namely,

(A.4)
$$(ZX)_{ij} = \begin{cases} y_{ij} & (j = 1, \ldots, r) \\ 0 & (j = r+1, \ldots, n) \end{cases}$$

This means that

(A.5)
$$ZX = \begin{bmatrix} Y & 0 \\ X_{1,\ldots,r}^{r+1,\ldots,n} & X_{r+1,\ldots,n}^{r+1,\ldots,n} \end{bmatrix}.$$

Since $\det Z = z_{11} \cdots z_{rr}$, we obtain

(A.6)
$$\det X \; (\det X_{r+1,\ldots,n}^{r+1,\ldots,n})^r = \det Y \; \det X_{r+1,\ldots,n}^{r+1,\ldots,n}.$$

This implies the polynomial identity

(A.7)
$$\det Y = \det X \; (\det X_{r+1,\ldots,n}^{r+1,\ldots,n})^{r-1}$$

in the variables $x_{ij}$ ($1 \le i, j \le n$). $\qquad\qquad\square$

When $(r, s) = (n-1, 1)$, (A.1) means that

(A.8)
$$\det \left( x_{ij} x_{nn} - x_{in} x_{nj} \right)_{i,j=1}^{n-1} = \det X \; x_{nn}^{n-2}.$$

Another extreme case $(r, s) = (2, n-2)$ implies

(A.9)
$$\det X_{1,3,\ldots,n}^{1,3,\ldots,n} \det X_{2,3,\ldots,n}^{2,3,\ldots,n} - \det X_{2,3,\ldots,n}^{1,3,\ldots,n} \det X_{1,3,\ldots,n}^{2,3,\ldots,n} = \det X_{1,2,\ldots,n}^{1,2,\ldots,n} \det X_{3,\ldots,n}^{3,\ldots,n},$$

which is equivalent to

(A.10)
$$\det X_{1,\ldots,n-1}^{1,\ldots,n-1} \det X_{2,\ldots,n}^{2,\ldots,n} - \det X_{2,\ldots,n}^{1,\ldots,n-1} \det X_{1,\ldots,n-1}^{2,\ldots,n} = \det X_{1,2,\ldots,n}^{1,2,\ldots,n} \det X_{2,\ldots,n-1}^{2,\ldots,n-1}.$$

These identities (A.9), (A.10) are often referred to as *Jacobi's formula* or *Lewis–Carroll's formula*.

The variant of Dodgson condensation that we use in this paper is the following identity due to Gasca–López-Carmona–Ramirez [**2**].

LEMMA A.2 (Condensation along a moving core). *Let* $X = (x_{ij})_{i,j=1}^{n}$ *an* $n \times n$ *matrix and set* $n = r + s$ $(r, s \ge 1)$. *We define an* $r \times r$ *matrix* $Y = (y_{ij})_{i,j=1}^{r}$ *by using the* $(s+1) \times (s+1)$ *minor determinants* $y_{ij} = \det X_{j,r+1,\ldots,n}^{i,i+1,\ldots,i+s}$ *of* $X$. *Then the determinant of* $Y$ *is expressed as*
(A.11)
$$\det Y = \det X \prod_{i=1}^{r-1} \det X_{r+1,\ldots,n}^{i+1,\ldots,i+s}; \quad Y = (y_{ij})_{i,j=1}^{r}, \quad y_{ij} = \det X_{j,r+1,\ldots,n}^{i,i+1,\ldots,i+s}.$$

PROOF. We define an $n \times n$ upper triangular matrix $Z = (z_{ij})_{i,j=1}^{n}$ as follows by using $s \times s$ minor determinants of $X$:

(A.12)
$$z_{ij} = \begin{cases} (-1)^{j-i} \det X_{r+1,\ldots,n}^{i,\ldots,\hat{j},\ldots,i+s} & (1 \le i \le r; i \le j \le i+s) \\ \delta_{ij} & \text{(otherwise)}. \end{cases}$$

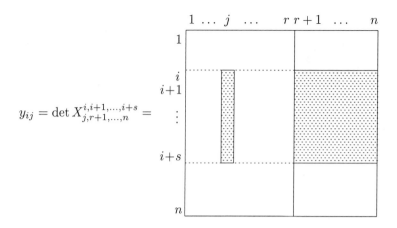

$$y_{ij} = \det X^{i,i+1,\ldots,i+s}_{j,r+1,\ldots,n} =$$

FIGURE 1. Condensation along a moving core

Then for $i = 1, \ldots, r$, we have

(A.13)
$$\left(ZX\right)_{ij} = \sum_{k=i}^{i+s} z_{ik}\, x_{kj} = \sum_{k=i}^{i+s} (-1)^{k-i} \det X^{i,i+1,\ldots,\widehat{k},\ldots,i+s}_{r+1,\ldots,n}\, x_{kj} = \det X^{i,\ldots,i+s}_{j,r+1,\ldots,n},$$

namely

(A.14)
$$\left(ZX\right)_{ij} = \begin{cases} y_{ij} & (1 \le j \le r) \\ 0 & (r+1 \le j \le n). \end{cases}$$

This means that

(A.15)
$$ZX = \begin{bmatrix} Y & 0 \\ X^{r+1,\ldots,n}_{1,\ldots,r} & X^{r+1,\ldots,n}_{r+1,\ldots,n} \end{bmatrix}.$$

Since $\det Z = z_{11} \cdots z_{rr}$, we obtain

(A.16)
$$\det X \prod_{i=1}^{r} \det X^{i,\ldots,i+s}_{r+1,\ldots,n} = \det Y \det X^{r+1,\ldots,n}_{r+1,\ldots,n}.$$

This implies the polynomial identity

(A.17)
$$\det Y = \det X \prod_{i=1}^{r-1} \det X^{i,\ldots,i+s}_{r+1,\ldots,n}$$

in the variables $x_{ij}$ $(1 \le i, j \le n)$.                              □

We remark that, if we renormalize the matrix $Y$ by setting

(A.18)
$$\widetilde{Y} = \left(\widetilde{y}_{ij}\right)^{r}_{i,j=1}, \quad \widetilde{y}_{ij} = \det X^{i,i+1,\ldots,i+s}_{j,r+1,\ldots,n} \left( \det X^{i+1,\ldots,i+s}_{r+1,\ldots,n} \right)^{-1} \quad (i,j = 1,\ldots,r),$$

then equality (A.11) is rewritten equivalently as

(A.19)
$$\det X = \det \widetilde{Y} \det X^{r+1,\ldots,n}_{r+1,\ldots,n}.$$

### Appendix B. Variations of Krattenthaler's determinant formula

In this Appendix B, we recall Krattenthaler's determinant formula [5] and its elliptic extension due to Warnaar [15]. Although these formulas can be proved in various ways, we remark here that they are consequences of Lemma B.3 below, which can be regarded as an abstract form of Krattenthaler's determinant formula (for recent works on the evaluation of determinants involving shifted factorials, see Normand [10]).

We first recall a typical form of Krattenthaler's determinant formula [5].

LEMMA B.1. *For any set of variables $x_i$ ($0 \le i \le m$) and parameters $\alpha_k, \beta_k, \gamma_k,$ $\delta_k$ ($0 \le k < m$), one has*

$$(\text{B.1}) \quad \det \left( \prod_{0 \le k < j} \frac{\alpha_k x_i + \beta_k}{\gamma_k x_i + \delta_k} \right)_{i,j=0}^m = \frac{\displaystyle\prod_{0 \le i < j \le m} (x_j - x_i) \prod_{0 \le k \le l < m} (\alpha_k \delta_l - \beta_k \gamma_l)}{\displaystyle\prod_{0 \le i \le m} \prod_{0 \le k < m} (\gamma_k x_i + \delta_k)}.$$

By specializing the parameters $\alpha_k, \beta_k, \gamma_k, \delta_k$, we obtain various determinant formulas. We quote below some of them.

(a)   Case where $\alpha_k = \gamma_k = 1$ and $\beta_k = a_k$, $\delta_k = b_k$:

$$(\text{B.2}) \quad \det \left( \prod_{0 \le k < j} \frac{x_i + a_k}{x_i + b_k} \right)_{i,j=0}^m = \frac{\displaystyle\prod_{0 \le i < j \le m} (x_j - x_i) \prod_{0 \le k \le l < m} (b_l - a_k)}{\displaystyle\prod_{0 \le i \le m} \prod_{0 \le k < m} (x_i + b_k)}.$$

In particular, by setting $a_k = a + k$, $b_k = b + k$ one obtains

$$(\text{B.3}) \quad \det \left( \frac{(a + x_i)_j}{(b + x_i)_j} \right)_{i,j=0}^m = \frac{\displaystyle\prod_{0 \le i < j \le m} (x_j - x_i) \prod_{k=1}^m (b - a)_k}{\displaystyle\prod_{0 \le i \le m} (b + x_i)_m}.$$

where $(a)_k = a(a + 1) \cdots (a + k - 1)$.

(b)   Case where $\beta_k = \delta_k = 1$, $\alpha_k = -a_k$, $\gamma_k = -b_k$:

$$(\text{B.4}) \quad \det \left( \prod_{0 \le k < j} \frac{1 - a_k x_i}{1 - b_k x_i} \right)_{i,j=0}^m = \frac{\displaystyle\prod_{0 \le i < j \le m} (x_j - x_i) \prod_{0 \le k \le l < m} (b_l - a_k)}{\displaystyle\prod_{0 \le i \le m} \prod_{0 \le k < m} (1 - b_k x_i)}.$$

By setting $a_k = p^k a$ and $b_k = q^k b$, one has

$$(\text{B.5}) \quad \det \left( \prod_{0 \le k < j} \frac{(ax_i; p)_j}{(bx_i; q)_j} \right)_{i,j=0}^m = \frac{\displaystyle\prod_{0 \le i < j \le m} (x_j - x_i) \prod_{0 \le k \le l < m} (q^l b - p^k a)}{\displaystyle\prod_{0 \le i \le m} (bx_i; q)_m}.$$

where $(a;p)_k = (1-a)(1-pa)\cdots(1-p^{k-1}a)$. In particular,

$$(\text{B.6})\quad \det\left(\prod_{0\le k<j}\frac{(ax_i;q)_j}{(bx_i;q)_j}\right)^m_{i,j=0} = \frac{a^{\binom{m+1}{2}}q^{\binom{m+1}{3}}\displaystyle\prod_{0\le i<j\le m}(x_i-x_j)\prod_{k=1}^m(b/a;q)_k}{\displaystyle\prod_{0\le i\le m}(bx_i;q)_m}.$$

(c)   Noting that

$$(\text{B.7})\qquad (az;q)_n(ac/z;q)_n = \prod_{0\le k<n}(1+a^2cq^{2k}-aq^k(z+c/z))$$

set

(B.8)
$$x_i = z_i+c/z_i;\quad \alpha_k = -ap^k,\quad \beta_k = 1+a^2cp^{2k},\quad \gamma_k = -bq^k,\quad \delta_k = 1+b^2cq^{2k}.$$

Then we have

$$\det\left(\frac{(az_i;p)_j(ac/z_i;p)_j}{(bz_i;q)_j(bc/z_i;q)_j}\right)^m_{i,j=0}$$

$$(\text{B.9})\qquad = \frac{\displaystyle\prod_{0\le i<j\le m}(z_j-z_i)(1-c/z_iz_j)\prod_{0\le k<l<m}(bq^l-ap^k)(1-p^kq^labc)}{\displaystyle\prod_{0\le i\le m}(bz_i;q)_m(bc/z_i;q)_m}.$$

In particular,

$$\det\left(\frac{(az_i;q)_j(ac/z_i;q)_j}{(bz_i;q)_j(bc/z_i;q)_j}\right)^m_{i,j=0}$$

(B.10)

$$= \frac{a^{\binom{m+1}{2}}q^{\binom{m+1}{3}}\displaystyle\prod_{0\le i<j\le m}(z_i-z_j)(1-c/z_iz_j)\prod_{k=1}^m(b/a;q)_k(q^{2(m-1-k)}abc;q)_k}{\displaystyle\prod_{0\le i\le m}(bz_i;q)_m(bc/z_i;q)_m}.$$

Let $[x]$ a nonzero entire function in $x\in\mathbb{C}$ and suppose that $[x]$ satisfies the so-called Riemann relation: For any $x,\alpha,\beta,\gamma\in\mathbb{C}$,

$$(\text{B.11})\qquad [x\pm\alpha][\beta\pm\gamma]+[x\pm\beta][\gamma\pm\alpha]+[x\pm\gamma][\alpha\pm\beta]=0,$$

where $[x\pm\alpha]=[x+\alpha][x-\alpha]$. This functional equation is equivalent to

$$(\text{B.12})\qquad [x\pm u][y\pm v]-[x\pm v][y\pm u]=[x\pm y][u\pm v].$$

The following lemma is the elliptic extension of Lemma B.1 due to Warnaar [15].

LEMMA B.2. *For any set of variables* $x_i$ $(0 \leq i \leq m)$ *and parameters* $a_k, b_k$ $(0 \leq k < m)$, *one has*

$$(B.13) \qquad \det \left( \prod_{0 \leq k < j} \frac{[a_k \pm x_i]}{[b_k \pm x_i]} \right)^m_{i,j=0} = \frac{\displaystyle\prod_{0 \leq i < j \leq m} [x_j \pm x_i] \prod_{0 \leq k \leq l < m} [a_k \pm b_l]}{\displaystyle\prod_{0 \leq i \leq m} \prod_{0 \leq k < m} [b_k \pm x_i]}.$$

As a special case where $a_k = a + k\delta$, $b_k = b + k\delta$ $(0 \leq k < m)$, we obtain

$$(B.14) \quad \det \left( \frac{[a \pm x_i]_j}{[b \pm x_i]_j} \right)^m_{i,j=0} = \frac{\displaystyle\prod_{0 \leq i < j \leq m} [x_i \pm x_j] \prod_{k=1}^{m} [b - a]_k [a + b + (k-1)\delta]_k}{\displaystyle\prod_{0 \leq i \leq m} [b \pm x_i]_m}.$$

where $[a]_k = [a][a + \delta] \cdots [a + (k-1)\delta]$ and $[a \pm b]_k = [a + b]_k [a - b]_k$.

Lemma B.1 and Lemma B.2 can be proved as consequences of the following abstract form of Krattenthaler's determinant formula.

Let $a_{ik}, b_{ik}$ $(0 \leq i \leq N; 0 \leq k < N)$ be elements a field $\mathbb{K}$ with $b_{ik} \neq 0$ for all $i, k$, and consider the matrix

$$(B.15) \qquad X_m = \left( \prod_{0 \leq k < j} \frac{a_{ik}}{b_{ik}} \right)^m_{i,j=0}.$$

for $m = 0, 1, \ldots, N$. Suppose that there exist elements $p_{ij}$ $(0 \leq i, j \leq N)$, $q_{kl}$ $(0 \leq k \leq l < N)$ of $\mathbb{K}$ such that

$$(B.16) \qquad a_{ik} b_{jl} - a_{jk} b_{il} = p_{ij} q_{kl}, \quad p_{ji} = -p_{ij}$$

for all $i, j \in \{0, 1 \ldots, N\}$ and $k, l \in \{0, 1, \ldots, N - 1\}$.

LEMMA B.3. *Under the assumption* (B.16), *the determinant* $\det X_m$ *is factorized as*

$$(B.17) \qquad \det X_m = \det \left( \prod_{0 \leq k < j} \frac{a_{ik}}{b_{ik}} \right)^m_{i,j=0} = \frac{\displaystyle\prod_{0 \leq i < j \leq m} p_{ji} \prod_{0 \leq k \leq l < m} q_{kl}}{\displaystyle\prod_{i=0}^{m} \prod_{k=0}^{m-1} b_{i,k}}.$$

*for* $m = 0, 1, \ldots, N$.

Set $\tau_m = \det X_m$ for $m = 0, 1, 2, \ldots, N$, so that

$$(B.18) \qquad \tau_0 = 1, \quad \tau_1 = \det \begin{bmatrix} 1 & \dfrac{a_{00}}{b_{00}} \\ 1 & \dfrac{a_{10}}{b_{10}} \end{bmatrix} = \frac{a_{10}}{b_{10}} - \frac{a_{00}}{b_{00}}.$$

The first nontrivial case is guaranteed by the assumption (B.16):

$$(B.19) \qquad \tau_1 = \frac{a_{10} b_{00} - a_{00} b_{10}}{b_{00} b_{10}} = \frac{p_{10} \, q_{00}}{b_{00} \, b_{10}}.$$

Then (B.17) can be proved by means of the Lewis–Carroll formula. In fact from (A.10), we obtain the bilinear identities

$$(B.20) \qquad \frac{a_{m+1,1}}{b_{m+1,1}} \tau_m \, T_C T_R(\tau_m) - \frac{a_{11}}{b_{11}} T_C(\tau_m) \, T_R(\tau_m) = \tau_{m+1} \, T_C T_R(\tau_{m-1}),$$

for $\tau_m$, where $T_R$ and $T_C$ stand for the symbolic shift operator for the row indices and the column indices:

$$T_R(a_{ij}) = a_{i+1,j}, \quad T_R(b_{ij}) = b_{i+1,j}, \quad T_R(p_{ij}) = p_{i+1,j+1}, \quad T_R(q_{kl}) = q_{kl},$$

$$(B.21) \quad T_C(a_{ij}) = a_{i,j+1}, \quad T_C(b_{ij}) = b_{i,j+1}, \quad T_C(p_{ij}) = p_{ij}, \quad T_C(q_{kl}) = q_{k+1,l+1}.$$

Thanks to the bilinear identities, one can prove

$$(B.22) \qquad \tau_m = \frac{\displaystyle\prod_{0 \le i < j \le m} p_{ji} \prod_{0 \le k \le l < m} q_{kl}}{\displaystyle\prod_{i=0}^{m} \prod_{k=0}^{m-1} b_{i,k}} \qquad (m = 0, 1, \ldots, N)$$

by the induction on $m$.

We remark that Lemma B.1 is the case where $a_{ik} = \alpha_k x_i + \beta_k$, $b_{ik} = \gamma_k x_i + \delta_k$. Since

$$(B.23) \quad (\alpha_k x_i + \beta_k)(\gamma_l x_j + \delta_l) - (\alpha_k x_j + \beta_k)(\gamma_l x_i + \delta_l) = (x_i - x_j)(\alpha_k \delta_l - \beta_k \gamma_l)$$

the factorization condition (B.16) is verified with $p_{ij} = x_i - x_j$ and $q_{kl} = \alpha_k \delta_l - \beta_k \gamma_l$. Lemma B.2 is the case where $a_{ik} = [a_k \pm x_i]$, $b_{ik} = [b_k \pm x_i]$. Since

$$(B.24) \qquad [a_k \pm x_i][b_l \pm x_j] - [a_k \pm x_j][b_l \pm x_i] = [a_k \pm b_l][x_i \pm x_j]$$

the condition (B.16) is satisfied with $p_{ij} = [x_i \pm x_j]$ and $q_{kl} = [a_k \pm b_l]$. One can prove in fact that *generic* solutions to the system of equations (B.16) reduce to the case of (B.23). It would be worthwhile, however, to recognize the role of bilinear equations (B.21) which lead to the factorization of determinants.

## References

[1] A. G. Akritas, E. K. Akritas, and G. I. Malaschonok, *Various proofs of Sylvester's (determinant) identity*, Math. Comput. Simulation **42** (1996), no. 4-6, 585–593, DOI 10.1016/S0378-4754(96)00035-3. Symbolic computation, new trends and developments (Lille, 1993). MR1430843 (98c:15021)

[2] M. Gasca, A. López-Carmona, and V. Ramírez, *A generalized Sylvester's identity on determinants and its applications to interpolation problems*, Multivariate approximation theory, II (Oberwolfach, 1982), Internat. Ser. Numer. Math., vol. 61, Birkhäuser, Basel-Boston, Mass., 1982, pp. 171–184. MR719906 (84k:41005)

[3] G. Gasper and M. Rahman, *Basic hypergeometric series*, 2nd ed., Encyclopedia of Mathematics and its Applications, vol. 96, Cambridge University Press, Cambridge, 2004. With a foreword by Richard Askey. MR2128719 (2006d:33028)

[4] Y. Kajihara and M. Noumi, *Multiple elliptic hypergeometric series. An approach from the Cauchy determinant*, Indag. Math. (N.S.) **14** (2003), no. 3-4, 395–421, DOI 10.1016/S0019-3577(03)90054-1. MR2083083 (2005g:33032)

[5] C. Krattenthaler, *Advanced determinant calculus*, Sém. Lothar. Combin. **42** (1999), Art. B42q, 67 pp. (electronic). The Andrews Festschrift (Maratea, 1998). MR1701596 (2002i:05013)

[6] C. Krattenthaler, *Advanced determinant calculus: a complement*, Linear Algebra Appl. **411** (2005), 68–166, DOI 10.1016/j.laa.2005.06.042. MR2178686 (2006g:05022)

[7] Y. L. Luke, *Mathematical functions and their approximations*, Academic Press, Inc. [Harcourt Brace Jovanovich, Publishers], New York-London, 1975. MR0501762 (58 #19039)

[8] Y. L. Luke, *Algorithms for the computation of mathematical functions*, Academic Press [Harcourt Brace Jovanovich, Publishers], New York-London, 1977. MR0494840 (58 #13624)

[9] G. Mühlbach and M. Gasca, *A generalization of Sylvester's identity on determinants and some applications*, Linear Algebra Appl. **66** (1985), 221–234, DOI 10.1016/0024-3795(85)90134-X. MR781303 (86h:15006)

[10] J.-M. Normand, *Calculation of some determinants using the s-shifted factorial*, J. Phys. A **37** (2004), no. 22, 5737–5762, DOI 10.1088/0305-4470/37/22/003. MR2066627 (2005h:33004)

[11] M. Noumi, S. Tsujimoto, and Y. Yamada, *Padé interpolation for elliptic Painlevé equation*, Symmetries, integrable systems and representations, Springer Proc. Math. Stat., vol. 40, Springer, Heidelberg, 2013, pp. 463–482, DOI 10.1007/978-1-4471-4863-0_18. MR3077695

[12] H. Rosengren and M. Schlosser, *Summations and transformations for multiple basic and elliptic hypergeometric series by determinant evaluations*, Indag. Math. (N.S.) **14** (2003), no. 3-4, 483–513, DOI 10.1016/S0019-3577(03)90058-9. MR2083087 (2005f:33033)

[13] P. E. Spicer, F. W. Nijhoff, and P. H. van der Kamp, *Higher analogues of the discrete-time Toda equation and the quotient-difference algorithm*, Nonlinearity **24** (2011), no. 8, 2229–2263, DOI 10.1088/0951-7715/24/8/006. MR2813585 (2012e:37156)

[14] V. P. Spiridonov and A. S. Zhedanov, *Elliptic grids, rational functions, and the Padé interpolation*, Ramanujan J. **13** (2007), no. 1-3, 285–310, DOI 10.1007/s11139-006-0253-1. MR2281167 (2007h:42031)

[15] S. O. Warnaar, *Summation and transformation formulas for elliptic hypergeometric series*, Constr. Approx. **18** (2002), no. 4, 479–502, DOI 10.1007/s00365-002-0501-6. MR1920282 (2003h:33018)

[16] Y. Yamada, *Padé method to Painlevé equations*, Funkcial. Ekvac. **52** (2009), no. 1, 83–92, DOI 10.1619/fesi.52.83. MR2538280 (2010f:34180)

[17] Y. Yamada, *A Lax formalism for the elliptic difference Painlevé equation*, SIGMA Symmetry Integrability Geom. Methods Appl. **5** (2009), Paper 042, 15, DOI 10.3842/SIGMA.2009.042. MR2506170 (2010h:37164)

DEPARTMENT OF MATHEMATICS, KOBE UNIVERSITY
*E-mail address*: noumi@math.kobe-u.ac.jp

Contemporary Mathematics
Volume **651**, 2015
http://dx.doi.org/10.1090/conm/651/13037

# A $q$-analogue of the Drinfeld-Sokolov Hierarchy of Type $A$ and $q$-Painlevé System

Takao Suzuki

ABSTRACT. In this article we derive a higher order generalization of the $q$-Painlevé VI equation from a $q$-analogue of the Drinfeld-Sokolov (DS) hierarchy of type $A$. We also discuss its particular solution in terms of the $q$-hypergeometric function $_n\phi_{n-1}$.

## 1. Introduction

A relationship between Painlevé systems and infinite-dimensional integrable hierarchies has been studied; e.g. [**4**, **5**, **12**, **16**, **22**]. In a recent work [**6**, **20**] we derive a coupled Painlevé VI system, which admits a particular solution in terms of the hypergeometric function $_nF_{n-1}$ [**19**, **23**], from the DS hierarchy of type $A$. In this article we give a $q$-difference analogue of the above result. Namely, we consider a $q$-analogue of the DS hierarchy of type $A$ and derive a $q$-Painlevé system which admits a $q$-hypergeometric solution.

The DS hierarchies are extensions of the KdV hierarchy for affine Lie algebras [**2**, **3**, **8**]. For type $A_{N-1}^{(1)}$, these hierarchies are characterized by partitions of a natural number $N$. In this article we give an explicit formula of a $q$-DS hierarchy corresponding to the partition $(n, \ldots, n)$ of $mn$. Note that its formulation is based on two preceding works: [**15**], equivalent to the case $m = 1$, and [**11**], equivalent to the case $n = 1$.

The coupled Painlevé VI system given in [**20**] (or [**24**]) is derived by a similarity reduction from the DS hierarchy corresponding to the partition $(n, n)$. Therefore, in this article, we investigate a similarity reduction of the $q$-DS hierarchy corresponding to the partition $(n, n)$; we denote it by $q$-$P_{(n,n)}$. The system $q$-$P_{(n,n)}$ can be regarded as a higher order generalization of the $q$-Painlevé VI equation. In fact, the system $q$-$P_{(2,2)}$ reduces to the $q$-Painlevé VI equation given in [**10**].

The $q$-Painlevé VI equation is a system of $q$-difference equations

(1.1)
$$\frac{x(t)x(q^{-1}t)}{\alpha_3\alpha_4} = \frac{(y(q^{-1}t) - t\beta_1)(y(q^{-1}t) - t\beta_2)}{(y(q^{-1}t) - \beta_3)(y(q^{-1}t) - \beta_4)},$$
$$\frac{y(t)y(q^{-1}t)}{\beta_3\beta_4} = \frac{(x(t) - t\alpha_1)(x(t) - t\alpha_2)}{(x(t) - \alpha_3)(x(t) - \alpha_4)},$$

---

2010 *Mathematics Subject Classification.* Primary 34M55; Secondary 37K10, 37K35, 39A13.

where the parameters satisfy

$$\frac{\beta_1\beta_2}{\beta_3\beta_4} = q^{-1}\frac{\alpha_1\alpha_2}{\alpha_3\alpha_4}.$$

It is the compatibility condition of a system of linear $q$-difference equations

(1.2)
$$Y(q^{-1}z, t) = \mathcal{A}(z, t)Y(z, t), \quad \mathcal{A}(z, t) = \mathcal{A}_0(t) + \mathcal{A}_1(t)z + \mathcal{A}_2(t)z^2,$$

$$Y(z, q^{-1}t) = \frac{z(zI + \mathcal{B}_0(t))}{(z - q^{-1}t\alpha_1)(z - q^{-1}t\alpha_2)}Y(z, t),$$

where the coefficient matrices $\mathcal{A}_i(t)$ $(i = 0, 1, 2)$ satisfy

$$\mathcal{A}_2(t) = \begin{bmatrix} \kappa_1 & 0 \\ 0 & \kappa_2 \end{bmatrix}, \quad \mathcal{A}_0(t) \text{ has eigenvalues } t\theta_1, t\theta_2,$$

$$\det \mathcal{A}(z, t) = \kappa_1\kappa_2(z - t\alpha_1)(z - t\alpha_2)(z - \alpha_3)(z - \alpha_4),$$

and

$$\kappa_1 = \frac{q}{\beta_3}, \quad \kappa_2 = \frac{1}{\beta_4}, \quad \theta_1 = \frac{\alpha_1\alpha_2}{\beta_1}, \quad \theta_2 = \frac{\alpha_1\alpha_2}{\beta_2}.$$

The dependent variables $x(t)$ and $y(t)$ are given by

$$x(t) = -\frac{(\mathcal{A}_0(t))_{12}}{(\mathcal{A}_1(t))_{12}}, \quad y(t) = \frac{q(x(t) - t\alpha_1)(x(t) - t\alpha_2)}{(\mathcal{A}(x(t), t))_{11}}.$$

In this article we derive the system (1.1) from $q$-$P_{(2,2)}$ with the aid of a $q$-Laplace transformation for a system of linear $q$-difference equations.

REMARK 1.1. The method used in this article was suggestd by the derivation of the $q$-Painlevé VI equation (1.1) from the $q$-$\mathfrak{gl}_3$ hierarchy via the $q$-Laplace transformation in [11]. Note also that (1.1) was derived from $q$-UC hierarchy in [25].

The $q$-hypergeometric function ${}_n\phi_{n-1}$ is defined by the power series

$${}_n\phi_{n-1}\left[\begin{matrix} \alpha_1, \ldots, \alpha_{n-1}, \alpha_n \\ \beta_1, \ldots, \beta_{n-1} \end{matrix}; q, t\right] = \sum_{k=0}^{\infty}\frac{(\alpha_1; q)_k \ldots (\alpha_{n-1}; q)_k(\alpha_n; q)_k}{(\beta_1; q)_k \ldots (\beta_{n-1}; q)_k(q; q)_k}t^k,$$

where $(\alpha; q)_k$ stands for the $q$-shifted factorial

$$(\alpha; q)_0 = 1, \quad (\alpha; q)_k = (1 - \alpha)(1 - q\alpha) \ldots (1 - q^{k-1}\alpha) \quad (k \geq 1).$$

Note that it reduces to the generalized hypergeometric function ${}_nF_{n-1}$ in the continuous limit $q \to 1$. We see that $x(t) = {}_n\phi_{n-1}$ satisfies an $n$-th order linear $q$-difference equation

$$\Big((1 - q^{-1}\beta_1 T_{q,t}) \ldots (1 - q^{-1}\beta_{n-1}T_{q,t})(1 - T_{q,t})$$

$$- t(1 - \alpha_1 T_{q,t}) \ldots (1 - \alpha_{n-1}T_{q,t})(1 - \alpha_n T_{q,t})\Big)x(t) = 0,$$

where $T_{q,t}$ stands for a $q$-shift operator, $T_{q,t}x(t) = x(qt)$. In this article, we show that the system $q$-$P_{(n,n)}$ admits a particular solution in terms of the $q$-hypergeometric function ${}_n\phi_{n-1}$.

This article is organized as follows. In Section 2, we formulate a $q$-DS hierarchy corresponding to the partition $(n, \ldots, n)$ of $mn$ and consider its similarity reduction. In Section 3, an explicit formula of $q$-$P_{(n,n)}$ is presented. We also discuss a group of symmetries and a Lax pair for $q$-$P_{(n,n)}$. In Section 4, the $q$-Painlevé VI equation is derived from $q$-$P_{(2,2)}$. In Section 5, we show that $q$-$P_{(n,n)}$ admits a particular solution in terms of $_n\phi_{n-1}$.

## 2. $q$-DS hierarchy

In this section we formulate a $q$-DS hierarchy corresponding to the partition $(n, \ldots, n)$ of $mn$ and consider its similarity reduction.

Let $\lambda$ be a grading parameter and $T_\lambda$ be the corresponding $q$-shift operator, $T_\lambda(\lambda) = q\lambda$. We define $mn \times mn$ matrices $e_j, f_j, h_j$ $(j = 1, \ldots, mn)$ by

$$
\begin{aligned}
e_j &= E_{j,j+1} & f_j &= E_{j+1,j} & h_j &= E_{j,j} & (j \notin m\mathbb{Z}), \\
e_j &= \lambda E_{j,j+1} & f_j &= \lambda^{-1} E_{j+1,j} & h_j &= E_{j,j} & (j \in m\mathbb{Z}; j \neq mn), \\
e_{mn} &= \lambda E_{mn,1} & f_{mn} &= \lambda^{-1} E_{1,mn} & h_{mn} &= E_{mn,mn},
\end{aligned}
$$

where $E_{i,j}$ stands for the matrix with 1 in the $(i,j)$-th entry and zeros elsewhere. We also set

$$
e_{j,k} = e_j e_{j+1} \cdots e_{j+k-1}, \quad f_{j,k} = f_{j+k-1} \cdots f_{j+1} f_j,
$$

where $e_{j+mn} = e_j$ and $f_{j+mn} = f_j$. Note that

$$
T_\lambda(e_{j,m}) = q e_{j,m}, \quad T_\lambda(f_{j,m}) = q^{-1} f_{j,m} \quad (j = 1, \ldots, mn).
$$

Let us consider a set of matrices $\{\Lambda_1, \ldots, \Lambda_m\}$ such that

$$
\Lambda_i \Lambda_j = \Lambda_j \Lambda_i = 0, \quad T_\lambda(\Lambda_i) = q\Lambda_i \quad (i, j = 1, \ldots, m; i \neq j).
$$

Explicitly, we can take

$$
\Lambda_i = \sum_{j=1}^{n} e_{i+(j-1)m,m} \quad (i = 1, \ldots, m).
$$

Note that this choice of matrices is suggested by the previous work [1, 13, 20]. In the following, we formulate a $q$-DS hierarchy by using such matrices.

Let $t_i$ and $T_i$ $(i = 1, \ldots, m)$ be independent variables and corresponding $q$-shift operators such that

$$
T_i(t_i) = q t_i, \quad T_i(t_j) = t_j \quad (j \neq i).
$$

Assuming $|q| > 1$, we consider a $q$-Sato equation

$$
(2.1) \qquad T_i(Z)Z^{-1} = T_i(W)(I - \varepsilon t_i \Lambda_i)W^{-1} \quad (i = 1, \ldots, m),
$$

with matrices of dependent variables

$$
Z = \sum_{j=1}^{mn} z_{j,0} h_j + \sum_{j=1}^{mn} \sum_{k=1}^{\infty} z_{j,k} e_{j,k}, \quad W = I + \sum_{j=1}^{mn} \sum_{k=1}^{\infty} w_{j,k} f_{j,k},
$$

where $I$ stands for the identity matrix and $\varepsilon = 1 - q$. We also assume that

$$
(2.2) \qquad z_{1,0} z_{2,0} \cdots z_{mn,0} = 1.
$$

If we consider the system (2.1) and set

$$
\Psi = W \prod_{i=1}^{m} \prod_{k=0}^{\infty} (I - q^{-k-1} \varepsilon t_i \Lambda_i),
$$

then we obtain a system of linear $q$-difference equations called *a Lax pair*

$$
(2.3) \qquad T_i(\Psi) = B_i \Psi \quad (i = 1, \ldots, m),
$$

where

$$B_i = T_i(W)(I - \varepsilon t_i \Lambda_i) W^{-1}.$$

The compatibility condition of system (2.3) is equivalent to

(2.4) $$T_i(B_j)B_i = T_j(B_i)B_j \quad (i, j = 1, \dots, m).$$

We call the system of $q$-difference equations (2.4) a $q$-DS hierarchy.

Next, we formulate a similarity reduction of the $q$-DS hierarchy. We start from system (2.4) and consider the following equation:

(2.5) $$T_\lambda(W) = q^\rho T_{1,\dots,m}(W)q^{-\rho}, \quad \rho = \sum_{i=1}^{m} \sum_{j=1}^{n} \rho_i h_{i+(j-1)m},$$

where $T_{i,\dots,m} = T_i T_{i+1} \dots T_m$ and $\rho_1 + \dots + \rho_m = 0$. Note that

$$\rho \Lambda_i = \Lambda_i \rho \quad (i = 1, \dots, m).$$

We set

$$\Psi = W \left( \prod_{i=1}^{m} \prod_{k=0}^{\infty} (I - q^{-k-1} \varepsilon t_i \Lambda_i) \right) \lambda^\rho.$$

Then we obtain a Lax pair

(2.6) $$T_\lambda(\Psi) = M\Psi, \quad T_i(\Psi) = B_i \Psi \quad (i = 1, \dots, m),$$

where

$$M = q^\rho T_{2,\dots,m}(B_1) T_{3,\dots,m}(B_2) \dots T_m(B_{m-1}) B_m.$$

Note that the matrix $M$ is also given by

$$M = T_\lambda(W) q^\rho (1 - \varepsilon t_1 \Lambda_1)(1 - \varepsilon t_2 \Lambda_2) \dots (1 - \varepsilon t_m \Lambda_m) W^{-1}.$$

The compatibility condition of system (2.6) is equivalent to

(2.7) $$T_i(M)B_i = T_\lambda(B_i)M, \quad T_i(B_j)B_i = T_j(B_i)B_j \quad (i, j = 1, \dots, m).$$

We call the system of $q$-difference equations (2.7) a similarity reduction of the $q$-DS hierarchy.

REMARK 2.1. Replacing $M \to I - \varepsilon M$ and $B_i \to I - \varepsilon t_i B_i$, we can rewrite the system (2.7) as

$$[D_\lambda - M, D_i - B_i] = 0, \quad [D_i - B_i, D_j - B_j] = 0 \quad (i, j = 1, \dots, m),$$

where

$$D_\lambda = \frac{1 - T_\lambda}{\varepsilon \lambda}, \quad D_i = \frac{1 - T_i}{\varepsilon t_i}.$$

In the continuous limit $q \to 1$ it reduces to the similarity reduction of the DS hierarchy of type $A_{mn-1}^{(1)}$ given in [20].

## 3. Higher order $q$-Painlevé system

In this section we present an explicit formula of the similarity reduction (2.7) corresponding to the partition $(n, n)$, where $n \geq 2$, in terms of dependent variables. We also discuss a group of symmetries and a Lax pair for $q$-$P_{(n,n)}$.

Consider the Lax pair

$$(3.1) \qquad T_\lambda(\Psi) = M\Psi, \quad T_i(\Psi) = B_i\Psi \quad (i = 1, 2),$$

with $2n \times 2n$ matrices

$$M = T_\lambda(W)q^\rho(1 - \varepsilon t_1\Lambda_1)(1 - \varepsilon t_2\Lambda_2)W^{-1},$$
$$B_i = T_i(W)(1 - \varepsilon t_i\Lambda_i)W^{-1} \quad (i = 1, 2),$$

where

$$\Lambda_1 = \sum_{j=1}^n e_{2j-1,2}, \quad \Lambda_2 = \sum_{j=1}^n e_{2j,2}, \quad \rho = \sum_{j=1}^n \rho_1(h_{2j-1} - h_{2j}),$$

and

$$W = I + \sum_{j=1}^{2n}\sum_{k=1}^\infty w_{j,k}f_{j,k}.$$

Then the compatibility condition of system (3.1) is equivalent to the similarity reduction

$$(3.2) \qquad T_1(B_2)B_1 = T_2(B_1)B_2, \quad T_i(M)B_i = T_\lambda(B_i)M \quad (i = 1, 2).$$

The matrices $B_1$, $B_2$ and $M$ can be expressed in terms of dependent variables $w_j = w_{j,1}$ and parameters $\kappa_j$ $(j = 1, \ldots, 2n)$. Equation (2.5) implies

$$T_{1,2}(w_{2j-1}) = q^{2\rho_1}w_{2j-1}, \quad T_{1,2}(w_{2j}) = q^{-2\rho_1-1}w_{2j} \quad (j = 1, \ldots, n),$$

from which we obtain

$$M = \sum_{j=1}^{2n}(1 - \varepsilon\kappa_j)h_j + \sum_{j=1}^n(q^{\rho_1}\varepsilon t_1 w_{2j} - q^{-\rho_1-1}\varepsilon t_2 w_{2j-2})e_{2j-1}$$

$$+ \sum_{j=1}^n(q^{-\rho_1}\varepsilon t_2 w_{2j+1} - q^{\rho_1}\varepsilon t_1 w_{2j-1})e_{2j} - q^{\rho_1}\varepsilon t_1\Lambda_1 - q^{-\rho_1}\varepsilon t_2\Lambda_2,$$

and

$$B_i = \sum_{j=1}^{2n}u_{i,j}h_j + \sum_{j=1}^{2n}v_{i,j}e_j - \varepsilon t_i\Lambda_i \quad (i = 1, 2),$$

where $w_0 = w_{2n}$, $w_{2n+1} = w_1$ and

$$u_{1,2j-1} = \frac{q^{-\rho_1}(1 - \varepsilon\kappa_{2j-1})}{1 + q^{-2\rho_1-1}\varepsilon t_2 w_{2j-2}T_1(w_{2j-1})}, \quad u_{1,2j} = 1 + \varepsilon t_1 T_1(w_{2j-1})w_{2j},$$

$$v_{1,2j-1} = \varepsilon t_1 w_{2j}, \quad v_{1,2j} = -\varepsilon t_1 T_1(w_{2j-1}),$$

$$u_{2,2j-1} = 1 + \varepsilon t_2 T_2(w_{2j-2})w_{2j-1}, \quad u_{2,2j} = \frac{q^{\rho_1}(1 - \varepsilon\kappa_{2j})}{1 + q^{2\rho_1}\varepsilon t_1 w_{2j-1}T_2(w_{2j})},$$

$$v_{2,2j-1} = -\varepsilon t_2 T_2(w_{2j-2}), \quad v_{2,2j} = \varepsilon t_2 w_{2j+1}.$$

Note that equation (2.2) implies

$$\prod_{j=1}^{2n}(1-\varepsilon\kappa_j)=1, \quad \prod_{j=1}^{2n}u_{i,j}=1 \quad (i=1,2).$$

We now assume that $t_2 = 1$. We also consider a change of variables and parameters

$$p=q^n, \quad t=q^{2n\rho_1}t_1^n, \quad z=q^{-n(4\rho_1-n+1)/2}(\epsilon\lambda)^n,$$

$$a_j=q^{-\rho_1+j-1}(1-\varepsilon\kappa_{2j-1}), \quad b_j=q^{-\rho_1+j-1}(1-\varepsilon\kappa_{2j}),$$

$$x_j(t)=q^{-2(j-1)\rho_1}t_1^{-j+1}w_{2j-1}, \quad y_j(t)=q^{(j-1)(2\rho_1+1)}t_1^j\varepsilon w_{2j},$$

for $j=1,\ldots,n$. Replacing $p \to q$, we can describe the similarity reduction (3.2) as follows.

THEOREM 3.1. *The dependent variables $x_j(t)$, $y_j(t)$ $(j=1,\ldots,n)$ satisfy a system of $q$-difference equations*

(3.3)
$$x_j(t)-x_{j-1}(t)=\frac{a_j x_j(qt)}{1+x_j(qt)y_{j-1}(t)}-\frac{b_{j-1}x_{j-1}(qt)}{1+x_{j-1}(qt)y_{j-1}(t)},$$

$$y_j(qt)-y_{j-1}(qt)=\frac{b_j y_j(t)}{1+x_j(qt)y_j(t)}-\frac{a_j y_{j-1}(t)}{1+x_j(qt)y_{j-1}(t)},$$

*for $j=1,\ldots,n$, where*

$$b_0=q^{-1}b_n, \quad x_0(t)=tx_n(t), \quad y_0(t)=q^{-1}t^{-1}y_n(t),$$

*with relations*

$$\prod_{j=1}^n a_j\frac{1+x_j(qt)y_j(t)}{1+x_j(qt)y_{j-1}(t)}=q^{(n-1)/2}.$$

We denote the $q$-Painlevé system (3.3) by $q\text{-}P_{(n,n)}$. As is seen in the next section, $q\text{-}P_{(2,2)}$ reduces to the $q$-Painlevé VI equation (1.1). Note that $q\text{-}P_{(n,n)}$ reduces to the coupled Painlevé VI system given in [20, 24] in the continuous limit $q \to 1$.

REMARK 3.2. In the previous work [21], the higher order generalizations of the $q$-Painlevé VI equation were presented. Its relationship to $q\text{-}P_{(n,n)}$ has not been clarified yet. However, we conjecture that $q\text{-}P_{(n,n)}$ coincides with the $q$-Painlevé equation of [21] with $(n,n)$ periodicity and $(|I|,|J|)=(2,2)$.

The system $q\text{-}P_{(n,n)}$ admits the affine Weyl group symmetry of type $A_{2n-1}^{(1)}$. We describe its action on the dependent variables and parameters. Recall that an extended affine Weyl group $\widetilde{W}(A_{2n-1}^{(1)})$ is generated by the transformations $r_0,\ldots,r_{2n-1}$ and $\pi$ with the fundamental relations

$$r_i^2=1, \quad (r_ir_j)^{2-a_{i,j}}=1 \quad (i,j=0,\ldots,2n-1;i\neq j),$$

$$\pi^{2n}=1, \quad \pi r_i=r_{i+1}\pi, \quad \pi r_{2n-1}=r_0\pi \quad (i=0,\ldots,2n-2),$$

where

$$a_{i,i}=2 \qquad\qquad\qquad\qquad (i=0,\ldots,2n-1),$$

$$a_{i,i+1}=a_{2n-1,0}=a_{i+1,i}=a_{0,2n-1}=-1 \quad (i=0,\ldots,2n-2),$$

$$a_{i,j}=0 \qquad\qquad\qquad\qquad\quad (\text{otherwise}).$$

THEOREM 3.3. *Let $r_j$ $(j = 0, \ldots, 2n-1)$ be birational transformations defined by*

$$r_{2j-2}(a_j) = b_{j-1}, \quad r_{2j-2}(b_{j-1}) = a_j,$$

$$r_{2j-2}(x_{j-1}(t)) = x_{j-1}(t), \quad r_{2j-2}(y_{j-1}(t)) = y_{j-1}(t) + \frac{b_{j-1} - a_j}{x_j(t) - x_{j-1}(t)},$$

$$r_{2j-2}(a_i) = a_i, \quad r_{2j-2}(b_{i-1}) = b_{i-1},$$

$$r_{2j-2}(x_{i-1}(t)) = x_{i-1}(t), \quad r_{2j-2}(y_{i-1}(t)) = y_{i-1}(t) \quad (i \neq j),$$

*and*

$$r_{2j-1}(a_j) = b_j \quad r_{2j-1}(b_j) = a_j,$$

$$r_{2j-1}(x_j(t)) = x_j(t) + \frac{a_j - b_j}{y_j(t) - y_{j-1}(t)}, \quad r_{2j-1}(y_j(t)) = y_j(t),$$

$$r_{2j-1}(a_i) = a_i, \quad r_{2j-1}(b_i) = b_i,$$

$$r_{2j-1}(x_i(t)) = x_i(t), \quad r_{2j-1}(y_i(t)) = y_i(t) \quad (i \neq j).$$

*Also, let $\pi$ be a birational transformation defined by*

$$a_i \to b_i, \quad b_i \to a_{i+1}, \quad a_n \to b_n, \quad b_n \to qa_1,$$

$$x_i(t) \to y_i(qt), \quad y_i(t) \to x_{i+1}(qt),$$

$$x_n(t) \to y_n(qt), \quad y_n(t) \to \frac{x_1(qt)}{qt}, \quad t \to \frac{1}{q^2 t} \quad (i \neq n).$$

*Then $q$-$P_{(n,n)}$ is invariant under actions of those transformations. Furthermore, the group of symmetries $\langle r_0, \ldots, r_{2n+1}, \pi \rangle$ is isomorphic to an extended affine Weyl group $\widetilde{W}(A_{2n+1}^{(1)})$.*

We rewrite the Lax pair (3.1) into a simpler form. Consider a gauge transformation

$$\widetilde{\Psi}(z, t) = \lambda^{-\rho_1} \sum_{j=1}^{n} q^{(j-1)(j-2)/2} \left( (\varepsilon t_1 \lambda)^{j-1} h_{2j-1} + (q^{-2\rho_1} \varepsilon \lambda)^{j-1} h_{2j} \right) \Psi.$$

Then we obtain a Lax pair for $q$-$P_{(n,n)}$,

(3.4)          $$\widetilde{\Psi}(qz, t) = \widetilde{M}(z, t)\widetilde{\Psi}(z, t), \quad \widetilde{\Psi}(z, qt) = \widetilde{B}(z, t)\widetilde{\Psi}(z, t),$$

with $2n \times 2n$ matrices

$$\widetilde{M}(z, t) = \sum_{j=1}^{n} a_j E_{2j-1, 2j-1} + \sum_{j=1}^{n} b_j E_{2j, 2j} + (y_1(t) - q^{-1}t^{-1}y_n(t))E_{1,2}$$

$$+ \sum_{j=1}^{n-1} (x_{j+1}(t) - x_j(t))E_{2j, 2j+1} + \sum_{j=2}^{n} (y_j(t) - y_{j-1}(t))E_{2j-1, 2j}$$

$$+ (x_1(t) - tx_n(t))zE_{2n,1} - \sum_{j=1}^{2n-2} E_{j, j+2} - tzE_{2n-1,1} - zE_{2n,2},$$

and

$$\widetilde{B}(z,t) = \sum_{j=1}^{n} \frac{a_j}{1 + x_j(qt)y_{j-1}(t)} E_{2j-1,2j-1} + \sum_{j=1}^{n}(1 + x_j(qt)y_j(t))E_{2j,2j}$$

$$+ \sum_{j=1}^{n} y_j(t)E_{2j-1,2j} - \sum_{j=1}^{n-1} x_j(qt)E_{2j,2j+1} - tx_n(qt)zE_{2n,1}$$

$$- \sum_{j=1}^{n-1} E_{2j-1,2j+1} - tzE_{2n-1,1}.$$

The group of symmetries defined above is derived from transformations for the Lax pair (3.4)

$$r_j(\widetilde{\Psi}(z,t)) = R_i(z,t)\widetilde{\Psi}(z,t) \quad (j = 0, \ldots, 2n-1),$$

where

$$R_0(z,t) = I + \frac{b_n - qa_1}{x_1(t) - tx_n(t)} z^{-1} E_{1,2n},$$

$$R_{2j-2}(z,t) = I + \frac{b_{j-1} - a_j}{x_j(t) - x_{j-1}(t)} E_{2j-1,2j-2} \quad (j = 2, \ldots, n),$$

$$R_{2j-1}(z,t) = I + \frac{a_j - b_j}{y_j(t) - y_{j-1}(t)} E_{2j,2j-1} \quad (j = 1, \ldots, n).$$

Note that a construction of those transformations is suggested by the previous works [14, 17].

## 4. $q$-Painlevé VI equation

The system $q$-$P_{(2,2)}$ is given as the compatibility condition of the Lax pair

(4.1)        $$\Psi_4(qz,t) = M_4(z,t)\Psi_4(z,t), \quad \Psi_4(z,qt) = B_4(z,t)\Psi_4(z,t),$$

where

$$M_4(z,t) = \begin{bmatrix} a_1 & y_1(t) - \frac{y_2(t)}{qt} & -1 & 0 \\ 0 & b_1 & x_2(t) - x_1(t) & -1 \\ -tz & 0 & a_2 & y_2(t) - y_1(t) \\ \{x_1(t) - tx_2(t)\}z & -z & 0 & b_2 \end{bmatrix},$$

$$B_4(z,t) = \begin{bmatrix} \frac{qta_1}{qt+x_1(qt)y_2(t)} & y_1(t) & -1 & 0 \\ 0 & 1 + x_1(qt)y_1(t) & -x_1(qt) & 0 \\ -tz & 0 & \frac{a_2}{1+x_2(qt)y_1(t)} & y_2(t) \\ -tx_2(qt)z & 0 & 0 & 1 + x_2(qt)y_2(t) \end{bmatrix}.$$

In this section we derive the system (1.2) from the Lax pair (4.1) with the aid of a $q$-Laplace transformation (cf. [9, 11]).

REMARK 4.1.  A $4 \times 4$ Lax pair of the $q$-Painlevé VI equation similar to (4.1) was already presented in [18].

First, we consider a gauge transformation

$$\Psi_4^*(z,t) = \tau_1(z,t)\Psi_4(z,t),$$

where $\tau_1(z,t)$ is a function such that

$$\frac{\tau_1(qz,t)}{\tau_1(z,t)} = \frac{1}{qa_1}, \quad \frac{\tau_1(z,qt)}{\tau_1(z,t)} = \frac{qt + x_1(qt)y_2(t)}{qta_1}.$$

Then the Lax pair (4.1) is transformed into

(4.2) $$\Psi_4^*(qz,t) = M_4^*(z,t)\Psi_4^*(z,t), \quad \Psi_4^*(z,qt) = B_4^*(z,t)\Psi_4^*(z,t),$$

where

$$M_4^*(z,t) = \frac{1}{qa_1} M_4(z,t), \quad B_4^*(z,t) = \frac{qt + x_1(qt)y_2(t)}{qta_1} B_4(z,t).$$

We set

$$M_4^*(z,t) = M_{4,0}^*(t) + z M_{4,1}^*(t), \quad B_4^*(z,t) = B_{4,0}^*(t) + z B_{4,1}^*(t).$$

Next, we consider a $q$-Laplace transformation

$$z\Psi_4^*(z) \to \frac{\Phi_4(\zeta) - \Phi_4(q^{-1}\zeta)}{\varepsilon\zeta}, \quad \Psi_4^*(qz) \to q^{-1}\Phi_4(q^{-1}\zeta),$$

where $\varepsilon = 1 - q$. Then the Lax pair (4.2) is transformed into

(4.3) $$\Phi_4(q^{-1}\zeta,t) = N_4(\zeta,t)\Phi_4(\zeta,t), \quad \Phi_4(\zeta,qt) = C_4(\zeta,t)\Phi_4(z,t).$$

where

$$N_4(\zeta,t) = \left(q^{-1}\varepsilon\zeta I + M_{4,1}^*(t)\right)^{-1}\left(\varepsilon\zeta M_{4,0}^*(t) + M_{4,1}^*(t)\right),$$
$$C_4(\zeta,t) = B_{4,0}^*(t) + \varepsilon^{-1}\zeta^{-1}B_{4,1}^*(t)\left(I - N_4(\zeta,t)\right).$$

Denoting $\zeta^{-1}$ by $z$, we can restrict the Lax pair (4.3) to the one with $3 \times 3$ matrices

(4.4) $$\Psi_3(qz,t) = M_3(z,t)\Psi_3(z,t), \quad \Psi_3(z,qt) = B_3(z,t)\Psi_3(z,t),$$

in view of the following lemma.

LEMMA 4.2. *For each of the matrices $N_4(\zeta,t)$ and $C_4(\zeta,t)$, the first column is equal to the fundamental vector ${}^t[1,0,0,0]$.*

REMARK 4.3. The Lax pair (4.4) coincides with the one for the $q$-$\mathfrak{gl}_3$ hierarchy given in [**11**].

In a similar way, we can reduce the Lax pair (4.4) to the one with $2 \times 2$ matrices. We consider a gauge transformation

$$\Psi_3^*(z,t) = \tau_2(z,t)\Psi_3(z,t),$$

where $\tau_2(z,t)$ is a function such that

$$\frac{\tau_2(qz,t)}{\tau_2(z,t)} = \frac{a_1}{qb_1}, \quad \frac{\tau_2(z,qt)}{\tau_2(z,t)} = \frac{qta_1}{(qt + x_1(qt)y_2(t))(1 + x_1(qt)y_1(t))}.$$

We also consider a $q$-Laplace transformation

$$z\Psi_3^*(z) \to \frac{\Phi_3(\zeta) - \Phi_3(q^{-1}\zeta)}{\varepsilon\zeta}, \quad \Psi_3^*(qz) \to q^{-1}\Phi_3(q^{-1}\zeta).$$

Then the Lax pair (4.4) is transformed into

(4.5) $$\Phi_3(q^{-1}\zeta,t) = N_3(\zeta,t)\Phi_3(\zeta,t), \quad \Phi_3(\zeta,qt) = C_3(\zeta,t)\Phi_3(z,t).$$

Denoting $\varepsilon^2 a_1 b_1 \zeta$ by $z$, we can restrict the Lax pair (4.5) to the one with $2 \times 2$ matrices

(4.6) $\qquad \Psi_2(q^{-1}z, t) = M_2(z, t)\Psi_2(z, t), \quad \Psi_2(z, qt) = B_2(z, t)\Psi_2(z, t),$

in view of the following lemma.

LEMMA 4.4. *For each of the matrices $N_3(\zeta, t)$ and $C_3(\zeta, t)$, the first column is equal to the fundamental vector ${}^t[1, 0, 0]$.*

The matrices $M_2(z, t)$ and $B_2(z, t)$ are of the form

$$M_2(z, t) = \frac{M_{2,0}(t) + zM_{2,1}(t) + z^2 M_{2,2}(t)}{(z - t)(z - 1)}, \quad B_2(z, t) = \frac{B_{2,0}(t) + zB_{2,1}(t)}{z - t},$$

where

$$M_{2,2}(t) = \frac{1}{b_1}\begin{bmatrix} a_2 & y_2(t) - y_1(t) \\ 0 & b_2 \end{bmatrix},$$

$$B_{2,1}(t) = \frac{1}{1 + x_1(qt)y_1(t)}\begin{bmatrix} \frac{a_2}{1 + x_2(qt)y_1(t)} & y_2(t) \\ 0 & 1 + x_2(qt)y_2(t) \end{bmatrix}.$$

In order to derive the system (1.2), we consider a gauge transformation

$$Y(z, t) = \frac{(q^{-1}tz^{-1}; q^{-1})_\infty (qz; q)_\infty^2}{(q^{-1}z^{-1}; q^{-1})_\infty}\begin{bmatrix} \tau_3(t) & 0 \\ 0 & \tau_4(t) \end{bmatrix}\begin{bmatrix} 1 & \frac{y_2(t) - y_1(t)}{a_2 - b_2} \\ 0 & 1 \end{bmatrix}\Psi_2(z, t).$$

where $\tau_3(z, t)$ and $\tau_4(z, t)$ are functions such that

$$\frac{\tau_3(q^{-1}z, t)}{\tau_3(z, t)} = b_1, \quad \frac{\tau_3(z, qt)}{\tau_3(z, t)} = \frac{(1 + x_1(qt)y_1(t))(1 + x_2(qt)y_1(t))}{a_2},$$

$$\frac{\tau_4(q^{-1}z, t)}{\tau_4(z, t)} = b_1, \quad \frac{\tau_4(z, qt)}{\tau_4(z, t)} = \frac{1 + x_1(qt)y_1(t)}{1 + x_2(qt)y_2(t)}.$$

Then the Lax pair (4.6) is transformed into

(4.7) $\qquad Y(q^{-1}z, t) = \widetilde{M}_2(z, t)Y(z, t), \quad Y(z, t) = \frac{\widetilde{B}_2(z, q^{-1}t)}{z}Y(z, q^{-1}t).$

Here the coefficient matrices satisfy

$$\widetilde{M}_2(z, t) = \widetilde{M}_{2,0}(t) + z\widetilde{M}_{2,1}(t) + z^2\widetilde{M}_{2,2},$$

$$\widetilde{M}_{2,2} = \begin{bmatrix} a_2 & 0 \\ 0 & b_2 \end{bmatrix}, \quad \widetilde{M}_{2,0}(t) \text{ has eigenvalues } ta_1, tb_1,$$

$$\det \widetilde{M}_2(z, t) = a_2 b_2(z - t)(z - a_1 b_1 q^{-1/2}t)(z - 1)(z - a_2^{-1}b_2^{-1}q^{1/2}),$$

and

$$\widetilde{B}_2(z, q^{-1}t) = \widetilde{B}_{2,0}(q^{-1}t) + zI_2,$$

$$\det \widetilde{B}_2(z, q^{-1}t) = (z - q^{-1}t)(z - q^{-1}a_1 b_1 q^{-1/2}t).$$

We can show this by direct computations; we omit the details here. By setting

$$\alpha_1 = 1, \quad \alpha_2 = \frac{a_1 b_1}{q^{1/2}}, \quad \alpha_3 = 1, \quad \alpha_4 = \frac{q^{1/2}}{a_2 b_2},$$

$$\beta_1 = \frac{b_1}{q^{1/2}}, \quad \beta_2 = \frac{a_1}{q^{1/2}}, \quad \beta_3 = \frac{q}{a_2}, \quad \beta_4 = \frac{1}{b_2},$$

we arrive at

THEOREM 4.5. *The matrix-valued function $Y(z,t)$ solves the system* (1.2).

COROLLARY 4.6. *If, in the system $q$-$P_{(2,2)}$, we set*

$$x(t) = \frac{t(x_2(t) - x_1(t))\xi_1(t)}{\xi_2(t)}, \quad y(t) = \frac{x_2(qt)(qt + x_1(qt)y_2(t))\psi_1(t)}{(1 + x_2(qt)y_2(t))\psi_2(t)},$$

*where*

$$\xi_1(t) = qtx_1(t)y_1(t) - x_1(t)y_2(t) - qtx_2(t)y_1(t) + x_2(t)y_2(t) - (b_1 - a_1)qt,$$

$$\xi_2(t) = (tx_2(t) - x_1(t))(x_2(t) - x_1(t))(y_2(t) - qty_1(t))$$
$$+ (b_1 - a_1)qtx_1(t) + \{(a_2 - b_1)t - (a_2 - a_1)\}qtx_2(t),$$

$$\psi_1(t) = (1 - a_1b_1q^{1/2}t)x_2(qt)y_2(t) + qt - a_1b_1q^{1/2}t,$$

$$\psi_2(t) = a_2(1 - a_1b_1q^{1/2}t)x_1(qt)x_2(qt)y_2(t)$$
$$+ a_1(qt - a_2b_1q^{1/2}t)x_1(qt) + (a_2 - a_1)qtx_2(qt),$$

*then those variables satisfy the $q$-Painlevé VI equation* (1.1).

Note that $q$-$P_{(2,2)}$ gives explicit formulas for $x_1(qt)$ and $x_2(qt)$,

$$x_1(qt) = \frac{-\xi_3(t)}{\xi_3(t)y_1(t) + (qt - a_1b_1q^{1/2}t)y_1(t) - (1 - a_1b_1q^{1/2}t)y_2(t)},$$

$$x_2(qt) = \frac{-\xi_4(t)}{\xi_4(t)y_1(t) + a_2(qt - a_1b_1q^{1/2}t)y_1(t) - a_2(1 - a_1b_1q^{1/2}t)y_2(t)},$$

where

$$\xi_3(t) = a_1q^{1/2}t(x_2(t) - x_1(t))(y_2(t) - y_1(t)) - qt + a_1a_2q^{1/2}t,$$

$$\xi_4(t) = (x_2(t) - x_1(t))(y_2(t) - qty_1(t)) - b_1(qt - a_1a_2q^{1/2}t).$$

## 5. $q$-Hypergeometric function $_n\phi_{n-1}$

In this section we show that $q$-$P_{(n,n)}$ admits a particular solution in terms of the $q$-hypergeometric function $_n\phi_{n-1}$.

PROPOSITION 5.1. *If, in the system $q$-$P_{(n,n)}$, we consider a specialization*

$$y_j(t) = 0 \quad (j = 1, \ldots, n), \quad \prod_{j=1}^{n} a_j = q^{(n-1)/2},$$

*then a vector of the variables $\mathbf{x}(t) = {}^t[x_1(t), \ldots, x_n(t)]$ satisfies a system of linear $q$-difference equations*

(5.1) $$\mathbf{x}(q^{-1}t) = \left(A_0 + \frac{A_1}{1 - q^{-1}t}\right)\mathbf{x}(t),$$

*with $n \times n$ matrices*

$$A_0 = \sum_{j=1}^{n} b_j E_{j,j} + \sum_{i=1}^{n} \sum_{j=i+1}^{n} (b_j - a_j)E_{i,j}, \quad A_1 = \sum_{i=1}^{n} \sum_{j=1}^{n} (a_j - b_j)E_{i,j}.$$

We always assume that

$$a_j \notin \mathbb{Z}, \quad a_i - a_j \notin \mathbb{Z} \quad (i, j = 1, \ldots, n; i \neq j).$$

Note that $A_0$ is an upper triangular matrix and

$$A_0 + A_1 = \sum_{j=1}^{n} a_j E_{j,j} + \sum_{i=1}^{n} \sum_{j=1}^{i-1} (a_j - b_j) E_{i,j},$$

is a lower triangular matrix.

We consider a formal power series of $\mathbf{x}(t)$ at $t = 0$,

$$\mathbf{x}(t) = t^{\log_q a_1} \sum_{k=0}^{\infty} (q^{-1}t)^k \mathbf{x}_k.$$

Recall that $|q| > 1$, and so $|q^{-1}| < 1$. Substituting it into system (5.1), we obtain

(5.2)    $(A_0 + A_1 - a_1 I)\mathbf{x}_0 = \mathbf{0}$,    $(A_0 + A_1 - a_1 q^{-k} I)\mathbf{x}_k = A_0 \mathbf{x}_{k-1}$    $(k \geq 1)$.

The matrices $A_0 + A_1 - a_1 I$ and $A_0 + A_1 - a_1 q^{-k}I$ are of rank $n - 1$ and $n$, respectively. Hence the recurrence formula (5.2) admits one parameter family of solutions. Its explicit formula is given by

$$\mathbf{x}_k = \left( \frac{b_1 \ldots b_n}{a_1 \ldots a_n} \right)^k \begin{bmatrix} x_{k,1} \\ \vdots \\ x_{k,n} \end{bmatrix} \quad (k \geq 0),$$

where

$$x_{0,1} = 1, \quad x_{0,j} = \prod_{i=1}^{j-1} \frac{b_i - a_1}{a_{i+1} - a_1} \quad (j = 2, \ldots, n),$$

and

$$x_{k,j} = \frac{(q\frac{a_1}{b_1};q)_k \ldots (q\frac{a_1}{b_{j-1}};q)_k (\frac{a_1}{b_j};q)_k \ldots (\frac{a_1}{b_{n-1}};q)_k (\frac{a_1}{b_n};q)_k}{(q\frac{a_1}{a_2};q)_k \ldots (q\frac{a_1}{a_j};q)_k (\frac{a_1}{a_{j+1}};q)_k \ldots (\frac{a_1}{a_n};q)_k (q;q)_k} x_{0,j},$$

for $k \geq 1$. Then we arrive at

THEOREM 5.2. *The system* (5.1) *admits a solution*

$$\mathbf{x}(t) = t^{\log_q a_1} \begin{bmatrix} c_1 \varphi_1(t) \\ \vdots \\ c_n \varphi_n(t) \end{bmatrix},$$

*where*

$$c_j = \prod_{i=1}^{j-1} \frac{b_i - a_1}{a_{i+1} - a_1},$$

$$\varphi_j(t) = {}_n\phi_{n-1} \left[ \begin{matrix} q\frac{a_1}{b_1}, \ldots, q\frac{a_1}{b_{j-1}}, \frac{a_1}{b_j}, \ldots, q\frac{a_1}{b_{n-1}}, \frac{a_1}{b_n} \\ q\frac{a_1}{a_2}, \ldots, q\frac{a_1}{a_j}, \frac{a_1}{a_{j+1}}, \ldots, \frac{a_1}{a_n} \end{matrix} ; q^{-1}, \frac{b_1 \ldots b_n}{a_1 \ldots a_n} q^{-1}t \right].$$

## Acknowledgement

The author would like to express his gratitude to Dr. Kenta Fuji for fruitful discussion. The author is also grateful to Professors Anton Dzhamay, Saburo Kakei, Tetsuya Kikuchi, Masatoshi Noumi, Teruhisa Tsuda and Yasuhiko Yamada for helpful comments and advices.

# References

[1] M. J. Bergvelt and A. P. E. ten Kroode, *Partitions, vertex operator constructions and multi-component KP equations*, Pacific J. Math. **171** (1995), no. 1, 23–88. MR1362978 (97c:58061)

[2] V. G. Drinfel'd and V. V. Sokolov, *Lie algebras and equations of Korteweg-de Vries type* (Russian), Current problems in mathematics, Vol. 24, Itogi Nauki i Tekhniki, Akad. Nauk SSSR, Vsesoyuz. Inst. Nauchn. i Tekhn. Inform., Moscow, 1984, pp. 81–180. MR760998 (86h:58071)

[3] L. Fehér, J. Harnad, and I. Marshall, *Generalized Drinfel'd-Sokolov reductions and KdV type hierarchies*, Comm. Math. Phys. **154** (1993), no. 1, 181–214. MR1220953 (94f:58056)

[4] H. Flaschka and A. C. Newell, *Monodromy- and spectrum-preserving deformations. I*, Comm. Math. Phys. **76** (1980), no. 1, 65–116. MR588248 (82g:35103)

[5] K. Fuji and T. Suzuki, *The sixth Painlevé equation arising from $D_4^{(1)}$ hierarchy*, J. Phys. A **39** (2006), no. 39, 12073–12082, DOI 10.1088/0305-4470/39/39/S04. MR2266212 (2008k:34343)

[6] K. Fuji and T. Suzuki, *Drinfeld-Sokolov hierarchies of type A and fourth order Painlevé systems*, Funkcial. Ekvac. **53** (2010), no. 1, 143–167, DOI 10.1619/fesi.53.143. MR2668518 (2011k:34199)

[7] K. Fuji and T. Suzuki, *Higher order Painlevé systems of type A, Drinfeld-Sokolov hierarchies and Fuchsian systems*, Progress in mathematics of integrable systems, RIMS Kôkyûroku Bessatsu, B30, Res. Inst. Math. Sci. (RIMS), Kyoto, 2012, pp. 181–208. MR2964442

[8] M. F. de Groot, T. J. Hollowood, and J. L. Miramontes, *Generalized Drinfel'd-Sokolov hierarchies*, Comm. Math. Phys. **145** (1992), no. 1, 57–84. MR1155284 (93d:58072)

[9] W. Hahn, *Beiträge zur Theorie der Heineschen Reihen. Die 24 Integrale der Hypergeometrischen q-Differenzengleichung. Das q-Analogon der Laplace-Transformation* (German), Math. Nachr. **2** (1949), 340–379. MR0035344 (11,720b)

[10] M. Jimbo and H. Sakai, *A q-analog of the sixth Painlevé equation*, Lett. Math. Phys. **38** (1996), no. 2, 145–154, DOI 10.1007/BF00398316. MR1403067 (97m:39002)

[11] S. Kakei and T. Kikuchi, *A q-analogue of $\widehat{\mathfrak{gl}}_3$ hierarchy and q-Painlevé VI*, J. Phys. A **39** (2006), no. 39, 12179–12190, DOI 10.1088/0305-4470/39/39/S11. MR2266219 (2007i:37124)

[12] S. Kakei and T. Kikuchi, *The sixth Painlevé equation as similarity reduction of $\widehat{\mathfrak{gl}}_3$ generalized Drinfel'd-Sokolov hierarchy*, Lett. Math. Phys. **79** (2007), no. 3, 221–234, DOI 10.1007/s11005-007-0144-4. MR2309777 (2008b:17050)

[13] F. ten Kroode and J. van de Leur, *Bosonic and fermionic realizations of the affine algebra $\widehat{\mathfrak{gl}}_n$*, Comm. Math. Phys. **137** (1991), no. 1, 67–107. MR1099256 (92f:17030)

[14] K. Kajiwara, M. Noumi, and Y. Yamada, *Discrete dynamical systems with $W(A_{m-1}^{(1)} \times A_{n-1}^{(1)})$ symmetry*, Lett. Math. Phys. **60** (2002), no. 3, 211–219, DOI 10.1023/A:1016298925276. MR1917133 (2003k:37123)

[15] K. Kajiwara, M. Noumi, and Y. Yamada, *q-Painlevé systems arising from q-KP hierarchy*, Lett. Math. Phys. **62** (2002), no. 3, 259–268, DOI 10.1023/A:1022216308475. MR1958118 (2003m:39029)

[16] M. Noumi and Y. Yamada, *Symmetries in the fourth Painlevé equation and Okamoto polynomials*, Nagoya Math. J. **153** (1999), 53–86. MR1684551 (2000c:34015)

[17] M. Noumi and Y. Yamada, *Birational Weyl group action arising from a nilpotent Poisson algebra*, Physics and combinatorics 1999 (Nagoya), World Sci. Publ., River Edge, NJ, 2001, pp. 287–319, DOI 10.1142/9789812810199_0010. MR1865042 (2003b:17031)

[18] V. G. Papageorgiou, F. W. Nijhoff, B. Grammaticos, and A. Ramani, *Isomonodromic deformation problems for discrete analogues of Painlevé equations*, Phys. Lett. A **164** (1992), no. 1, 57–64, DOI 10.1016/0375-9601(92)90905-2. MR1162062 (93c:58092)

[19] T. Suzuki, *A particular solution of a Painlevé system in terms of the hypergeometric function $_{n+1}F_n$*, SIGMA Symmetry Integrability Geom. Methods Appl. **6** (2010), Paper 078, 11. MR2769937 (2012a:33014)

[20] T. Suzuki, *A class of higher order Painlevé systems arising from integrable hierarchies of type A*, Algebraic and geometric aspects of integrable systems and random matrices, Contemp. Math., vol. 593, Amer. Math. Soc., Providence, RI, 2013, pp. 125–141, DOI 10.1090/conm/593/11875. MR3087953

[21] T. Tsuda, *On an integrable system of q-difference equations satisfied by the universal characters: its Lax formalism and an application to q-Painlevé equations*, Comm. Math. Phys. **293** (2010), no. 2, 347–359, DOI 10.1007/s00220-009-0913-2. MR2563787 (2011d:39004)

[22] T. Tsuda, *From KP/UC hierarchies to Painlevé equations*, Internat. J. Math. **23** (2012), no. 5, 1250010, 59, DOI 10.1142/S0129167X11007537. MR2914651

[23] T. Tsuda, *Hypergeometric solution of a certain polynomial Hamiltonian system of isomon-odromy type*, Q. J. Math. **63** (2012), no. 2, 489–505, DOI 10.1093/qmath/haq040. MR2925302

[24] T. Tsuda, *UC hierarchy and monodromy preserving deformation*, J. Reine Angew. Math. **690** (2014), 1–34, DOI 10.1515/crelle-2012-0022. MR3200332

[25] T. Tsuda and T. Masuda, *q-Painlevé VI equation arising from q-UC hierarchy*, Comm. Math. Phys. **262** (2006), no. 3, 595–609, DOI 10.1007/s00220-005-1461-z. MR2202304 (2007e:32014)

DEPARTMENT OF MATHEMATICS, KINKI UNIVERSITY, 3-4-1, KOWAKAE, HIGASHI-OSAKA, OSAKA 577-8502, JAPAN

*E-mail address*: suzuki@math.kindai.ac.jp

Contemporary Mathematics
Volume **651**, 2015
http://dx.doi.org/10.1090/conm/651/13045

# Fractional Calculus of Quantum Painlevé Systems of Type $A_l^{(1)}$

## Hajime Nagoya

ABSTRACT. In this paper, considering the Schrödinger equations obtained
from the quantum Painlevé systems of type $A_l^{(1)}$ introduced by the author
previously, we realize the affine Weyl group symmetries of type $A_l^{(1)}$ on the
Schrödinger equations. As a result, we obtain several integral representations
as particular solutions to the Schrödinger equations.

## 1. Introduction

One of the important problems of mathematics in the ninetieth century was to
find *good transcendental functions* defined by nonlinear algebraic differential equations. In the beginning of the twentieth century, P. Painlevé [**20**] and B. Gambier
[**3**] classified second-order nonlinear ordinary differential equations without movable singular points in their solutions, and discovered the Painlevé equations $P_J$
($J = I, \dots, VI$). Since then, many studies on the Painlevé equations have been
done (see, for example, [**5**], [**2**] and references therein) and the Painlevé equations
are regarded as the most important nonlinear ordinary differential equations.

Among properties on the Painlevé equations, we emphasize that the Painlevé
equations can be written as Hamiltonian equations and the Painlevé equations $P_{VI}$,
$P_V$, $P_{IV}$, $P_{III}$, $P_{II}$ admit affine Weyl group actions of type $D_4^{(1)}$, $A_3^{(1)}$, $A_2^{(1)}$, $C_2^{(1)}$,
$A_1^{(1)}$, respectively, as Bäcklund transformation groups [**18**].

It is natural to study a canonical quantization of the Painlevé equations. Quantizations of the Painlevé equations preserving the affine Weyl group symmetries was
proposed in the Heisenberg picture in [**9**], [**6**], [**11**]. Later, it was shown in [**12**] that
the (confluent) Knizhnik-Zamolodchikov equations [**6**] in the conformal field theory are equivalent to quantizations of the Painlevé equations in the Schrödinger
picture, which we call quantum Painlevé equations in what follows. Realizing the
affine Weyl group symmetries on the quantum Painlevé equations, as Bäcklund
transformations, we obtained an infinite number of hypergeometric solutions to
them [**13**].

In the classical case, higher order analogues of the Painlevé equations have
been studied by several authors. For example, Noumi and Yamada constructed
nonlinear differential systems with the affine Weyl group symmetry of type $A_l^{(1)}$

2010 *Mathematics Subject Classification.* Primary 17B80, 33C70, 34M55, 81R12.
*Key words and phrases.* Affine Weyl groups, Painlevé equations, Integral representation.

$(l \geq 2)$, as higher order analogues of the fourth and fifth Painlevé equations [15]. Quantization of those systems preserving the affine Weyl group symmetries was proposed in [9].

In this paper, considering the Schrödinger equations obtained from the quantization introduced in [9], we realize the affine Weyl group symmetries of type $A_l^{(1)}$ on the Schrödinger equations, in the same manner as in [13]. As a result, we obtain several integral representations as particular solutions to the Schrödinger equations.

Let us explain the affine Weyl symmetries on the quantum Painlevé equations, taking the quantum sixth Painlevé equation $QP_{VI}(\alpha)$

$$\kappa z(z-1)\frac{\partial}{\partial z}\Psi(q,z) = H_{VI}\left(q,\frac{\partial}{\partial q},z,\alpha\right)\Psi(q,z)$$

as an example. Here $H_{VI}(q,\partial/\partial q, z, \alpha)$ is a canonical quantization of the Hamiltonian of $P_{VI}$ (see Section 3.1) and $\alpha = (\alpha_0, \alpha_1, \alpha_2, \alpha_3, \alpha_4)$, $(\alpha_i \in \mathbb{C})$ such that $\alpha_0 + \alpha_1 + 2\alpha_2 + \alpha_3 + \alpha_4 = -\kappa$. Let $\Psi(q,z)$ be a solution to $QP_{VI}(\alpha)$. Then, gauge transformations of $\Psi(q,z)$

$$(q-z)^{-\alpha_0}\Psi(q,z), \quad \Psi(q,z), \quad (q-1)^{-\alpha_3}\Psi(q,z), \quad q^{-\alpha_4}\Psi(q,z)$$

are solutions to $QP_{VI}(s_0(\alpha))$, $QP_{VI}(s_1(\alpha))$, $QP_{VI}(s_3(\alpha))$, $QP_{VI}(s_4(\alpha))$, respectively and the Euler transformation

$$\partial^{-\alpha_2}\Psi(q,z) = \int_\Delta (q-t)^{\alpha_2-1}\Psi(t,z)dt$$

with an appropriate cycle $\Delta$, is a solution to $QP_{VI}(s_2(\alpha))$. Here, $s_i$ $(i=0,1,2,3,4)$ act on the parameters $\alpha_j$ $(j=0,1,2,3,4)$ as $s_i(\alpha_j) = \alpha_j - \alpha_i a_{ij}$ where $(a_{ij})_{i,j=0}^4$ is the Cartan matrix of type $D_4^{(1)}$. Namely, $\langle s_0, s_1, s_2, s_3, s_4 \rangle$ give a representation of the affine Weyl group of type $D_4^{(1)}$. It is easy to see that $QP_{VI}(\alpha)$ has the constant solution 1 at $\alpha_2 = -1$. Applying the gauge transformations and the Euler transformation to the seed solution 1 repeatedly, we obtain an infinite number of integral representations as particular solutions to the quantum sixth Painlevé equation.

We note that since $s_4 s_2 s_4 = s_2 s_4 s_2$, compositions of the gauge transformation and the Euler transformation $q^{-\alpha_2}\partial^{-\alpha_2-\alpha_4}q^{-\alpha_4}$ and $\partial^{-\alpha_4}q^{-\alpha_2-\alpha_4}\partial^{-\alpha_2}$ transform solutions of $QP_{VI}(\alpha)$ to solutions of the same equation $QP_{VI}(s_2 s_4 s_2(\alpha))$. Hence, we expect that the integral transformation formula

$$q^{-\alpha_2}\partial^{-\alpha_2-\alpha_4}q^{-\alpha_4}\Psi(q,z) = \partial^{-\alpha_4}q^{-\alpha_2-\alpha_4}\partial^{-\alpha_2}\Psi(q,z)$$

holds if we take the paths of integrations suitably. In appendix, we also show that the above observation is true, that is, we give integral transformation formulas arising from the affine Weyl group symmetries on the quantum Painlevé equations. As an example of applications, we show that, the integral transformation formula corresponding to the braid relation for the quantum sixth Painlevé equation, yields a three term relation and a contiguity relation for the Gauss hypergeometric function. It can be said that integral transformation formulas obtained by symmetries of the quantum Painlevé equations give a new understanding of relations of special functions, such as the Gauss hypergeometric function.

In section 2, we recall quantum Painlevé systems of type $A_l^{(1)}$ introduced in [9] and reformulate the affine Weyl group symmetries on the Schrödinger operators.

Then, we realize the symmetries on the Schrödinger equations as Bäcklund transformations. Finally, we give examples of particular solutions obtained by applying the Bäcklund transformations to a seed solution. In appendix, we recall definition of the Euler transformation and prepare some facts. Then, we give integral transformation formulas arising from the affine Weyl group symmetries on the quantum Painlevé equations.

## 2. Fractional calculus of quantum Painlevé systems of type $A_l^{(1)}$

**2.1. Preliminary.** We recall results on quantum Painlevé systems of type $A_l^{(1)}$ in [9].

Let $\mathcal{K}_l$ $(l \in \mathbb{Z}_{\geq 2})$ be the skew field over $\mathbb{C}$ with generators $f_i$, $\alpha_i$ $(0 \leq i \leq l)$, and the fundamental relations

$$[f_i, f_{i+1}] = 1, \quad [f_i, f_j] = 0 \quad (j \neq i \pm 1), \quad [f_i, \alpha_j] = 0,$$
$$[\alpha_i, \alpha_j] = 0, \quad (0 \leq i, j \leq l),$$

where the indices are understood as elements in $\mathbb{Z}/(l+1)\mathbb{Z}$.

For each $i = 1, \ldots, l$, we denote by $\varpi_i$ the $i$-th fundamental weight of the finite root system of type $A_l$,

$$\varpi_i = \frac{1}{l+1}\{(l+1-i)\sum_{r=1}^{i} r\alpha_r + i \sum_{r=i+1}^{l}(l+1-r)\alpha_r\}$$

(2.1)
$$= \sum_{r=1}^{l}(\min\{i, r\} - \frac{ir}{l+1})\alpha_r$$

and set $\varpi_0 = 0$.

Put $\Gamma = \mathbb{Z}/(l+1)\mathbb{Z}$. For each subset $C_{i,m} := \{i, i+1, \ldots, i+m-1\}$ $(i = 0, 1, \ldots, l, \ m \in \mathbb{Z}_{>0}, \ m \leq l)$ of $\Gamma$, we define $\chi(C_{i,m})$ by

(2.2)
$$\chi(C_{i,m}) := \varpi_i - \varpi_{i+1} + \cdots + (-1)^{m-1}\varpi_{i+m-1}.$$

For each proper subset $C = \coprod_i C_{i,m_i}$ (disjoint union) of $\Gamma$, we define $\chi(C)$ by

(2.3)
$$\chi(C) := \sum_i \chi(C_{i,m_i}),$$

where we assume that the intersection of $C_{i,m_i+1}$ and $C_{j,m_j}$ is empty for $i \neq j$. Then, we call each $C_{i,m_i}$ a connected component of $C$ with length $m_i$.

For each $d = 1, \ldots, l+1$, let $S_d$ be the set consisting of the subset $K \subset \Gamma$, such that $|K| = d$, and the length of each connected component of $\Gamma \backslash K$ is even.

For $C_{i,m}$, we set

(2.4)
$$f_{C_{i,m}} = f_i f_{i+1} \cdots f_{i+m-1}.$$

For each $K = \coprod_i C_{i,m_i} \in S_d$ $(d = 1, \ldots, l)$, we define $f_K$ by

(2.5)
$$f_K = \prod_i f_{C_{i,m_i}},$$

where $C_{i,m_i}$ is a connected component of $K$. Note that we do not define $f_K$ for $K \in S_{l+1}$.

DEFINITION 2.1. *We define the Hamiltonian $H$ for the quantum Painlevé systems of type $A_l^{(1)}$ as follows:*
    *(1) For $l = 2n$:*

$$(2.6) \qquad H = \begin{cases} f_0 f_1 f_2 + f_1 + \sum_{K \in S_1} \chi(\Gamma \backslash K) f_K & (l = 2), \\[2mm] \sum_{K \in S_3} f_K + \sum_{K \in S_1} \chi(\Gamma \backslash K) f_K & (l = 2n, n \geq 2). \end{cases}$$

    *(2) For $l = 2n + 1$:*

$$(2.7) \quad zH = \begin{cases} f_0 f_1 f_2 f_3 + f_1 f_2 + \sum_{K \in S_2} \chi(\Gamma \backslash K) f_K + \left( \sum_{i=1}^{3} (-1)^{i-1} \varpi_i \right)^2 \\ (l = 3), \\[2mm] \sum_{K \in S_4} f_K + \sum_{K \in S_2} \chi(\Gamma \backslash K) f_K + \left( \sum_{i=1}^{l} (-1)^{i-1} \varpi_i \right)^2 \\ (l = 2n + 1, n \geq 2). \end{cases}$$

Denote by $\widetilde{W}(A_l^{(1)}) = \langle s_0, s_1, \dots, s_l \rangle \rtimes \langle \pi \rangle$, the extended affine Weyl group of type $A_l^{(1)}$.

Let $A = (a_{ij})_{i,j=0}^{l}$ be the generalized Cartan matrix of type $A_l^{(1)}$, namely,

$$(2.8) \qquad a_{ii} = 2, \quad a_{i,i \pm 1} = -1, \quad a_{ij} = 0 \ (j \neq i, i \pm 1).$$

PROPOSITION 2.2. *Let $s_i$ $(0 \leq i \leq l)$ and $\pi$ be the actions on the generators of $\mathcal{K}_l$ defined as*

$$s_i(f_j) = f_j + \frac{\alpha_i}{f_i}[f_i, f_j], \quad s_i(\alpha_j) = \alpha_j - \alpha_i a_{ij},$$

$$\pi(f_j) = f_{j+1}, \quad \pi(\alpha_j) = \alpha_{j+1}, \quad (if \ l = 2n, \ n \in \mathbb{Z}_{\geq 1}),$$

$$\pi(f_{2i}) = \frac{z}{w} f_{2i+1}, \quad \pi(f_{2i-1}) = \frac{w}{z} f_{2i}, \quad (if \ l = 2n + 1, \ n \in \mathbb{Z}_{\geq 1}),$$

*where $z = -\sum_{i=0}^{n} f_{2i}$ and $w = -\sum_{i=0}^{n} f_{2i+1}$. Then, $s_i$ $(0 \leq i \leq l)$ and $\pi$ preserve the fundamental relations of $\mathcal{K}_l$, and hence, they give rise to automorphisms on $\mathcal{K}_l$.*

The action of $\pi$ in the case of $l = 2n+1$ is modified from the one in [**9**], although $\pi$ acts on $H$ as same as the previous one.

THEOREM 2.3 ([**9**], Theorem 2.6). *The automorphisms $s_i$ $(0 \leq i \leq l)$ and $\pi$ defines a representation of $\widetilde{W}(A_l^{(1)})$ on $\mathcal{K}_l$.*

Put $\kappa = -\sum_{i=0}^{l} \alpha_i$.

PROPOSITION 2.4 ([**9**], Proposition 2.7 and A.2). *The automorphisms $s_i$ $(0 \leq i \leq l)$ and $\pi$ act on the Hamiltonian $H$ as follows.*
    *(1) For $l = 2n$,*

$$(2.9) \quad s_i(H) = H - \delta_{i,0} \kappa \frac{\alpha_0}{f_0} \quad (0 \leq i \leq l), \qquad \pi(H) = H - \kappa \sum_{i=1}^{n} f_{2i} + \frac{n\kappa}{2n+1} z,$$

*where $z = \sum_{i=0}^{l} f_i$.*
    *(2) For $l = 2n + 1$,*

$$s_i(H) = H + \delta_{i,0} \kappa \frac{\alpha_0}{f_0} z \quad (0 \leq i \leq l),$$

(2.10) $\quad \pi(H) = H - \kappa \sum_{1 \leq i \leq j \leq n} f_{2i} f_{2j+1} + \frac{n\kappa}{(2n+1)} \sum_{K \in S_2} f_K - \frac{\kappa}{4} \sum_{i=0}^{l} (-1)^i \alpha_i.$

Let us introduce a canonical coordinate as

$$p_i = f_{2i}, \quad q_i = -\sum_{j=1}^{i} f_{2j-1}, \quad (1 \leq i \leq n).$$

Then, the symmetric variables $f_i$ $(0 \leq i \leq l)$ are completely solved by $z$, $w$, $q_i$, $p_i$ $(1 \leq i \leq n)$ as follows. For $l = 2n$,

$$f_0 = z - \sum_{i=1}^{n} p_i + q_n, \quad f_{2i} = p_i, \quad f_{2i-1} = q_{i-1} - q_i, \quad (1 \leq i \leq n),$$

where $q_0 = 0$. For $l = 2n + 1$,

$$f_0 = -z - \sum_{i=1}^{n} p_i, \quad f_{2i} = p_i, \quad f_{2i-1} = q_{i-1} - q_i, \quad (1 \leq i \leq n),$$

$$f_{2n+1} = q_n - w,$$

where $q_0 = 0$. Note that $z$, $w$ are invariant under the actions of $s_i$ $(0 \leq i \leq l)$ and $\pi$.

We consider a skew field $\mathcal{K}_l(d)$ modified from $\mathcal{K}_l$, adding a generator $d$ and fundamental relations

$$[d, z] = 1, \quad [d, q_i] = [d, p_i] = 0 \quad (1 \leq i \leq n), \quad [d, \alpha_i] = 0,$$

and $w = 1$ for $l = 2n + 1$.

Let the actions of $s_i$ $(0 \leq i \leq l)$ and $\pi$ on $d$ be defined as follows. For $l = 2n$,

$$s_0(d) = d + \frac{\alpha_0}{\sum_{i=1}^{n} p_i - q_n - z}, \quad s_i(d) = d \quad (1 \leq i \leq l),$$

$$\pi(d) = d - \sum_{i=1}^{n} p_i + \frac{n}{2n+1} z.$$

For $l = 2n + 1$,

$$s_0(d) = d - \frac{\alpha_0}{z + \sum_{i=1}^{n} p_i}, \quad s_i(d) = d \quad (1 \leq i \leq l),$$

$$\pi(d) = d - \frac{1}{z} \sum_{i=1}^{n} p_i (q_i - 1) + \frac{n(n+z)}{(2n+1)z} - \frac{1}{4z} \sum_{i=0}^{l} (-1)^i \alpha_i.$$

Then, $s_i$ $(0 \leq i \leq l)$ and $\pi$ preserve the commutation relations of $\mathcal{K}_l(d)$ and give rise to automorphisms on $\mathcal{K}_l(d)$, although they do not satisfy the fundamental relations of $\widetilde{W}(A_l^{(1)})$ on $\mathcal{K}_l(d)$.

THEOREM 2.5. *The automorphisms $s_i$ $(0 \leq i \leq l)$ and $\pi$ preserve the Schrödinger operator $\kappa d - H$, $\kappa z d - H$ for $l = 2n$, $l = 2n + 1$, respectively, that is, we have, for $l = 2n$,*

$$s_i(kd - H) = kd - H \quad (0 \leq i \leq l), \quad \pi(\kappa d - H) = \kappa d - H,$$

*and for $l = 2n + 1$,*

$$s_i(kzd - H) = kzd - H \quad (0 \leq i \leq l), \quad \pi(\kappa z d - H) = \kappa z d - H.$$

PROOF. It follows from Proposition 2.4 that the automorphisms $s_i$ $(i = 0, 1, \ldots, l)$ preserve the Schrödinger operator.

The automorphism $\pi$ act on $d$ as

$$\pi(d) = d - \sum_{i=1}^{n} f_{2i} + \frac{n}{2n+1} z.$$

for $l = 2n$, and

$$\pi(d) = d - \sum_{1 \le i \le j \le n} f_{2i} f_{2j+1} + \frac{n}{(2n+1)} \sum_{K \in S_2} f_K - \frac{1}{4} \sum_{i=0}^{l} (-1)^i \alpha_i$$

for $l = 2n + 1$, because, for $l = 2n$, we have

$$\sum_{i=1}^{n} f_{2i} = \sum_{i=1}^{n} p_i$$

and for $l = 2n + 1$, we have

$$\sum_{1 \le i \le j \le n} f_{2i} f_{2j+1} = \sum_{i=1}^{n} p_i(q_i - 1), \qquad \sum_{K \in S_2} f_K = z + n.$$

Hence, from Proposition 2.4, the automorphism $\pi$ preserve the Schrödinger operator. □

### 2.2. Realization.

2.2.1. *Affine Weyl group symmetry.* Let $\mathcal{L}_i$ $(i = 1, \ldots, n)$ be the Laplace transformations on $\mathcal{K}_l(d)$ with respect to $q_i$ defined by

$$\mathcal{L}_i(p_i) = q_i, \quad \mathcal{L}_i(q_i) = -p_i.$$

For linear combinations $g$ of $q_i$ $(1 \le i \le n)$ and $t$, and $\alpha$ of $\alpha_i$ $(0 \le i \le 2n)$, let $\mathrm{Ad}(g^\alpha)$ be the gauge transformation on $\mathcal{K}_l(d)$ defined by

$$\mathrm{Ad}(g^\alpha)(f) = f - \frac{\alpha}{g}[f, g]$$

for $f \in \{q_1, \ldots, q_n, p_1, \ldots, p_n, t, d, \alpha_0, \ldots, \alpha_{2n}\}$. For a polynomial $g$ of $q_i$ $(1 \le i \le n)$ and $t$ with the degree 1, let $\mathrm{Ad}(\exp(g))$ be the gauge transformation on $\mathcal{K}_l(d)$ defined by

$$\mathrm{Ad}(\exp(g))(f) = f - [f, g]$$

for $f \in \{q_1, \ldots, q_n, p_1, \ldots, p_n, t, d, \alpha_0, \ldots, \alpha_{2n}\}$. Moreover, denote by $(q_i \to q_i/z)$ $(i = 1, \ldots, n)$, transformations of variables, precisely, $(q_i \to q_i/z)$ is an automorphism on $\mathcal{K}_l(d)$ defined by

$$q_i \to \frac{q_i}{z}, \quad p_i \to z p_i, \quad d \to d + \frac{q_i p_i}{z}.$$

DEFINITION 2.6. *We define transformations $R_s$ for $s \in \{s_0, \ldots, s_l, \pi\}$ as*

$$R_{s_{2i-1}} = \mathrm{Ad}\left((q_{i-1} - q_i)^{-\alpha_{2i-1}}\right), \quad R_{s_{2i}} = \mathcal{L}_i^{-1} \circ \mathrm{Ad}\left(q_i^{-\alpha_{2i}}\right) \circ \mathcal{L}_i \quad (1 \le i \le n),$$

*and for $l = 2n$,*

$$R_{s_0} = \mathrm{Ad}\left(\exp\left(\frac{q_n^2}{2}\right)\right) \circ \prod_{i=1}^{n} \mathcal{L}_i^{-1} \circ \mathrm{Ad}\left(\left(\sum_{i=1}^{n} q_i - z\right)^{-\alpha_0}\right)$$

$$\circ \prod_{i=1}^{n} \mathcal{L}_i \circ \mathrm{Ad}\left(\exp\left(-\frac{q_n^2}{2}\right)\right),$$

$$R_\pi = \mathcal{L}_1 \circ \mathrm{Ad}\left(\exp\left(q_1 q_2\right)\right) \circ \cdots \circ \mathcal{L}_{n-1} \circ \mathrm{Ad}\left(\exp\left(q_{n-1} q_n\right)\right) \circ \mathcal{L}_n$$

$$\circ \mathrm{Ad}\left(\exp\left(-z q_n - \frac{q_n^2}{2}\right)\right) \circ \mathrm{Ad}\left(\exp\left(\frac{-n z^2}{2(2n+1)}\right)\right),$$

and for $l = 2n + 1$,

$$R_{s_0} = \prod_{i=1}^{n} \mathcal{L}_i^{-1} \circ \mathrm{Ad}\left(\left(z + \sum_{i=1}^{n} q_i\right)^{-\alpha_0}\right) \circ \prod_{i=1}^{n} \mathcal{L}_i,$$

$$R_{s_{2n+1}} = \mathrm{Ad}\left(\left(q_n - 1\right)^{-\alpha_{2n+1}}\right),$$

$$R_\pi = \mathcal{L}_1 \circ \mathrm{Ad}\left(\exp\left(q_1 q_2\right)\right) \circ \left(q_1 \to \frac{q_1}{z}\right) \circ \cdots \circ \mathcal{L}_n \circ \mathrm{Ad}\left(\exp\left(q_n q_{n+1}\right)\right)$$

$$\circ \left(q_n \to \frac{q_n}{z}\right) \circ \mathrm{Ad}\left(\exp\left(\frac{nz}{2n+1}\right) z^{\frac{n^2}{2n+1} - \frac{1}{4}\sum_{i=0}^{l}(-1)^i \alpha_i}\right),$$

where $q_{n+1} = 1$.

Let $\tilde{s}$ for $s \in \{s_0, \ldots, s_l, \pi\}$ be defined by

$$\tilde{s}(\alpha_i) = s(\alpha_i) \quad (0 \leq i \leq l), \quad \tilde{s}(y) = y \quad (y \in \{q_1, \ldots, q_n, p_1, \ldots, p_n, z, d\}).$$

THEOREM 2.7 (cf. [**7**], [**10**]). *For $s \in \{s_0, \ldots, s_l, \pi\}$, we have*

$$s = \tilde{s} \circ R_s.$$

PROOF. It is sufficient to check that the actions of $\tilde{s}_i \circ R_i(\alpha_i)$ and $\tilde{\pi} \circ R_\pi$ on the generators of $\mathcal{K}$ are equal to those of $s_i$ and $\pi$, respectively. For the case of $R_i$, $i = 1, \ldots, l$, it is easy to see equality. We compute the action of $\tilde{\pi} \circ R_\pi$ on the generators of $\mathcal{K}$ as follows. Let us consider the case of $l = 2n$. For $q_i$ $(i = 1, \ldots, n)$, we have

$$\tilde{\pi} \circ R_\pi(q_i) = \mathcal{L}_1 \circ \mathrm{Ad}\left(\exp\left(q_1 q_2\right)\right) \circ \cdots \circ \mathcal{L}_{i-1} \circ \mathrm{Ad}\left(\exp\left(q_{i-1} q_i\right)\right)(-p_i)$$

$$= \mathcal{L}_1 \circ \mathrm{Ad}\left(\exp\left(q_1 q_2\right)\right) \circ \cdots \circ \mathcal{L}_{i-1}(-p_i + q_{i-1})$$

$$= -\sum_{j=1}^{i} p_j.$$

For $p_i$ $(i = 1, \ldots, n-1)$, we have

$$\tilde{\pi} \circ R_\pi(p_i) = \mathcal{L}_1 \circ \mathrm{Ad}\left(\exp\left(q_1 q_2\right)\right) \circ \cdots \circ \mathcal{L}_{i-1} \circ \mathrm{Ad}\left(\exp\left(q_{i-1} q_i\right)\right) \circ \mathcal{L}_i(p_i - q_{i+1})$$

$$= q_i - q_{i+1}.$$

For $p_n$, we have

$$\tilde{\pi} \circ R_\pi(p_n) = \mathcal{L}_1 \circ \mathrm{Ad}\left(\exp\left(q_1 q_2\right)\right) \circ \cdots \circ \mathcal{L}_{n-1} \circ \mathrm{Ad}\left(\exp\left(q_{n-1} q_n\right)\right) \circ \mathcal{L}_n(t + q_n)$$

$$= z - \sum_{i=1}^{n} p_i + q_n.$$

For $d$, we have

$$\tilde{\pi} \circ R_\pi(p_n) = \mathcal{L}_1 \circ \mathrm{Ad}\left(\exp\left(q_1 q_2\right)\right) \circ \cdots \circ \mathcal{L}_{n-1} \circ \mathrm{Ad}\left(\exp\left(q_{n-1} q_n\right)\right)$$

$$\circ \mathcal{L}_n \left( d + q_n + \frac{nz}{2n+1} \right)$$

$$= d - \sum_{i=1}^{n} p_i + \frac{nz}{2n+1}.$$

The actions of $R_\pi$ on the other generators are trivial. A fundamental relation $s_0 = \pi^{-1} s_1 \pi$ shows that $s_0 = \tilde{s}_0 \circ R_0(\alpha_0)$. We omit computation of the case of $l = 2n + 1$. ☐

PROPOSITION 2.8. *The transformations $R_s$ ($s \in \{s_0, \ldots, s_l, \pi\}$) are Bäcklund transformations of the Schrödinger operator, that is, it holds*

$$R_{s_i} (\kappa d - H) = \kappa d - \tilde{s}_i(H) \quad (0 \le i \le l), \quad R_\pi (\kappa d - H) = \kappa d - \tilde{\pi}^{-1}(H)$$

*for $l = 2n$, and*

$$R_{s_i} (\kappa z d - H) = \kappa z d - \tilde{s}_i(H) \quad (0 \le i \le l), \quad R_\pi (\kappa z d - H) = \kappa z d - \tilde{\pi}^{-1}(H)$$

*for $l = 2n + 1$.*

2.2.2. *Schrödinger equation.* We consider the following partial differential equation

$$(2.11) \qquad \kappa \frac{\partial}{\partial z} \Psi(q, z, \alpha) = H \left( q, \frac{\partial}{\partial q}, z, \alpha \right) \Psi(q, z, \alpha)$$

in the complex domain. Here, the Hamiltonian $H(q, \partial/\partial q, z, \alpha)$ is obtained by substituting $\partial/\partial q_i$ into $p_i$ ($i = 1, \ldots, n$) of the Hamiltonian $H$ defined in the skew field $\mathcal{K}_l$, and we regard $q_1, \ldots, q_n, z$ as variables and $\alpha_0, \alpha_1, \ldots, \alpha_l$ as complex parameters. We denote by $\mathrm{Sch}(\alpha)$, the Schrödinger equation above.

Proposition 2.8 is rewritten as

PROPOSITION 2.9. *We have*

$$R_{s_i} \left( \kappa \frac{\partial}{\partial z} - H \left( q, \frac{\partial}{\partial q}, z, \alpha \right) \right) = \kappa \frac{\partial}{\partial z} - H \left( q, \frac{\partial}{\partial q}, z, s_i(\alpha) \right) \quad (0 \le i \le l),$$

$$R_\pi \left( \kappa \frac{\partial}{\partial z} - H \left( q, \frac{\partial}{\partial q}, z, \alpha \right) \right) = \kappa \frac{\partial}{\partial z} - H \left( q, \frac{\partial}{\partial q}, z, \pi^{-1}(\alpha) \right)$$

*for $l = 2n$, and*

$$R_{s_i} \left( \kappa z \frac{\partial}{\partial z} - H \left( q, \frac{\partial}{\partial q}, z, \alpha \right) \right) = \kappa z \frac{\partial}{\partial z} - H \left( q, \frac{\partial}{\partial q}, z, s_i(\alpha) \right) \quad (0 \le i \le l),$$

$$R_\pi \left( \kappa z \frac{\partial}{\partial z} - H \left( q, \frac{\partial}{\partial q}, z, \alpha \right) \right) = \kappa z \frac{\partial}{\partial z} - H \left( q, \frac{\partial}{\partial q}, z, \pi^{-1}(\alpha) \right)$$

*for $l = 2n + 1$.*

DEFINITION 2.10. *For a complex number $\beta$, we define transformations $\tilde{R}_i(\beta)$ ($0 \le i \le l$) and $\tilde{R}_\pi$ as*

$$\tilde{R}_{2i-1}(\beta) = (q_{i-1} - q_i)^{-\beta}, \quad \tilde{R}_{2i}(\beta) = \mathcal{L}_i^{-1} \circ q_i^{-\beta} \circ \mathcal{L}_i \quad (1 \le i \le n),$$

*and for $l = 2n$,*

$$\tilde{R}_0(\beta) = \exp\left( \frac{q_n^2}{2} \right) \circ \prod_{i=1}^{n} \mathcal{L}_i^{-1} \circ \left( \sum_{i=1}^{n} q_i - z \right)^{-\beta} \circ \prod_{i=1}^{n} \mathcal{L}_i \circ \exp\left( -\frac{q_n^2}{2} \right),$$

$$\widetilde{R}_\pi = \mathcal{L}_1 \circ \exp(q_1 q_2) \circ \cdots \circ \mathcal{L}_{n-1} \circ \exp(q_{n-1} q_n) \circ \mathcal{L}_n \circ \exp\left(-z q_n - \frac{q_n^2}{2}\right)$$

$$\circ \exp\left(-\frac{nz^2}{2(2n+1)}\right),$$

and for $l = 2n+1$,

$$\widetilde{R}_0 = \prod_{i=1}^n \mathcal{L}_i^{-1} \circ \left(z + \sum_{i=1}^n q_i\right)^{-\alpha_0} \circ \prod_{i=1}^n \mathcal{L}_i, \quad \widetilde{R}_{2n+1} = (q_n - 1)^{-\alpha_{2n+1}},$$

$$\widetilde{R}_\pi = \mathcal{L}_1 \circ \exp(q_1 q_2) \circ \left(q_1 \to \frac{q_1}{z}\right) \circ \cdots \circ \mathcal{L}_n \circ \exp(q_n q_{n+1}) \circ \left(q_n \to \frac{q_n}{z}\right)$$

$$\circ \exp\left(\frac{nz}{2n+1}\right) z^{\frac{n^2}{2n+1} - \frac{1}{4}\sum_{i=0}^l (-1)^i \alpha_i},$$

where $q_{n+1} = 1$.

Here, Laplace transformations $\mathcal{L}_i$ for a function $f(q_i)$ satisfy

$$\mathcal{L}_i\left[\frac{\partial}{\partial q_i} f\right] = q_i \mathcal{L}_i[f], \quad \mathcal{L}_i[q_i f] = -\frac{\partial}{\partial q_i} \mathcal{L}_i[f].$$

It may be realized by the integral transformation

$$\mathcal{L}_i[f(q_i)] = \int_\gamma \exp(-t q_i) f(t) dt$$

with an appropriate cycle $\gamma$. The transformations $\widetilde{R}_{2i}(\beta)$ $(1 \le i \le n)$ for a function $f(q_i)$ are equivalent to the following integral transformations

$$\int_\gamma (q_i - t)^{\beta-1} f(t) dt,$$

which are called Euler transformations. We use the Euler transformations above as $\widetilde{R}_{2i}(\beta)$ in what follows.

Proposition 2.9 implies

COROLLARY 2.11. *If $\Psi(q, z, \alpha)$ is a solution to the Schrödinger equation* Sch($\alpha$), *then*

$$\widetilde{R}_i(\alpha_i)\Psi(q, z, \alpha) \quad (i = 0, 1, \ldots, l), \quad \widetilde{R}_\pi \Psi(z, q, \alpha)$$

*are solutions to the Schrödinger equation* Sch($s_i(\alpha)$), Sch($\pi^{-1}(\alpha)$), *respectively.*

From Corollary 2.11, if there is a known solution to the Schrödinger equation Sch($\alpha$), then, we obtain new solutions by applying Bäcklund transformations $\widetilde{R}_i(\alpha_i)$, $\widetilde{R}_\pi$ to the known solution successively.

PROPOSITION 2.12. *The Hamiltonian $H(q, \partial/\partial q, z, \alpha)$ acts on 1 as follows:*
*(1) For $l = 2n$:*

$$H\left(q, \frac{\partial}{\partial q}, z, \alpha\right) \cdot 1 = -\sum_{i=1}^n (\alpha_{2i} + 1) q_i + c_l,$$

where

$$c_l = z \sum_{i=1}^{2n} (-1)^{i-1} \varpi_i.$$

*(2) For $l = 2n + 1$:*

$$H\left(q, \frac{\partial}{\partial q}, z, \alpha\right) \cdot 1 = z \sum_{i=1}^{n} (\alpha_{2i} + 1) q_i + c_l z,$$

*where*

$$c_l z = 2n - 1 + \frac{(n-2)(n-1)}{2} + \sum_{i=0}^{n} \sum_{j=0}^{2n-1} (-1)^{j-1} \varpi_{2i+j} + z \sum_{i=1}^{2n} (-1)^{i-1} \varpi_i$$

$$+ \left(\sum_{i=1}^{l} (-1)^{i-1} \varpi_i\right)^2.$$

PROOF. It follows from direct computations.                                  □

COROLLARY 2.13. *The Schrödinger equation* $\mathrm{Sch}(\alpha)$ *with* $\alpha_{2i} = -1$ *($i = 1, \ldots, n$) admits a function independent on $q_j$ ($j = 1, \ldots, n$) as a solution expressed by elementary functions.*

PROOF. Put $l = 2n$. When $\alpha_{2i} = -1$ ($i = 1, \ldots, n$), from Proposition 2.12, the Schrödinger equation is equal to

$$(2.12) \qquad\qquad \kappa \frac{\partial}{\partial z} \Psi(z, \alpha) = c_l \Psi(z, \alpha).$$

Hence, $\Psi(z, \alpha)$ is solved by elementary functions.                  □

If we apply Bäcklund transformations $\widetilde{R}_i(\alpha_i)$, $\widetilde{R}_\pi$ to a solution of (2.12) successively, then we obtain integral representations as particular solutions to the Schrödinger equation $\mathrm{Sch}(\alpha)$. In the next subsection, we give integral representations as examples.

2.2.3. *Examples.* Fix complex parameters $\kappa$, $\beta_{2i-1}$ ($i = 1, \ldots, n$) for the case of $l = 2n$ and $\beta_{2i-1}$ ($i = 1, \ldots, n, n+1$) for the case of $l = 2n+1$. Note that for any complex parameters $\kappa$, $\beta_{2i-1}$, the Schrödinger equation obtained by substituting $-1$ into all $\alpha_{2i}$ has a solution due to Corollary 2.13. Let us denote by $f_l(z)$, a solution of (2.12), that is, a solution of the Schrödinger equation $\mathrm{Sch}(\alpha)$ with

$$(2.13) \qquad \alpha_{2i-1} = \beta_{2i-1} \quad (1 \le i \le n), \quad \alpha_{2i} = -1 \quad (1 \le i \le n)$$

for $l = 2n$, and

$$(2.14) \qquad \alpha_{2i-1} = \beta_{2i-1} \quad (1 \le i \le n+1), \quad \alpha_{2i} = -1 \quad (1 \le i \le n)$$

for $l = 2n + 1$.

Applying all gauge transformations to the seed solution $f_l(z)$, we have a solution

$$(2.15) \qquad\qquad \prod_{i=1}^{n} (q_{i-1} - q_i)^{-\beta_{2i-1}} f_l(z)$$

to the Schrödinger equation for $l = 2n$ such that

$$\alpha_{2i-1} = -\beta_{2i-1} \quad (1 \le i \le n), \quad \alpha_{2i} = -1 + \beta_{2i-1} + \beta_{2i+1} \quad (1 \le i \le n-1),$$
$$\alpha_{2n} = -1 + \beta_{2n-1},$$

and we have a solution

$$(2.16) \qquad\qquad \prod_{i=1}^{n+1} (q_{i-1} - q_i)^{-\beta_{2i-1}} f_l(z)$$

to the Schrödinger equation for $l = 2n + 1$ such that

$$\alpha_{2i-1} = -\beta_{2i-1} \quad (1 \leq i \leq n+1), \quad \alpha_{2i} = -1 + \beta_{2i-1} + \beta_{2i+1} \quad (1 \leq i \leq n).$$

Note that the paramters $-\beta_{2i-1}$ in the solutions (2.15), (2.16) are determined by Corollary 2.11.

Applying Euler transformations $\widetilde{R}_{2i}(\beta)$ $(1 \leq i \leq n)$ to these solutions (2.15), (2.16), we obtain integral representations as solutions.

PROPOSITION 2.14. *The integral representation*

$$\int_\Delta \prod_{i=1}^n (t_{i-1} - t_i)^{-\beta_{2i-1}} (q_i - t_i)^{\beta_{2i-1}+\beta_{2i+1}-2} dt_1 \cdots dt_n,$$

*where $\Delta$ is an appropriate cycle, $t_0 = 0$, and $\beta_{2n+1} = 0$, is a solution to the Schrödiger equation for $l = 2n$ such that*

$$\alpha_1 = \beta_3 - 1, \quad \alpha_{2i-1} = \beta_{2i-3} + \beta_{2i-1} + \beta_{2i+1} - 2 \quad (2 \leq i \leq n),$$
$$\alpha_{2i} = 1 - \beta_{2i-1} - \beta_{2i+1} \quad (1 \leq i \leq n),$$

*and the integral representation*

$$\int_\Delta \prod_{i=1}^{n+1} (t_{i-1} - t_i)^{-\beta_{2i-1}} (q_i - t_i)^{\beta_{2i-1}+\beta_{2i+1}-2} dt_1 \wedge \cdots \wedge dt_n,$$

*where $\Delta$ is an appropriate cycle, $t_0 = 0$, and $t_{n+1} = 1$ is a solution to the Schrödiger equation for $l = 2n + 1$ such that*

$$\alpha_1 = \beta_3 - 1, \quad \alpha_{2i-1} = \beta_{2i-3} + \beta_{2i-1} + \beta_{2i+1} - 2 \quad (2 \leq i \leq n),$$
$$\alpha_{2n+1} = \beta_{2n-1} - 1, \quad \alpha_{2i} = 1 - \beta_{2i-1} - \beta_{2i+1} \quad (1 \leq i \leq n).$$

When $l = 2$, the Schrödinger equation Sch$(\alpha)$ is the quantum fourth Painlevé equation. In [**13**], the integral representation

$$(2.17) \qquad \int_\Delta (q_1 - t)^{-\kappa} t^{-\beta_1} \exp\left(\frac{t^2}{2} + tz\right) dt$$

was given as a particular solution to the quantum fourth Painlevé equation with parameters $\alpha_1 = 1 - \kappa - \beta_1$, $\alpha_2 = -1$. The next proposition is a generalization of (2.17).

PROPOSITION 2.15. *The integral representation*

$$\Phi(q, z) = f(z) \exp\left(\frac{nz^2}{2(2n+1)}\right) \int_\Delta \prod_{i=1}^n (q_i - t)^{\beta_{2i+1}-1} t^{\beta_1-1} \exp\left(zt + \frac{t^2}{2}\right) dt,$$

*where $\beta_{2n+1} = n - \kappa - \sum_{i=2}^n \beta_{2i-1}$, is a solution to Sch$(\alpha)$ for $l = 2n$ such that*

$$(2.18) \qquad \alpha_{2i-1} = \beta_{2i-1} + \beta_{2i+1} - 1 \quad (1 \leq i \leq n), \quad \alpha_{2i} = -\beta_{2i+1} \quad (1 \leq i \leq n).$$

PROOF. The integral representation $\Phi(q, z)$ is obtained by applying

$$\widetilde{R}_2(\beta_3) \circ \widetilde{R}_4(\beta_5) \circ \cdots \circ \widetilde{R}_{2n}(\beta_{2n+1}) \circ \widetilde{R}_\pi^{-1} \circ \widetilde{R}_1(\beta_1)$$

to $f(z)$. First, we compute $\widetilde{R}_\pi^{-1} \circ \widetilde{R}_1(\beta_1)(\exp(-nz^2/2(2n+1)))$ as follows:

$$\widetilde{R}_\pi^{-1} \circ \widetilde{R}_1(\beta_1)\left(\exp\left(-\frac{nz^2}{2(2n+1)}\right)\right)$$

$$= \exp\left(zq_n + \frac{q_n^2}{2}\right)\left(\prod_{i=2}^{n}\mathcal{L}_i^{-1}\circ\exp\left(-q_iq_{i-1}\right)\right)(q_1^{\beta_1-1})$$

$$(2.19) \qquad = q_1^{\beta_1-1}\exp\left(zq_n + \frac{q_n^2}{2}\right)\prod_{i=2}^{n}\delta(q_i - q_{i-1}),$$

where $\delta(x)$ is the delta function such that

$$\int_\gamma \delta(x-a)\varphi(x)dx = \varphi(a).$$

Here we used $\mathcal{L}_i^{-1}[1] = \delta(q_i)$.

Because the function (2.19) is a solution to the Schrödinger equation such that

$$(2.20) \qquad \alpha_1 = \beta_1 - 1, \quad \alpha_{2i} = \beta_{2i+1} \ (1\le i\le n), \quad \alpha_{2i-1} = -1 \ (2\le i\le n),$$

the following transformations

$$\widetilde{R}_2(\beta_3), \widetilde{R}_4(\beta_5),\ldots,\widetilde{R}_{2n}(\beta_{2n+1})$$

are Bäcklund transformations of the Schrödinger equation satifying (2.20). Applying $\widetilde{R}_{2n}(\beta_{2n+1})$ to (2.19), we obtain

$$\widetilde{R}_{2n}(\beta_{2n+1})\left(q_1^{\beta_1-1}\exp\left(zq_n + \frac{q_n^2}{2}\right)\prod_{i=2}^{n}\delta(q_i - q_{i-1})\right)$$

$$= q_1^{\beta_1-1}\int_\Delta (q_n - t)^{\beta_{2n+1}-1}\exp\left(zt + \frac{t^2}{z}\right)\delta(t - q_{n-1})dt\prod_{i=2}^{n-1}\delta(q_i - q_{i-1})$$

$$= q_1^{\beta_1-1}(q_n - q_{n-1})^{\beta_{2n+1}-1}\exp\left(zq_{n-1} + \frac{q_{n-1}^2}{z}\right)\prod_{i=2}^{n-1}\delta(q_i - q_{i-1}).$$

Similarly, we have

$$\widetilde{R}_4(\beta_5)\circ\cdots\circ\widetilde{R}_{2n}(\beta_{2n+1})\left(q_1^{\beta_1-1}\exp\left(zq_n + \frac{q_n^2}{2}\right)\prod_{i=2}^{n}\delta(q_i - q_{i-1})\right)$$

$$= q_1^{\beta_1-1}\exp\left(zq_1 + \frac{q_1^2}{2}\right)\prod_{i=2}^{n}(q_i - q_1)^{\beta_{2i+1}-1}.$$

Finally, applying $\widetilde{R}_2(\beta_3)$ to the solution above, we obtain

$$\int_\Delta \prod_{i=1}^{n}(q_i - t)^{\beta_{2i+1}-1}t^{\beta_1-1}\exp\left(zt + \frac{t^2}{2}\right)dt.$$

$\square$

The cycle $\Delta$ of $\Phi(q,z)$ is, for example, the double loop winding about any two points in $\{0, q_1,\ldots,q_n\}$ or a contour starting at infinity, encircling any one point in $\{0, q_1,\ldots,q_n\}$ and returning to infinity, or a contour starting at infinity in some region and returning to at infinity in another region.

Proposition 2.15 can be verified by direct computation using the explicit expression of the Hamiltonian for $l = 2n$

(2.21)
$$
H = \sum_{i=1}^{n} H_{\mathrm{IV}}\left(q_i, \frac{\partial}{\partial q_i}, z, \alpha_{2i}, \sum_{j=1}^{i} \alpha_{2j-1}\right) + \sum_{1 \le i < j \le n}\left(2q_i \frac{\partial}{\partial q_i} + 1\right)\frac{\partial}{\partial q_j} + g(z),
$$

where $g(z)$ is some function depending on $z$ only and

$$
H_{\mathrm{IV}}\left(q, \frac{\partial}{\partial q}, z, a, b\right) = q\frac{\partial^2}{\partial q^2} - q(q+z)\frac{\partial}{\partial q} - (a+1)q - (b-1)\frac{\partial}{\partial q}.
$$

We remark that Sasano pointed out that the classical Hamiltonians given by Noumi and Yamada [15] were written in terms of the Hamiltonian of the fourth Painlevé equation or the fifth Painlevé equation. See, for example, [22]. The expression (2.21) of the Hamiltonian for $l = 2n$ is quantum version of the coupled form of the Hamiltonian.

**Another proof of Proposition 2.15.** Denote by $\langle \varphi(t) \rangle$, the integral representation

$$
\int_\Delta \varphi(t) \prod_{i=1}^{n}(q_i - t)^{\beta_{2i+1}-1} t^{\beta_1-1} \exp\left(zt + \frac{t^2}{2}\right) dt.
$$

for convenience. For a cycle $\Delta$ chosen appropriately, it holds

(2.22)
$$
0 = \langle \nabla 1 \rangle = \left\langle \sum_{i=1}^{n} \frac{1 - \beta_{2i+1}}{q_i - t} + \frac{\beta_1 - 1}{t} + z + t \right\rangle,
$$

and

$$
0 = \left\langle \nabla \frac{1}{q_i - t} \right\rangle = \left\langle \frac{2 - \beta_{2i+1}}{(q_i - t)^2} + \sum_{j \ne i} \frac{\beta_{2j+1} - 1}{q_i - q_j}\left(\frac{1}{q_i - t} - \frac{1}{q_j - t}\right) \right.
$$

(2.23)
$$
\left. + \frac{q_i + z}{q_i - t} - 1 + \frac{\beta_1 - 1}{q_i}\left(\frac{1}{q_i - t} + \frac{1}{t}\right) \right\rangle.
$$

where $\nabla = \partial/\partial t + \partial(\log(U(t)))/\partial t$ with the integrand $U(t)$ of the integral representation $\Phi(q, z)$. Since $H\langle 1 \rangle$ is computed as follows:

$$
H\langle 1 \rangle = \sum_{i=1}^{n} q_i \left\langle \frac{(\beta_{2i+1} - 1)(\beta_{2i+1} - 2)}{(q_i - t)^2} \right\rangle - \sum_{i=1}^{n} q_i(q_i + z)\left\langle \frac{\beta_{2i+1} - 1}{q_i - t} \right\rangle
$$
$$
- \sum_{i=1}^{n}(\alpha_{2i} + 1)q_i\langle 1 \rangle - \sum_{i=1}^{n}\left(\sum_{j=1}^{i} \alpha_{2j-1} - 1\right)\left\langle \frac{\beta_{2i+1} - 1}{q_i - t} \right\rangle
$$
$$
+ \sum_{1 \le i < j \le n} \frac{2(\beta_{2i+1} - 1)(\beta_{2j+1} - 1)q_i}{q_j - q_i}\left\langle \frac{1}{q_i - t} - \frac{1}{q_j - t} \right\rangle
$$
$$
+ \sum_{j=2}^{n}(j - 1)(\beta_{2j+1} - 1)\left\langle \frac{1}{q_j - t} \right\rangle + g(z)\langle 1 \rangle,
$$

by (2.22) and (2.23), we obtain

$$
\left(\kappa \frac{\partial}{\partial z} - H\right)\langle 1 \rangle
$$

$$= \left( \kappa \frac{\partial}{\partial z} - H \right) \langle 1 \rangle - \kappa \langle \nabla 1 \rangle - \sum_{i=1}^{n} (\beta_{2i+1} - 1) q_i \left\langle \nabla \frac{1}{q_i - t} \right\rangle$$

$$= \sum_{i=1}^{n} (\alpha_{2i} + \beta_{2i+1}) q_i \langle 1 \rangle - (\kappa z + g(z)) \langle 1 \rangle$$

$$+ \sum_{i=1}^{n} (\beta_{2i+1} - 1) \left( \sum_{j=1}^{i} \alpha_{2j-1} - 2 \sum_{j=1}^{i-1} \beta_{2j+1} - \beta_{2i+1} + i - \beta_1 \right) \left\langle \frac{1}{q_i - t} \right\rangle.$$

Hence, it holds that

$$\left( \kappa \frac{\partial}{\partial z} - H \right) \langle 1 \rangle = -(\kappa z + g(z)) \langle 1 \rangle$$

if the parameters $\alpha_i$ $(1 \le i \le 2n)$ satisfy the condition (2.18).    $\square$

When $l = 3$, the Schrödinger equation Sch($\alpha$) is the quantum fifth Painlevé equation. In [**13**], the integral representation

$$(2.24) \qquad \int_{\Delta} (q_1 - t)^{-\kappa} t^{\beta_1 - 1} (t - 1)^{\beta_3 - 1 - \kappa} \exp(-zt) dt$$

was given as a particular solution to the quantum fifth Painlevé equation with parameters $\alpha_1 = \beta_1 - \kappa$, $\alpha_2 = -1 + \kappa$, $\alpha_3 = \beta_3$. Similarly to the case of $l = 2n$, we have the next proposition as a generalization of (2.24).

PROPOSITION 2.16. *The integral representation*

$$\Phi(q, z) = f(z) \exp \left( \frac{-nz}{2n+1} \right) z^{\frac{-n^2}{2n+1} - \frac{1}{4} \left( 2 \sum_{i=1}^{n+1} \beta_{2i-1} + \kappa \right) - (n+1) + \beta_1}$$

$$\times \int_{\Delta} \prod_{i=1}^{n-1} (q_i - t)^{\beta_{2i+1} - 1} (q_n - t)^{-(\kappa + n + 1 + \sum_{i=2}^{n} \beta_{2i-1})}$$

$$\cdot t^{\beta_1 - 1} (t - 1)^{n + \kappa + \sum_{i=1}^{n} \beta_{2i+1}} \exp (-zt) \, dt$$

*is a solution to* Sch($\alpha$) *for* $l = 2n + 1$ *such that*

$$\alpha_{2i-1} = \beta_{2i-1} + \beta_{2i+1} - 1 \quad (1 \le i \le n - 2), \quad \alpha_{2n-1} = -\kappa - n - 1 - \sum_{i=2}^{n-1} \beta_{2i-1},$$

$$\alpha_{2n+1} = \beta_{2n+1}, \quad \alpha_{2i} = -\beta_{2i+1} \quad (1 \le i \le n - 1) \quad \alpha_{2n} = \kappa + n + \sum_{i=2}^{n} \beta_{2i-1}.$$

PROOF. The integral representation $\Phi(q, z)$ is obtained by applying the following Bäcklund transformation

$$\widetilde{R}_2(\beta_3) \circ \widetilde{R}_4(\beta_5) \circ \cdots \circ \widetilde{R}_{2n-2}(\beta_{2n-1}) \circ \widetilde{R}_{2n}(-(\kappa + n + \sum_{i=2}^{n} \beta_{2i-1}))$$

$$\circ \widetilde{R}_{2n+1}(-(\kappa + n + \sum_{i=2}^{n+1} \beta_{2i-1})) \circ \widetilde{R}_{\pi}^{-1} \circ \widetilde{R}_1(\beta_1)$$

to $f(z)$. We omit the computation.    $\square$

In order to construct an infinite number of integral representation as particular solutions to the Schrödinger equation, the following shift operators are useful:

$$T_1 = \pi s_l s_{l-1} \cdots s_1,$$
$$T_2 = s_1 \pi s_l s_{l-1} \cdots s_2,$$
$$\vdots$$
$$T_{l+1} = s_l s_{l-1} \cdots s_1 \pi.$$

The shift operators $T_i$ $(i = 1, \ldots, l+1)$ act on the parameters as follows.

$$T_i(\alpha_{i-1}) = \alpha_{i-1} - \kappa, \quad T_i(\alpha_i) = \alpha_i + \kappa, \quad T_i(\alpha_j) = \alpha_j \quad (j \neq i, i-1).$$

Denote by $\widetilde{T}_i(\alpha)$, the Bäcklund transformations for $\mathrm{Sch}(\alpha)$ corresponding to $T_i$, that is,

$$\widetilde{T}_1(\alpha) = \widetilde{R}_1 \left( \sum_{i=1}^l \alpha_{i+1} \right) \circ \widetilde{R}_2 \left( \sum_{i=2}^l \alpha_{i+1} \right) \circ \cdots \circ \widetilde{R}_{l-1}(\alpha_l + \alpha_0) \circ \widetilde{R}_l(\alpha_0) \circ \widetilde{R}_\pi^{-1},$$

$$\widetilde{T}_2(\alpha) = \widetilde{R}_2 \left( \sum_{i=2}^{l+1} \alpha_{i+1} \right) \circ \cdots \circ \widetilde{R}_{l-1}(\alpha_l + \alpha_0 + \alpha_1) \circ \widetilde{R}_l(\alpha_0 + \alpha_1) \circ \widetilde{R}_\pi^{-1} \circ \widetilde{R}_1(\alpha_1),$$

$$\vdots$$

$$\widetilde{T}_{l+1}(\alpha) = \widetilde{R}_\pi^{-1} \circ \widetilde{R}_1 \left( \sum_{i=1}^l \alpha_i \right) \circ \widetilde{R}_2 \left( \sum_{i=2}^l \alpha_i \right) \circ \cdots \circ \widetilde{R}_{l-1}(\alpha_{l-1} + \alpha_l) \circ \widetilde{R}_l(\alpha_l).$$

If $\Psi(q, z)$ is a solution to the Schröinger equation $\mathrm{Sch}(\alpha)$, then $\widetilde{T}_i(\alpha)\Psi(q, z)$ is a solution to the Schrödinger equation $\mathrm{Sch}(T_i(\alpha))$.

Since the seed solution $f(z)$ is a solution to the Schrödinger equation $\mathrm{Sch}(\alpha)$ such that $\alpha_{2i} = -1$ $(i = 1, \ldots, n)$, the Bäcklund transformations $\widetilde{T}_{2i-1}(\alpha)$, $\widetilde{T}_{2i}^{-1}(\alpha)$ do not generate non-trivial solutions from $f(z)$. We present integral representations of $\widetilde{T}_{2i}(\alpha)(f(z))$ $(1 \leq i \leq n)$.

PROPOSITION 2.17. *For $i = 1, \ldots, n$, the integral representation*

$$f(z) \exp \left( \frac{nz^2}{2(2n+1)} \right) \int_\Delta (q_i - t)^{-\kappa} t^{\sum_{j=1}^i \beta_{2j-1} - i} \exp \left( zt + \frac{t^2}{z} \right) dt$$

*is a solution to the Schrödinger equation for $l = 2n$ such that*

$$\alpha_{2i} = \kappa - 1, \quad \alpha_{2j} = -1 \quad (j \neq i, 1 \leq j \leq n),$$
$$\alpha_{2i-1} = \beta_{2i-1} - \kappa, \quad \alpha_{2j-1} = \beta_{2j-1} \quad (j \neq i, 1 \leq j \leq n).$$

PROOF. An explicit form of $\widetilde{T}_{2i}(\alpha)$ is that

$$\widetilde{T}_{2i}(\alpha)$$

$$= \widetilde{R}_{2i} \left( \sum_{j=2i+1}^l \alpha_j + \sum_{j=0}^{2i-1} \alpha_j \right) \circ \widetilde{R}_{2i+1} \left( \sum_{j=2i+2}^l \alpha_j + \sum_{j=0}^{2i-1} \alpha_j \right) \circ \cdots \circ \widetilde{R}_{l-1} \left( \sum_{j=-1}^{2i-1} \alpha_j \right)$$

$$\circ \widetilde{R}_l \left( \sum_{j=0}^{2i-1} \alpha_j \right) \circ \widetilde{R}_\pi^{-1}$$

$$\circ \widetilde{R}_1 \left( \sum_{j=1}^{2i-1} \alpha_j \right) \circ \widetilde{R}_2 \left( \sum_{j=2}^{2i-1} \alpha_j \right) \circ \cdots \circ \widetilde{R}_{2i-2}(\alpha_{2i-2} + \alpha_{2i-1}) \circ \widetilde{R}_{2i-1}(\alpha_{2i-1}).$$

First, we apply $\widetilde{R}_1(\sum_{j=1}^{i} \beta_{2j-1} - i + 1) \circ \cdots \circ \widetilde{R}_{2i-1}(\beta_{2i-1})$ to $f(z)$, then we have

$$\widetilde{R}_1 \left( \sum_{j=1}^{i} \beta_{2j-1} - i + 1 \right) \circ \widetilde{R}_2 \left( \sum_{j=2}^{i} \beta_{2j-1} - i + 1 \right) \circ \cdots \circ \widetilde{R}_{2i-2}(\beta_{2i-1} - 1)$$

$$\circ \widetilde{R}_{2i-1}(\beta_{2i-1})(f(z))$$

$$= C q_1^{-\sum_{j=1}^{i} \beta_{2j-1} + i - 1} \prod_{j=2}^{i} \delta(q_{j-1} - q_j),$$

where $C \in \mathbb{C} - \{0\}$. Here, we used a relation

$$\mathcal{L}_k \circ q_k^{1-a} \circ \mathcal{L}_k^{-1} \left( (q_k - q_{k+1})^{-a} \right) = C_k \delta(q_k - q_{k+1}),$$

where $C_k \in \mathbb{C} - \{0\}$. Second, we apply $\widetilde{R}_\pi^{-1}$ to $q_1^{-\sum_{j=1}^{i} \beta_{2j-1} + i - 2} \prod_{j=2}^{i} \delta(q_{j-1} - q_j)$ as follows. Using a relation

$$\mathcal{L}_k^{-1} \left( q_k^a \delta(q_k - q_{k+1}) \right) = \exp(q_k q_{k+1}) q_{k+1}^a,$$

we have

$$\widetilde{R}_\pi^{-1} \left( q_1^{-\sum_{j=1}^{i} \beta_{2j-1} + i - 1} \prod_{j=2}^{i} \delta(q_{j-1} - q_j) \right)$$

$$= \exp \left( \frac{nz^2}{2(2n+1)} \right) \exp \left( zq_n + \frac{q_n^2}{2} \right) \mathcal{L}_n^{-1} \circ \prod_{k=2}^{n-1} \exp(-q_k q_{k+1})$$

$$\circ \mathcal{L}_k^{-1} \left( q_2^{-\sum_{j=1}^{i} \beta_{2j-1} + i - 1} \prod_{j=3}^{i} \delta(q_{j-1} - q_j) \right)$$

$$= \exp \left( \frac{nz^2}{2(2n+1)} \right) \exp \left( zq_n + \frac{q_n^2}{2} \right) \mathcal{L}_n^{-1} \circ \prod_{k=i+1}^{n-1} \exp(-q_k q_{k+1})$$

$$\circ \mathcal{L}_k^{-1} \left( q_i^{\sum_{j=1}^{i} \beta_{2j-1} - i} \exp(-q_i q_{i+1}) \right)$$

$$= \exp \left( \frac{nz^2}{2(2n+1)} \right) \exp \left( zq_n + \frac{q_n^2}{2} \right) q_i^{\sum_{j=1}^{i} \beta_{2j-1} - i} \prod_{j=i+1}^{n} \delta(q_j - q_{j-1}).$$

Finally, we obtain

$$\widetilde{T}_{2i}(\beta)(1) = \exp \left( \frac{nz^2}{2(2n+1)} \right) \widetilde{R}_{2i}(1 - \kappa) \left( \exp \left( zq_i + \frac{q_i^2}{2} \right) q_i^{\sum_{j=1}^{i} \beta_{2j-1} - i} \right)$$

$$= \exp \left( \frac{nz^2}{2(2n+1)} \right) \int_\Delta (q_i - t)^{-\kappa} t^{\sum_{j=1}^{i} \beta_{2j-1} - i} \exp \left( zt + \frac{t^2}{z} \right) dt.$$

$\square$

Similarly to Proposition 2.17, we have the

PROPOSITION 2.18. *For $i = 1, \ldots, n$, the integral representation*

$$f(z) \exp\left(\frac{-nz}{2n+1}\right) z^{\frac{-n^2}{2n+1} - \frac{1}{4}\left(\sum_{i=1}^{n+1} \beta_{2i-1} + \kappa\right) + \delta_{i,n}\left(\sum_{j=1}^{i} \beta_{2j-1} - i + 1\right)}$$

$$\times \int_{\Delta} (q_i - t)^{-\kappa} t^{\sum_{j=1}^{i} \beta_{2j-1} - i} (t-1)^{\sum_{j=i}^{n} \beta_{2j+1} + \kappa - n + i - 1} \exp\left(-zt\right) dt,$$

*where $\delta_{i,j}$ is the Kronecker's delta, is a solution to the Schrödinger equation for $l = 2n + 1$ such that*

$$\alpha_{2i} = \kappa - 1, \quad \alpha_{2j} = -1 \quad (j \neq i, 1 \leq j \leq n),$$
$$\alpha_{2i-1} = \beta_{2i-1} - \kappa, \quad \alpha_{2j-1} = \beta_{2j-1} \quad (j \neq i, 1 \leq j \leq n+1).$$

The Bäcklund transformations $\widetilde{T}_{2i}(\alpha)$ act on a function $\Psi(q, z)$ as follows.

PROPOSITION 2.19. *We have*

$$\widetilde{T}_{2i}(\alpha)\left(\Psi(q,z)\right) = g_l(z) \int_{\Delta} \prod_{k=i}^{n} (q_k - w_k)^{\sum_{j=2k+1}^{2i+l} \alpha_j - 1} \prod_{k=i}^{n-1} (w_k - q_{k+1})^{-\sum_{j=2k+2}^{2i+l} \alpha_j}$$

$$\times \exp((w_i - q_{i-1})u_{i,l}) \prod_{k=i+1}^{n} \exp((w_k - w_{k-1})u_{k,l})$$

$$\times \prod_{k=1}^{i-1} \exp((q_k - q_{k-1})u_{k,l})$$

$$\times \prod_{k=1}^{i-1} (u_k - t_k)^{\sum_{j=2k}^{2i-1} \alpha_j - 1} \prod_{k=0}^{i-1} (t_k - u_{k+1})^{-\sum_{j=2k+1}^{2i-1} \alpha_j}$$

$$\times \Psi(t_1, \ldots, t_{i-1}, u_i, \ldots, u_n, z) G_l(w_n) dt du dw,$$

*where $t_0 = 1$, $u_{k,2n} = u_k$, $u_{k,2n+1} = zu_k$, and*

$$g_{2n}(z) = \exp\left(\frac{nz^2}{2(2n+1)}\right), \quad g_{2n+1}(z) = \exp\left(\frac{-nz}{2n+1}\right) z^{\frac{-n^2}{2n+1} + \frac{1}{4}\sum_{i=0}^{l} (-1)^i \alpha_i},$$

$$G_{2n}(x) = \exp\left(zx + \frac{x^2}{2}\right), \quad G_{2n+1}(x) = \exp(-zx)(x-1)^{-\sum_{j=0}^{2i-1} \alpha_j}.$$

PROOF. Using a realization of the inverse Laplace transformations:

$$\mathcal{L}_i^{-1}\left(\Phi(q_i)\right) = \int_{\Delta} \exp(q_i t) \Phi(t) dt,$$

we obtain the integral representation above by straightforward calculations. □

From this proposition, we would obtain an infinite number of explicit forms of integral representations as solutions to the Schrödinger equation by applying the Bäcklund transformations $\widetilde{T}_2, \ldots, \widetilde{T}_{2n}$ repeatedly to the seed solution.

Finally, we present an explicit form of the integral representation of

$$\widetilde{T}_2(\widetilde{T}_4 \cdots \widetilde{T}_{2n}(\alpha)) \circ \widetilde{T}_4(\widetilde{T}_6 \cdots \widetilde{T}_{2n}(\alpha)) \circ \cdots \circ \widetilde{T}_{2n-2}(\widetilde{T}_{2n}(\alpha)) \circ \widetilde{T}_{2n}(\alpha)(f(z))$$

PROPOSITION 2.20. *The integral representation*

$$f(z) \exp\left(\frac{n^2 z^2}{2(2n+1)}\right)$$

$$\times \int_\Delta \prod_{k=1}^n (q_k - w_k^{(1)})^{\sum_{j=2k+1}^{2n+2} \beta_j - 1} \prod_{k=1}^{n-1} (w_k^{(1)} - q_{k+1})^{-\sum_{j=2k+2}^{2n+2} \beta_j - \kappa}$$

$$\times \prod_{i=2}^n \left( \prod_{k=i}^n (u_k^{(i-1)} - w_k^{(i)})^{\sum_{j=2k+1}^{2i+l} \beta_j - 1} \prod_{k=i}^{n-1} (w_k^{(i)} - u_{k+1}^{(i-1)})^{-\sum_{j=2k+2}^{2i+l} \beta_j - \kappa} \right)$$

(2.25)

$$\times \prod_{i=1}^n \left( (w_i^{(i)})^{\sum_{j=1}^{2i-1} \beta_j - 1} \exp\left( zw_n^{(i)} + \frac{(w_n^{(i)})^2}{2} \right) \prod_{k=i+1}^n \exp\left( (w_k^{(i)} - w_{k-1}^{(i)}) u_k^{(i)} \right) \right) du\,dw,$$

*where $\beta_{2i} = -1$ $(1 \le i \le n)$, is a solution to the Schrödinger equation for $l = 2n$ such that*

$$\alpha_{2i-1} = \beta_{2i-1} - \kappa \quad (1 \le i \le n), \quad \alpha_{2i} = \kappa - 1 \quad (1 \le i \le n).$$

PROOF. We note that for any $i = 1, \ldots, n$, $\widetilde{T}_{2i+2}(T_{2i+4} \cdots T_{2n}(\alpha)) \circ \cdots \circ \widetilde{T}_{2n}(\alpha)(f(z))$, where $\alpha_{2j} = -1$ $(1 \le j \le n)$, is a function depending on the variables $q_{i+1}, \ldots, q_n$ only. Using the action of $\widetilde{T}_{2i}(\alpha)$ on $\Psi(q_{i+1}, \ldots, q_n)$:

$$\widetilde{T}_{2i}(\alpha)(\Psi(q_{i+1}, \ldots, q_n))$$

$$= g_{2n}(z) \int_\Delta \prod_{k=i}^n (q_k - w_k)^{\sum_{j=2k+1}^{2i+l} \alpha_j - 1} \prod_{k=i}^{n-1} (w_k - q_{k+1})^{-\sum_{j=2k+2}^{2i+l} \alpha_j}$$

$$\times \prod_{k=i+1}^n \exp((w_k - w_{k-1})u_k) \exp\left( zw_n + \frac{w_n^2}{2} \right)$$

$$\times w_i^{\sum_{j=1}^{2i-1} - 1} \Psi(u_{i+1}, \ldots, u_n)\,du\,dw,$$

we obtain the integral representation (2.25). □

Similarly to Proposition 2.20, we have the

PROPOSITION 2.21. *The integral representation*

$$f(z) \exp\left(\frac{-n^2 z}{2n+1}\right) z^{\frac{-n^3}{2n+1} + \frac{n}{4}\sum_{i=0}^l (-1)^i \beta_i + \frac{n\kappa}{2} + \sum_{i=1}^n \left(\sum_{j=1}^{2i-1} \beta_j - 1\right)}$$

$$\times \int_\Delta \prod_{k=1}^n (q_k - w_k^{(1)})^{\sum_{j=2k+1}^{2n+2} \beta_j - 1} \prod_{k=1}^{n-1} (w_k^{(1)} - q_{k+1})^{-\sum_{j=2k+2}^{2n+2} \beta_j - \kappa}$$

$$\times \prod_{i=2}^n \left( \prod_{k=i}^n (u_k^{(i-1)} - w_k^{(i)})^{\sum_{j=2k+1}^{2i+l} \beta_j - 1} \prod_{k=i}^{n-1} (w_k^{(i)} - u_{k+1}^{(i-1)})^{-\sum_{j=2k+2}^{2i+l} \beta_j - \kappa} \right)$$

$$\times \prod_{i=1}^n \left( (w_i^{(i)})^{\sum_{j=1}^{2i-1} \beta_j - 1} \exp\left( -zw_n^{(i)} \right) (w_n^{(i)} - 1)^{-\sum_{j=0}^{2i-1} \beta_j} \right.$$

$$\times \prod_{k=i+1}^n \exp\left( z(w_k^{(i)} - w_{k-1}^{(i)}) u_k^{(i)} \right) \Big)\,du\,dw,$$

*where $\beta_{2i} = -1$ $(1 \leq i \leq n)$, is a solution to the Schrödinger equation for $l = 2n+1$ such that*

$$\alpha_{2i-1} = \beta_{2i-1} - \kappa \quad (1 \leq i \leq n), \quad \alpha_{2n+1} = \beta_{2n+1}, \quad \alpha_{2i} = \kappa - 1 \quad (1 \leq i \leq n).$$

## Appendix A. Integral transformation formulas

**A.1. Preliminary.** In this subsection, we recall the Euler transformation, following [5], Section 2.3.3. See also [8]. Denote the Gamma function by $\Gamma(x)$.

DEFINITION A.1 (The Euler transformation). *The Euler transformation of $f(x)$ is the integral:*

$$(A.1) \qquad \partial_a^{-\alpha} f(x) = \frac{1}{A(\alpha, \mu)} \int_\gamma (x - t)^{\alpha-1} f(t) dt,$$

*where $a \in \mathbb{C}$ and $\alpha \notin \mathbb{Z}_{\geq 1}$,*

$$A(\alpha, \mu) = (1 - e^{2\pi\sqrt{-1}\alpha})(1 - e^{2\pi\sqrt{-1}\mu})\Gamma(\alpha),$$

*and the path $\gamma$ of integration is the Pochhammer cycle turning around $a$ and $x$, and $f(x)$ is of the form*

$$(A.2) \qquad f(x) = (x - a)^\mu g(x), \quad \mu \in \mathbb{C} - \mathbb{Z},$$

*where $g(x)$ is holomorphic in a neighborhood of $\gamma$ and $g(a) \neq 0$. We fix branches of the multi-valued functions $\arg(x - t)$ and $\arg(t - a)$ along the cycle $\gamma$.*

Throughout the appendix, we assume that $f(x)$ is of the form (A.2), because we have the following Proposition.

PROPOSITION A.2 ([5], Proposition 3.3.3). *Let $\alpha \notin \mathbb{Z}_{\geq 1}$, $\mu \notin \mathbb{Z}$ and $\mu + \alpha \notin \mathbb{Z}_{\leq -1}$. Then, we have*

$$\partial_a^{-\alpha} f(x) = (x - a)^{\mu+\alpha} \tilde{g}(x),$$

*where $\tilde{g}(x)$ is holomorphic in a neighborhood of $a$.*

PROPOSITION A.3 ([5], Proposition 3.3.4). *Let $\alpha \notin \mathbb{Z}_{\geq 1}$, $\mu \notin \mathbb{Z}$. Then, we have*

$$\partial_a^{-\beta} \partial_a^{-\alpha} f(x) = \partial_a^{-\alpha-\beta} f(x).$$

REMARK A.4. *If we replace the path of integration appropriately, then the Euler transformation of $f(x, \alpha, \mu)$ is continued analytically to an entire function in $\alpha$ and a meromorphic function in $\mu$ (see [5], Section 2.3.3).*

We can take an arc $C$ with initial point $a$ and terminal point $x$ as the path of integration of the Euler transformation of $f(x)$. Namely, we have

PROPOSITION A.5 ([5], Proposition 3.3.7). *If $\alpha \notin \mathbb{Z}_{\geq 1}$, $\mu \notin \mathbb{Z}$, $\Re(\alpha) > 0$ and $\Re(\mu) > -1$, then, we have*

$$\partial_a^{-\alpha} f(x) = \frac{1}{\Gamma(\alpha)} \int_C (x - t)^{\alpha-1} f(t) dt.$$

**A.2. $P_{VI}$ case.** The affine Weyl group $W(D_4^{(1)})$ is defined by the generators $s_0, s_1, s_2, s_3, s_4$ and the fundamental relations:

$$s_i^2 = 1 \quad (i = 0, 1, 2, 3, 4), \quad s_i s_2 s_i = s_2 s_i s_2, \quad s_i s_j = s_j s_i \quad (i, j = 0, 1, 3, 4).$$

It was shown in [**13**] that the quantum sixth Painlevé equation $QP_{VI}(\alpha)$:

$$\kappa z(z-1)\frac{\partial}{\partial z}\Psi(x, z) = H_{VI}\left(x, \frac{\partial}{\partial x}, z, \alpha\right)\Psi(x, z),$$

where

$$H_{VI} = x(x-1)(x-t)\left(\frac{\partial}{\partial x} - \frac{\alpha_4 - 1}{x} - \frac{\alpha_3 - 1}{x-1} - \frac{\alpha_0 - 1 + \kappa}{x-t}\right)\frac{\partial}{\partial x}$$
$$+ (\alpha_2 + 1)(\alpha_1 + \alpha_2 + 1)x,$$

has the affine Weyl group symmetry of type $D_4^{(1)}$. Here $\alpha_0, \alpha_1, \alpha_2, \alpha_3, \alpha_4$ are complex parameters such that $\alpha_0 + \alpha_1 + 2\alpha_2 + \alpha_3 + \alpha_4 = -\kappa$. Let us explain this symmetry further. The generators of $W(D_4^{(1)})$ act on the complex parameters $\alpha_0, \alpha_1, \alpha_2, \alpha_3, \alpha_4$ as $s_i(\alpha_j) = \alpha_j - \alpha_i a_{ij}$ $(i, j = 0, 1, 2, 3, 4)$, where $(a_{ij})$ is the Cartan matrix of type $D_4^{(1)}$. We define transformations $R_0, R_1, R_2, R_3, R_4$ for solutions as

$$R_0(\alpha_0) = (x-z)^{-\alpha_0}, \quad R_1(\alpha_1) = 1, \quad R_2(\alpha_2) = \partial^{-\alpha_2},$$
$$R_3(\alpha_3) = (x-1)^{-\alpha_3}, \quad R_4(\alpha_4) = q^{-\alpha_4},$$

where

$$\partial^{-\alpha_2}\Psi(x, z) = \int_\Delta (x-t)^{\alpha_2 - 1}\Psi(t, z)dt$$

with an appropriate cycle $\Delta$. Then, we have

THEOREM A.6 ([**13**]). *If $\Psi(x, z)$ is a solution to $QP_{VI}(\alpha)$, then $R_i(\alpha_i)\Psi(x, z)$ is a solution to $QP_{VI}(s_i(\alpha))$, up to a gauge factor of $t$.*

REMARK A.7. *The Euler transformation produces both the classical and quantum version of the Okamoto transformation of the sixth Painlevé equation [**18**], since the Euler transformation for certain Fuchsian equation yields the classical Okamoto transformation (see [**16**], [**4**], [**23**], for example). This is because the Lax equations for the sixth Painlevé equation give the quantum sixth Painlevé equation [**21**], [**17**], [**24**].*

Theorem A.6 implies that $R_i(-\alpha_2)R_2(-\alpha_i-\alpha_2)R_i(-\alpha_i)\Psi(x, z)$ and $R_2(-\alpha_i)R_i(-\alpha_i-\alpha_2)R_2(-\alpha_2)\Psi(x, z)$ are solutions to $QP_{VI}(s_i s_2 s_i(\alpha))$. Hence, we expect that if we take the path of integration appropriately, both solutions are equal. This observation leads to the following theorem.

THEOREM A.8. *Let $\alpha, \beta, \alpha + \beta \notin \mathbb{Z}_{\geq 1}$, $\mu \notin \mathbb{Z}$ and $b \in \mathbb{C}$ such that $b \neq a$ and $\mu + \alpha \notin \mathbb{Z}$, or $b = a$ and $\mu + 2\alpha + \beta \notin \mathbb{Z}$. Then, we have*

$$(x-b)^{-\alpha}\partial_a^{-\alpha-\beta}(x-b)^{-\beta}f(x) = \partial_a^{-\beta}(x-b)^{-\alpha-\beta}\partial_a^{-\alpha}f(x).$$

PROOF. It is sufficient to prove that

$$\frac{1}{\Gamma(\alpha + \beta)}(x-b)^{-\alpha}\int_a^x (x-t)^{\alpha+\beta-1}(t-b)^{-\beta}(t-a)^\mu g(t)dt$$

(A.3)
$$= \frac{1}{\Gamma(\alpha)\Gamma(\beta)} \int_a^x (x - t_2)^{\beta-1}(t_2 - b)^{-\alpha-\beta} \left( \int_a^{t_2} (t_2 - t_1)^{\alpha-1}(t_1 - a)^{\mu} g(t_1)dt_1 \right) dt_2$$

for $a, b, x \in \mathbb{R}$ such that $a < x$, when the integrals of both sides converge, because of Proposition A.5. We fix the branches of the multivalued functions of the integrand of both sides as $\arg(x - t) = 0$, $\arg(t - b) = 0$, $\arg(t - a) = 0$, $\arg(x - t_2) = 0$, $\arg(t_2 - b) = 0$, $\arg(t_2 - t_1) = 0$, and $\arg(t_1 - a) = 0$.

Changing the variable $t_2$ in the integrand of the right-hand side of (A.3) as

$$z = \frac{x(t_2 - t_1)}{t_2(x - t_1)},$$

we obtain

$$\text{R.H.S of (A.3)} = \frac{1}{\Gamma(\alpha)\Gamma(\beta)}(x - b)^{-\alpha} \int_a^x (x - t_1)^{\alpha+\beta-1}(t_1 - b)^{-\beta}(t_1 - a)^{\mu} g(t_1)dt_1$$
$$\times \int_0^1 z^{\alpha-1}(1 - z)^{\beta-1}dz.$$

Notice that the last term above is the beta function $B(\alpha, \beta)$. Thanks to the famous formula

$$B(\alpha, \beta) = \frac{\Gamma(\alpha)\Gamma(\beta)}{\Gamma(\alpha + \beta)},$$

the right-hand side of (A.3) becomes the left-hand side of (A.3).    □

The above theorem corresponds to $s_i s_2 s_i = s_2 s_i s_2$ for $i = 0, 3, 4$. The braid relation $s_1 s_2 s_1 = s_2 s_1 s_2$ corresponds to Proposition A.3.

EXAMPLE A.9 (Three term ralation). *Applying Threorem A.8 to the case of* $\alpha = -1$, $a = b = 0$, *and* $f(x) = x^{\mu}(1 - x)^{\eta}$, *we obtain*
(A.4)
$$(\beta - 1)x \int_0^x (x - t)^{\beta-2}t^{\mu-\beta}(1 - t)^{\eta}dt = \int_0^x (x - t)^{\beta-1}(\mu t - \eta(1 - t))t^{\mu-\beta}(1 - t)^{\eta-1}dt.$$

*Since the Gauss hypergeometric function* $F(p, q, r; x)$ *has the integral representation*

(A.5)    $$F(p, q, r; x) = \frac{\Gamma(r)}{\Gamma(p)\Gamma(r - p)} \int_0^1 u^{p-1}(1 - u)^{r-p-1}(1 - xu)^{-q}du,$$

*where* $\Re(p) > 0$, $\Re(r - p) > 0$, *the relation* (A.4) *is equivalent to a three term relation of the Gauss hypergeometric function*

$$F(p, q, r; x) = F(p, q, r + 1; x) + \frac{pq}{r(r + 1)}xF(p + 1, q + 1, r + 2; x),$$

*where* $p = \mu - \beta + 1$, $q = -\eta$ *and* $r = \mu$.

EXAMPLE A.10 (Contiguity relation). *Applying Threorem A.8 to the case of* $\beta = -1$, $a = b = 0$, *and* $f(x) = x^{\mu}(1 - x)^{\eta}$, *we obtain*
(A.6)
$$(\alpha - 1) \int_0^x (x - t)^{\alpha-2}t^{\mu+1}(1 - t)^{\eta}dt = \left( x\frac{\partial}{\partial x} + 1 - \alpha \right) \int_0^x (x - t)^{\alpha-1}t^{\mu+1}(1 - t)^{\eta}dt.$$

*Using* (A.5), *the relation* (A.6) *is equivalent to a contiguity relation of the Gauss hypergeometric function*

$$pF(p+1, q, r; x) = \left(x\frac{\partial}{\partial x} + p\right) F(p, q, r; x),$$

*where* $p = \mu + 1$, $q = -\eta$ *and* $r = \alpha + \mu + 1$.

REMARK A.11. *It is known that fractional calculus is useful to obtain relations of special functions. We refer to* [1] *and* [19] *as examples.*

**A.3. $P_V$ case.** The affine Weyl group $W(A_3^{(1)})$ is defined by the generators $s_0, s_1, s_2, s_3$ and the fundamental relations:

$$s_i^2 = 1 \quad (i = 0, 1, 2, 3), \quad s_i s_{i+1} s_i = s_{i+1} s_i s_{i+1}, \quad s_i s_{i+2} = s_{i+2} s_i \quad (i = 0, 1, 2, 3).$$

Here the indices are understood as elements in $\mathbb{Z}/4\mathbb{Z}$. It was shown in [13] that the quantum fifth Painlevé equation $QP_V(\alpha)$:

$$\kappa z\frac{\partial}{\partial z}\Psi(x, z) = H_V\left(x, \frac{\partial}{\partial x}, z, \alpha\right)\Psi(x, z),$$

where

$$H_V = (x - 1)(y + z)xy - (\alpha_1 + \alpha_3 - 1)xy + \alpha_1 y + (\alpha_2 + 1)zx, \quad y = \frac{\partial}{\partial x},$$

has the affine Weyl group symmetry of type $A_3^{(1)}$. Here $\alpha_0, \alpha_1, \alpha_2, \alpha_3$ are complex parameters such that $\alpha_0 + \alpha_1 + \alpha_2 + \alpha_3 = -\kappa$. Let us explain this symmetry further. The generators of $W(A_3^{(1)})$ act on the complex parameters $\alpha_0, \alpha_1, \alpha_2, \alpha_3$ as $s_i(\alpha_j) = \alpha_j - \alpha_i a_{ij}$ $(i, j = 0, 1, 2, 3)$, where $(a_{ij})$ is the Cartan matrix of type $A_3^{(1)}$. We define transformations $R_0, R_1, R_2, R_3$ for solutions as

$$R_0(\alpha_0) = e^{-xz}\partial^{-\alpha_0}e^{xz}, \quad R_1(\alpha_1) = x^{-\alpha_1}, \quad R_2(\alpha_2) = \partial^{-\alpha_2}, \quad R_3(\alpha_3) = (x-1)^{-\alpha_3}.$$

Then, we have

THEOREM A.12 ([13]). *If $\Psi(x, z)$ is a solution to $QP_V(\alpha)$, then $R_i(\alpha_i)\Psi(x, z)$ is a solution to $QP_V(s_i(\alpha))$, up to a gauge factor of $t$.*

The fundamental relations of $W(A_3^{(1)})$ and Theorem A.12 lead to the following theorem.

THEOREM A.13. *Let $\alpha, \beta \notin \mathbb{Z}_{\geq 1}$, $\mu, \mu + \alpha, \mu + \beta \notin \mathbb{Z}$. Then, we have*

$$e^{-xz}\partial_a^{-\alpha}e^{xz}\partial_a^{-\beta}f(x) = \partial_a^{-\beta}e^{-xz}\partial_a^{-\alpha}e^{xz}f(x).$$

PROOF. It is sufficient to prove that

$$e^{-xz}\int_a^x (x - t_2)^{\alpha-1}e^{t_2 z}\left(\int_a^{t_2} (t_2 - t_1)^{\beta-1}(t_1 - a)^\mu g(t_1)dt_1\right)dt_2$$

$$\text{(A.7)} \quad = \int_a^x (x - s_2)^{\beta-1}e^{-s_2 z}\left(\int_a^{s_2} (s_2 - s_1)^{\alpha-1}e^{s_1 z}(s_1 - a)^\mu g(s_1)ds_1\right)ds_2$$

for $a, x \in \mathbb{R}$ such that $a < x$, when the integrals of both sides converge, because of Proposition A.5. We fix branches of the multivalued functions of the integrand of both sides as $\arg(x - t_2) = 0$, $\arg(t_2 - t_1) = 0$, $\arg(t_1 - a) = 0$, $\arg(x - s_2) = 0$, $\arg(s_2 - s_1) = 0$, and $\arg(s_1 - a) = 0$.

Changing the variables $s_1$, $s_2$ in the integrand of the right-hand side of (A.7) as

$$s_2 = x - t_2 + t_1, \quad s_1 = t_1,$$

we see that the right-hand side of (A.7) becomes the left-hand side of (A.7).     $\square$

**A.4. $P_{IV}$ case.** The affine Weyl group $W(A_2^{(1)})$ is defined by the generators $s_0, s_1, s_2$ and the fundamental relations:

$$s_i^2 = 1 \quad (i = 0, 1, 2), \quad s_i s_{i+1} s_i = s_{i+1} s_i s_{i+1} \quad (i = 0, 1, 2).$$

Here the indices are understood as elements in $\mathbb{Z}/3\mathbb{Z}$. It was shown in [**13**] that the quantum fourth Painlevé equation $QP_{IV}(\alpha)$:

$$\kappa z \frac{\partial}{\partial z} \Psi(x, z) = H_{IV}\left( x, \frac{\partial}{\partial x}, z, \alpha \right) \Psi(x, z),$$

where

$$H_{IV} = yxy - xyx - zxy - \alpha_2 x - \alpha_1 y, \quad y = \frac{\partial}{\partial x},$$

has the affine Weyl group symmetry of type $A_2^{(1)}$. Here $\alpha_0, \alpha_1, \alpha_2$ are complex parameters such that $\alpha_0 + \alpha_1 + \alpha_2 = -\kappa$. Let us explain this symmetry further. The generators of $W(A_2^{(1)})$ act on the complex parameters $\alpha_0, \alpha_1, \alpha_2$ as $s_i(\alpha_j) = \alpha_j - \alpha_i a_{ij}$ $(i, j = 0, 1, 2)$, where $(a_{ij})$ is the Cartan matrix of type $A_2^{(1)}$. We define transformations $R_0, R_1, R_2$ for solutions as

$$R_0(\alpha_0) = e^{\frac{x^2}{2} + xz} \partial^{-\alpha_0} e^{-\frac{x^2}{2} - xz}, \quad R_1(\alpha_1) = x^{-\alpha_1}, \quad R_2(\alpha_2) = \partial^{-\alpha_2}.$$

Then, we have

THEOREM A.14 ([**13**]). *If $\Psi(x, z)$ is a solution to $QP_{IV}(\alpha)$, then $R_i(\alpha_i)\Psi(x, z)$ is a solution to $QP_{IV}(s_i(\alpha))$, up to a gauge factor of $t$.*

The fundamental relations of $W(A_2^{(1)})$ and Theorem A.14 lead to the following theorem.

THEOREM A.15. *Let $\alpha, \beta, \alpha+\beta \notin \mathbb{Z}_{\geq 1}$, $\mu, \mu+\alpha, \mu+\beta, \mu+\alpha+2\beta, \mu+2\alpha+\beta \notin \mathbb{Z}$. Then, we have*

$$\partial_a^{-\alpha} e^{\frac{x^2}{2} + xz} \partial_a^{-\alpha-\beta} e^{-\frac{x^2}{2} - xz} \partial_a^{-\beta} f(x)$$

$$= e^{\frac{x^2}{2} + xz} \partial_a^{-\beta} e^{-\frac{x^2}{2} - xz} \partial_a^{-\alpha-\beta} e^{\frac{x^2}{2} + xz} \partial_a^{-\alpha} e^{-\frac{x^2}{2} - xz} f(x).$$

PROOF. It is sufficient to prove that

$$\int_a^x dt_3 (x - t_3)^{\alpha-1} e^{\frac{t_3^2}{2} + t_3 z} \int_a^{t_3} dt_2 (t_3 - t_2)^{\alpha+\beta-1} e^{-\frac{t_2^2}{2} - t_2 z}$$

$$\times \int_a^{t_2} dt_1 (t_2 - t_1)^{\beta-1} (t_1 - a)^\mu g(t_1)$$

$$= e^{\frac{x^2}{2} + xz} \int_a^x ds_3 (x - s_3)^{\beta-1} e^{-\frac{s_3^2}{2} - s_3 z} \int_a^{s_3} ds_2 (s_3 - s_2)^{\alpha+\beta-1} e^{\frac{s_2^2}{2} + s_2 z}$$

(A.8)     $$\cdot \int_a^{s_2} ds_1 (s_2 - s_1)^{\alpha-1} e^{-\frac{s_1^2}{2} - s_1 z} (s_1 - a)^\mu g(s_1)$$

for $a, x \in \mathbb{R}$ such that $a < x$, when the integrals of both sides converge, because of Proposition A.5. We fix branches of the multivalued functions of the integrand of both sides as $\arg(x - t_3) = 0$, $\arg(t_3 - t_2) = 0$, $\arg(t_2 - t_1) = 0$, $\arg(t_1 - a) = 0$, $\arg(x - s_3) = 0$, $\arg(s_3 - s_2) = 0$, $\arg(s_2 - s_1) - 0$, and $\arg(s_1 - a) = 0$.

Changing the variables $s_1$, $s_2$, $s_3$ in the integrand of the right-hand side of (A.8) as

$$s_3 = x - \frac{(t_3 - t_2)(t_2 - t_1)}{x - t_3 + t_2 - t_1}, \quad s_2 = \frac{(t_3 - t_2)(x - t_3)}{x - t_3 + t_2 - t_1} - t_1, \quad s_1 = t_1,$$

we see that the right-hand side of (A.8) becomes the left-hand side of (A.8).    □

**A.5. $\mathrm{P_{III}}$ case.** The affine Weyl group $W(C_2^{(1)})$ is defined by the generators $s_0, s_1, s_2$ and the fundamental relations:

$$s_i^2 = 1 \quad (i = 0, 1, 2), \quad s_i s_2 s_i s_2 = s_2 s_i s_2 s_i \quad (i = 0, 2).$$

It was shown in [**13**] that the quantum third Painlevé equation $\mathrm{QP_{III}}(\alpha)$:

$$\kappa z \frac{\partial}{\partial z} \Psi(x, z) = H_{\mathrm{III}} \left( x, \frac{\partial}{\partial x}, z, \alpha \right) \Psi(x, z),$$

where

$$H_{\mathrm{III}} = xyxy - xyx + (\alpha_0 + \alpha_2 + 1)xy - \alpha_2 x + zy, \quad y = \frac{\partial}{\partial x}$$

has the affine Weyl group symmetry of type $C_2^{(1)}$. Here $\alpha_0, \alpha_1, \alpha_2$ are complex parameters such that $\alpha_0 + 2\alpha_1 + \alpha_2 = -\kappa$. Let us explain this symmetry further. The generators of $W(C_2^{(1)})$ act on the complex parameters $\alpha_0, \alpha_1, \alpha_2$ as $s_i(\alpha_j) = \alpha_j - \alpha_i a_{ij}$ $(i, j = 0, 1, 2)$, where $(a_{ij})$ is the Cartan matrix of type $C_2^{(1)}$. We define transformations $R_0, R_1, R_2$ for solutions as

$$R_0(\alpha_0) = e^x \partial^{-\alpha_0} e^{-x}, \quad R_1(\alpha_1) = (z \mapsto -z) e^{-\frac{z}{x}} x^{-2\alpha_1}, \quad R_2(\alpha_2) = \partial^{-\alpha_2}.$$

Then, we have

THEOREM A.16 ([**13**]). *If $\Psi(x, z)$ is a solution to $\mathrm{QP_{III}}(\alpha)$, then $R_i(\alpha_i)\Psi(x, z)$ is a solution to $\mathrm{QP_{III}}(s_i(\alpha))$, up to a gauge factor of $t$.*

The fundamental relations of $W(C_2^{(1)})$ and Theorem A.16 lead to the following theorem.

THEOREM A.17. *Let $\alpha, \alpha + 2\beta \notin \mathbb{Z}_{\geq 1}$, $\mu, \mu + \alpha, \mu + \alpha + 2\beta, \mu + 2\alpha + 2\beta \notin \mathbb{Z}$, and $a \neq 0$. Then, we have*

$$\partial_a^{-\alpha} e^{\frac{z}{x}} x^{-2(\alpha+\beta)} \partial_a^{-\alpha-2\beta} e^{-\frac{z}{x}} x^{-2\beta} f(x) = e^{\frac{z}{x}} x^{-2\beta} \partial_a^{-\alpha-2\beta} e^{-\frac{z}{x}} x^{-2(\alpha+\beta)} \partial_a^{-\alpha} f(x).$$

PROOF. It is sufficient to prove that

$$\int_a^x (x - t_2)^{\alpha-1} e^{\frac{z}{t_2}} t_2^{-2(\alpha+\beta)} \left( \int_a^{t_2} dt_1 (t_2 - t_1)^{\alpha+2\beta-1} e^{-\frac{z}{t_1}} t_1^{-2\alpha} (t_1 - a)^\mu g(t_1) \right) dt_2$$

(A.9)

$$= \int_a^x (x - s_2)^{\beta-1} e^{-s_2 z} \left( \int_a^{s_2} (s_2 - s_1)^{\alpha-1} e^{s_1 z} (s_1 - a)^\mu g(s_1) ds_1 \right) ds_2$$

for $a, x \in \mathbb{R}$ such that $0 < a < x$, when the integrals of both sides converge, because of Proposition A.5. We fix branches of the multivalued functions of the integrand of both sides as $\arg(x - t_2) = 0$, $\arg(t_2 - t_1) = 0$, $\arg(t_1) = 0$, $\arg(t_1 - a) = 0$, $\arg(x - s_2) = 0$, $\arg(s_2 - s_1) = 0$, and $\arg(s_1 - a) = 0$.

Changing the variables $s_1$, $s_2$ in the integrand of the right-hand side of (A.9) as

$$s_2 = \left( \frac{1}{x} - \frac{1}{t_2} + \frac{1}{t_1} \right)^{-1}, \quad s_1 = t_1,$$

we see that the right-hand side of (A.9) becomes the left-hand side of (A.9).    □

## References

[1] R. Askey, *Orthogonal polynomials and special functions*, Society for Industrial and Applied Mathematics, Philadelphia, Pa., 1975. MR0481145 (58 #1288)

[2] R. Conte (ed.), *The Painlevé property: One century later*, CRM Series in Mathematical Physics, Springer-Verlag, New York, 1999. MR1713574 (2000e:34001)

[3] B. Gambier, *Sur les équations différentielles du second ordre et du premier degré dont l'intégrale générale est a points critiques fixes* (French), Acta Math. **33** (1910), no. 1, 1–55, DOI 10.1007/BF02393211. MR1555055

[4] Y. Haraoka and G. Filipuk, *Middle convolution and deformation for Fuchsian systems*, J. Lond. Math. Soc. (2) **76** (2007), no. 2, 438–450, DOI 10.1112/jlms/jdm064. MR2363425 (2009e:34262)

[5] K. Iwasaki, H. Kimura, S. Shimomura, and M. Yoshida, *From Gauss to Painlevé: A modern theory of special functions*, Aspects of Mathematics, E16, Friedr. Vieweg & Sohn, Braunschweig, 1991. MR1118604 (92j:33001)

[6] M. Jimbo, H. Nagoya, and J. Sun, *Remarks on the confluent KZ equation for $\mathfrak{sl}_2$ and quantum Painlevé equations*, J. Phys. A **41** (2008), no. 17, 175205, 14, DOI 10.1088/1751-8113/41/17/175205. MR2451670 (2009h:32023)

[7] G. Kuroki, *Quantum groups and quantization of Weyl group symmetries of Painlevé systems*, Exploring new structures and natural constructions in mathematical physics, Adv. Stud. Pure Math., vol. 61, Math. Soc. Japan, Tokyo, 2011, pp. 289–325. MR2867150 (2012k:81140)

[8] K. S. Miller and B. Ross, *An introduction to the fractional calculus and fractional differential equations*, A Wiley-Interscience Publication, John Wiley & Sons, Inc., New York, 1993. MR1219954 (94e:26013)

[9] H. Nagoya, *Quantum Painlevé systems of type $A_l^{(1)}$*, Internat. J. Math. **15** (2004), no. 10, 1007–1031, DOI 10.1142/S0129167X0400265X. MR2106261 (2005j:37126)

[10] H. Nagoya, *Quantum Painlevé systems of the type $A_{n-1}^{(1)}$ with higher degree Lax operators*, Internat. J. Math. **18** (2007), no. 7, 839–868, DOI 10.1142/S0129167X07004321. MR2337159 (2008g:34238)

[11] H. Nagoya, *A quantization of the sixth Painlevé equation*, Noncommutativity and singularities, Adv. Stud. Pure Math., vol. 55, Math. Soc. Japan, Tokyo, 2009, pp. 291–298. MR2463505 (2010i:37177)

[12] H. Nagoya, *Hypergeometric solutions to Schrödinger equations for the quantum Painlevé equations*, J. Math. Phys. **52** (2011), no. 8, 083509, 16, DOI 10.1063/1.3620412. MR2858065 (2012j:81090)

[13] H. Nagoya, *Realizations of affine Weyl group symmetries on the quantum Painlevé equations by fractional calculus*, Lett. Math. Phys. **102** (2012), no. 3, 297–321, DOI 10.1007/s11005-012-0557-6. MR2989486

[14] H. Nagoya, *Integral formulas for quantum isomonodromic systems*, Publ. Res. Inst. Math. Sci. **49** (2013), no. 4, 651–678, DOI 10.4171/PRIMS/115. MR3141719

[15] M. Noumi and Y. Yamada, *Higher order Painlevé equations of type $A_l^{(1)}$*, Funkcial. Ekvac. **41** (1998), no. 3, 483–503. MR1676885 (2000i:34019)

[16] D. P. Novikov, *Integral transformation of solutions of a Fuchs-class equation that corresponds to the Okamoto transformation of the Painlevé VI equation* (Russian, with Russian summary), Teoret. Mat. Fiz. **146** (2006), no. 3, 355–364, DOI 10.1007/s11232-006-0040-6; English transl., Theoret. and Math. Phys. **146** (2006), no. 3, 295–303. MR2253624 (2008e:34215)

[17] D. P. Novikov, *The Schlesinger system with $2 \times 2$ matrices and the Belavin-Polyakov-Zamolodchikov equation* (Russian, with Russian summary), Teoret. Mat. Fiz. **161** (2009), no. 2, 191–203, DOI 10.1007/s11232-009-0135-y; English transl., Theoret. and Math. Phys. **161** (2009), no. 2, 1485–1496. MR2667345 (2011j:34297)

[18] K. Okamoto, *Studies on the Painlevé equations. IV. Third Painlevé equation $P_{III}$*, Funkcial. Ekvac. **30** (1987), no. 2-3, 305–332. MR927186 (88m:58065)

[19] T. Oshima, *Fractional calculus of Weyl algebra and Fuchsian differential equations*, MSJ Memoirs, vol. 28, Mathematical Society of Japan, Tokyo, 2012. MR2986408

[20] P. Painlevé, *Mémoire sur les équations différentielles dont l'intégrale générale est uniforme* (French), Bull. Soc. Math. France **28** (1900), 201–261. MR1504376

[21] B. I. Suleĭmanov, *The Hamilton property of Painlevé equations and the method of isomonodromic deformations* (Russian, with Russian summary), Differentsial'nye Uravneniya **30** (1994), no. 5, 791–796, 917; English transl., Differential Equations **30** (1994), no. 5, 726–732. MR1306348 (95h:35226)

[22] Y. Sasano and Y. Yamada, *Symmetry and holomorphy of Painlevé type systems*, Algebraic, analytic and geometric aspects of complex differential equations and their deformations. Painlevé hierarchies, RIMS Kôkyûroku Bessatsu, B2, Res. Inst. Math. Sci. (RIMS), Kyoto, 2007, pp. 215–225. MR2310032 (2008c:34191)

[23] K. Takemura, *Integral representation of solutions to Fuchsian system and Heun's equation*, J. Math. Anal. Appl. **342** (2008), no. 1, 52–69, DOI 10.1016/j.jmaa.2007.11.015. MR2440779 (2009e:34260)

[24] A. Zabrodin and A. Zotov, *Quantum Painlevé-Calogero correspondence for Painlevé VI*, J. Math. Phys. **53** (2012), no. 7, 073508, 19, DOI 10.1063/1.4732534. MR2985248

DEPARTMENT OF MATHEMATICS, RIKKYO UNIVERSITY, TOKYO 171-8501, JAPAN
*E-mail address*: nagoya.hajime@rikkyo.ac.jp

Contemporary Mathematics
Volume **651**, 2015
http://dx.doi.org/10.1090/conm/651/13046

# Spectral Curves and Discrete Painlevé Equations

## Christopher M. Ormerod

ABSTRACT. It is well known that isomonodromic deformations admit a Hamiltonian description. These Hamiltonians appear as coefficients of the characteristic equations of their Lax matrices, which define spectral curves for linear systems of differential and difference systems. The characteristic equations in the case of the associated linear problems for various discrete Painlevé equations is biquadratic in the Painlevé variables. We show that the discrete isomonodromic deformations that define the discrete Painlevé equations may be succinctly described in terms of the characteristic equation of their Lax matrices.

## 1. Introduction

This article concerns Lax pairs for the discrete Painlevé equations [**36**]. These Lax pairs are pairs of differential or difference operators in two variables; a spectral variable, $x$, and an independent variable, $n$. The operators may be written in matrix form as

$$(1.1a) \qquad (\Delta_x - A_n(x))Y_n(x) = 0,$$

$$(1.1b) \qquad (\Delta_n - R_n(x))Y_n(x) = 0,$$

where $A_n(x)$ and $R_n(x)$ are meromorphic matrices in $x$ and $\Delta_x$ is one of three cases

$$(1.2) \qquad \Delta_x = \frac{d}{dx} : f_n(x) \mapsto \frac{df_n(x)}{dx},$$

$$(1.3) \qquad \Delta_x = \sigma_h : f_n(x) \mapsto f_n(x + h),$$

$$(1.4) \qquad \Delta_x = \sigma_q : f_n(x) \mapsto f_n(qx),$$

and $\Delta_n : f_n(x) \to f_{n+1}(x)$ [**2, 21, 28, 36, 41, 56**]. Computing the compatibility between (1.1a) and (1.1b) induces a transformation of the form

$$(1.5) \qquad A_n(x) \to A_{n+1}(x),$$

which we call a discrete isomonodromic deformation.

Given a particular operator of the form (1.1a) there is an algorithmic method for obtaining an operator of the form (1.1b) compatible with (1.1a). When $\Delta_x$ is a differential operator these deformations are known as Schlesinger transformations

---

2010 *Mathematics Subject Classification.* Primary 34M55, 34M56, 39A13, 39A10.

*Key words and phrases.* Discrete Painlevé, integrable, Lax pairs, spectral curve, isomonodromy.

[**20**, **47**]. When $\Delta_x$ is a difference operator these transformations are called connection (matrix) preserving deformations [**7**, **21**, **32**, **46**]. In fact for any given (1.1a) there is a finitely generated lattice of operators of the form of (1.1b), which we call a system of discrete isomonodromic deformations [**32**] or Schlesinger system [**43**].

Information such as the number and multiplicity of poles of $A_n(x)$ and asymptotic behavior of the solutions of (1.1a) determine which systems arises as discrete isomonodromic deformations. We could say that this information defines the "type" of a linear system. For example, the associated linear problem for the sixth Painlevé equation is determined by four Fuchsian singularities. Once an equations type been ascertained, it is a simple matter of parameterizing (1.1a) in the right way. This idea has been incredibly useful in applications such as reductions of partial differential and difference equations [**13**, **34**, **35**] and semiclassical orthogonal polynomials [**14**, **27**, **31**].

In the language of sheaves a system of linear differential equations defines a connection on a vector bundle, which coincides with a matrix presentation when specialized to a trivial bundle [**26**]. A discrete version of this framework is the $d$-connection, which was considered by Arinkin and Borodin [**2**]. In this setting, the type of a system of linear equations lends itself naturally to the idea of moduli spaces of ($d$-)connections. The Painlevé variables parameterize these moduli spaces of ($d$-)connections on vector bundles. In fact, the minimal compactification of these moduli spaces may be identified with the rational surfaces of initial conditions for the Painlevé equations [**2**, **45**]. Using this approach, it is possible to show that Lax pairs of a certain form exist without necessarily providing a parameterization [**6**].

The identification of an integrable system on the cotangent bundle of the moduli space of connections is the subject of Hitchin systems [**17**]. This paper has been inspired by of the analogies one can draw in the discrete setting, which have been called generalized Hitchin systems [**42**]. The key observation in Hitchin's framework is that the characteristic equation that defines the spectral curve gives a set of Hamiltonians whose flows are linear on the Jacobian of the spectral curve [**17**]. This may be extended in the non-autonomous case for Painlevé equations, giving a Hamiltonian formulation for isomonodromic deformations [**26**, **30**].

If we turn our attention to (1.5) in the autonomous setting (where $A_{n+1}(x) = A_n(x)$) we expect that the characteristic equation gives invariants [**55**]. In the case of Lax pairs for QRT mappings the invariants that appear in the characteristic equation are biquadratics [**38**, **39**]. Since QRT mappings are defined by the addition law on a biquadratic [**51**] the QRT maps are linear on the Jacobian of the spectral curve [**42**]. A similar geometric setting for the discrete Painlevé equations may be posed in terms of the addition law on a moving biquadratic curve [**22**]. The way in which these systems are defined and related suggests that the discrete isomonodromic deformations in the case of the QRT mappings and the discrete Painlevé equations admit a description of the form

$$(1.6\mathrm{a}) \qquad\qquad \tilde{\Gamma}(\lambda, x, y_{n+1}, z_n) = \tilde{\Gamma}(\lambda, x, y_n, z_n),$$

$$(1.6\mathrm{b}) \qquad\qquad \tilde{\Gamma}(\lambda, x, y_{n+1}, z_{n+1}) = \tilde{\Gamma}(\lambda, x, y_{n+1}, z_n),$$

where $\Gamma = \det(\lambda - A_n(x))$ is the characteristic equation for $A_n(x)$ and $(y_n, z_n)$ parameterize the biquadratic spectral curve [**16**, **18**] (the trivial solutions $y_{n+1} = y_n$ and $z_{n+1} = z_n$ are discarded). We use the notation $\tilde{\Gamma}$ to mean the characteristic equation with some intermediate parameter values that do not necessarily correspond to $A_n(x)$ or $A_{n+1}(x)$. For differential operators (1.6) coincides with treating the Hamiltonian as a QRT-type invariant, which is considered a method for obtaining integrable discretizations of biquadratic Hamiltonian systems, such as the discrete Painlevé equations [29]. Our contribution is that the characteristic equation for difference operators is also tied to the geometry of the discrete Painlevé equations.

In §2, we will review some of the theory regarding the Hamiltonian description of isomonodomic deformations for differential equations and the geometry of the QRT mappings and discrete Painlevé equations. In §3 we consider how this applies to contiguity relations for the second and sixth Painlevé equations and two discrete analogues of the sixth Painlevé equation.

## 2. Spectral curves and isomonodromic deformations

We wish to explain why the role of the characteristic equation in the Hamiltonian description of isomonodromic deformations. To relate this to discrete isomonodromic deformations, we require the formal series solutions of (1.1a) in each of the cases, which will give us a way of computing (1.1b) to compare against (1.6).

**2.1. Hamiltonian description of isomondromic deformations.** We start with a linear problem of the form of (1.1a) with (1.2) where $A_n(x)$ is rational. Let $A_n(x)$ have a finite collection of poles, $\{a_1, \ldots, a_N\}$ (and possibly $\{\infty\}$) where the order of the pole at $x = a_\nu$ is $r_\nu$ $(r_\infty)$. The matrix $A_n(x)$ takes the general form

$$(2.1) \qquad A_n(x) = \sum_{\nu=1}^{N} \sum_{k=0}^{r_\nu} \frac{A_{\nu,k}}{(x - a_\nu)^{k+1}} - \sum_{k=1}^{r_\infty} A_{\infty,-k} x^{k-1}.$$

We should assume that the leading coefficients, $A_{\nu,r_\nu}$, are semisimple with matrices $C_\nu$ such that

$$A_{\nu,r_\nu} = C_\nu T_\nu C_\nu^{-1},$$

where $T_\nu = \mathrm{diag}(t_{\nu,1}, \ldots t_{\nu,m})$. We may normalize the system so that $C_\infty = I$. We also require the technical conditions (see [19]) that

$$t_{\nu,i} \neq t_{\nu,j} \quad \text{if} \quad r_\nu \geq 1, \quad i \neq j,$$
$$t_{\nu,i} - t_{\nu,j} \notin \mathbb{Z} \quad \text{if} \quad r_\nu = 0, \quad i \neq j.$$

When prolonging a solution along a path around any collection of the poles, we obtain a relation

$$(2.2) \qquad Y_n(\gamma(1)) = Y_n(\gamma(0)) M_{[\gamma]},$$

where $[\gamma]$ denotes the equivalence class of paths under homotopy and $M_{[\gamma]}$ is called a monodromy matrix [54]. If $X$ denotes the punctured sphere $\mathbb{P}_1 \backslash \{a_1, \ldots, a_N, \infty\}$, for any element $[\gamma] \in \pi_1(X)$ we obtain a matrix representation

$$\Pi : \pi_1(X) \to \mathrm{GL}_m(\mathbb{C}).$$

We may choose a set of generators of $\pi_1(X)$, denoted $[\gamma_i]$, so that the images, $\Pi([\gamma_i]) = M_i$, satisfy

$$M_1 M_2 \ldots M_N M_\infty = I,$$

which is equivalent to $[\gamma_1 \ldots \gamma_N \gamma_\infty] = 1$.

It will be useful to specify a formal solution, which we write as

$$(2.3) \qquad Y_n(x) = C_\nu \hat{Y}_{n,\nu}(x) \exp \hat{T}_\nu(x),$$

where $\hat{Y}_{n,\nu}(x)$ is just some series expansion in $(x - a_\nu)$ such that the constant term $\hat{Y}_{n,\nu}(a_\nu)$ is $I$ and $T_\nu(x)$ is an expansion of the form

$$(2.4) \qquad \hat{T}_\nu(x) = \sum_{k=1}^{r_\nu} T_{\nu,k} \frac{(x - a_\nu)^{-k}}{-k} + T_{\nu,0} \log (x - a_\nu).$$

Generally $\hat{Y}_{n,\nu}(x)$ is not necessarily convergent. Given a point, $z$, in some neighborhood of $x = a_\nu$, there is a basis of solutions that is convergent in some neighborhood of $z$. Let us denote the matrix containing the basis of meromorphic solutions by $Y_{n,\nu}^{(i)}$. The collection of points in which $Y_{n,\nu}^{(i)}$ is convergent defines a Stokes sector. This divides a neighborhood of $a_\nu$ into a collection of precisely $2r_\nu$ Stokes sectors.

Given the columns of $Y_{n,\nu}^{(i)}$ and $Y_{n,\nu}^{(i+1)}$ both constitute a basis for formal solutions to (1.1a), we may express the solution, $Y_{n,\nu}^{(i+1)}$ as a linear combination of $Y_{n,\nu}^{(i)}$, which means that there exists a relation of the form

$$(2.5) \qquad Y_{n,\nu}^{(i+1)} = Y_{n,\nu}^{(i)} S_\nu^{(i)},$$

where $S_\nu^{(i)}$ is a constant matrix called the Stokes matrix (here $S_\nu^{(2r_\nu)}$ relates $S_\nu^{(2r_\nu)}$ to $S_\nu^{(1)}$). This gives us a collection of constants that govern the asymptotic behavior of the solutions around the poles of $A_n(x)$ and the Stokes matrices for irregular singularities [19]. In forming the monodromy matrix, every path around $x = a_\nu$ passes through each of the Stokes sectors, collecting a contribution from each of the Stokes matrices. We may specify the monodromy matrices in terms of this data as

$$(2.6) \qquad M_\nu = C_\nu \exp \left(2\pi i T_{\nu,0}\right) S_j^{(2r_\nu)} \ldots S_\nu^{(2)} S_\nu^{(1)} C_\nu^{-1}.$$

The Pfaffian system describing monodromy preserving deformations is specified by the following theorem.

THEOREM 2.1 (Theorem 1 of [19]). *The monodromy matrices are preserved if and only if there exists a matrix of 1-forms $\Omega_n(x)$ depending rationally on $x$ and a matrix of 1-forms $\Theta_\nu$ such that*

$$(2.7a) \qquad \mathrm{d}A_n(x) = \frac{\partial \Omega_n(x)}{\partial x} + \Omega_n(x)A_n(x) - A_n(x)\Omega_n(x),$$

$$(2.7b) \qquad \mathrm{d}C_\nu = \Theta_\nu C_\nu,$$

*where these 1-forms, $\Omega_n(x)$ and $\Theta_\nu$, are calculable by a rational procedure from $A(x)$ and d denotes exterior differentiation with respect to some deformation parameters.*

We have specified the process is rational, however, as the precise formulation is not the emphasis of this paper. We leave the reader with a reference to the work of Jimbo et al. [19]. A remarkable consequence of [19, 20] is the general integrability of the resulting system of partial differential equations defined by (2.7).

THEOREM 2.2 (Theorem 2 of [19]). *The non-linear differential equations are completely integrable in the sense of Frobenius in each of the variables*

$$\left\{ \begin{array}{c} a_1, \ldots, a_n \\ t_{1,1}, \ldots, t_{1,n} \\ t_{\nu,1}, \ldots, t_{\mu,n} \end{array} \right\}.$$

This means there is a continuous deformation in each of the $t_{i,j}$ for $j \geq 1$. For Painlevé equations we have an ideal of one-forms in the ring of differentials in one varible that is closed under external differentiation on isomonodromic deformations.

This Pfaffian system is defined by the collection of 1-forms

$$(2.8) \qquad \omega = \sum_i \omega_i,$$

$$(2.9) \qquad \omega_i = -\text{Res}_{x=a_i} \text{Tr} \hat{Y}_i(x) \frac{\partial \hat{Y}_i(x)}{\partial x}(x) \mathrm{d} T_i(x),$$

which is closed on solutions of the isomonodromic deformations. A succinct form for the Hamiltonian is due to Krichever [23].

THEOREM 2.3 (Theorem 2.1 in [23]). *The non-linear equations isomonodromic deformations are Hamiltonian with respect to the Hamiltonians defined by*

$$(2.10) \qquad H_{n,t_p} := -\frac{1}{n+1} \text{Tr} A(x)^{n+1} \Big|_{x=t_p}.$$

These Hamiltonians also appear as the coefficients of the characteristic equations. This theorem is reminiscent of the theory of invariants for discrete autonomous integrable mappings arising as reductions of partial difference equations [55]. When $r_\nu = 0$ for $\nu = 1, \ldots, N$ (and $r_\infty = 0$), (2.1) defines a Fuchsian system whose isomonodromic deformations is a Hamiltonian system with respect to the Hamiltonians

$$(2.11) \qquad H_j = \sum_{k \neq j \neq \infty} \frac{\text{Tr}(A_{j,0} A_{k,0})}{a_j - a_k}.$$

This description is due to Okamoto [30]. A simple expansion shows how these Hamiltonians appear in the coefficients of $\lambda$ in the characteristic equation

$$\Gamma(\lambda) = \lambda^m - \lambda^{m-1} \text{Tr} A(x)$$

$$+ \left( \sum_j \frac{1}{x - a_j} \sum_{k \neq j \neq \infty} \frac{\text{Tr}(A_{j,0} A_{k,0})}{a_j - a_k} \right) \lambda^{m-2} + O(\lambda^{m-3}).$$

More generally, the coefficients of the characteristic equations are expressible in terms of the determinants and traces of the $A_{i,j}$ and these Hamiltonians.

**2.2. Schlesinger transformations and spectral curves.** The aim of this section is to provide a way computing (1.1b). For systems of differential equations, from (2.6) it is easy to see that an integer shift in any collection of the entries of the $T_{\nu,0}$ results in the same monodromy matrices. If we identify an collection of integer shifts in the entries of $T_{\nu,0}$ with the shift $n \to n + 1$, we may use (2.3) to compute $R_n(x) = Y_{n+1}(x) Y_n(x)^{-1}$ [20, 47]. We need to specify what the discrete analogue of the formal solutions in (2.3) to calculate $R_n(x)$ for systems of difference equations.

If $\Delta_x$ is specified by (1.3) and $A_n(x)$ is rational, multiplying $Y_n(x)$ by gamma functions allows us to express $A_n(x)$ in polynomial form as

$$(2.12) \qquad A_n(x) = A_0 + A_1 x + \ldots + A_N x^N,$$

where the $A_i$ are constant in $x$. For systems of difference equations, we may use the formal solution specified by the following theorem.

THEOREM 2.4. *If $A_N = \mathrm{diag}(\kappa_1, \ldots, \kappa_m)$ where*

$$\kappa_i \neq 0, \quad i = 1, \ldots, m, \qquad\qquad \kappa_i/\kappa_j \notin \mathbb{R}, \quad i \neq j,$$

*then there exists unique fundamental solutions of* (1.1a), $Y_{-\infty}(x)$ *and* $Y_{\infty}(x)$, *of the form*

$$(2.13) \qquad Y_{\pm\infty}(x) = x^{Nx} e^{-Nx} \left( Y_0 + \frac{Y_1}{x} + \frac{Y_2}{x^2} + \ldots \right) \mathrm{diag}\left( \kappa_1^x x^{r_1}, \ldots, \kappa_m^x x^{r_m} \right)$$

*such that*

(1) $Y_{\infty}(x)$ *and* $Y_{-\infty}(x)$ *are analytic throughout the complex plane, except at possibly integer multiples of $h$ to the left and right of the roots of $A_n(x)$ respectively.*

(2) $Y_{\infty}(x)$ *and* $Y_{-\infty}(x)$ *are asymptotically represented by* (2.12).

For systems of $q$-difference equation, we may use $q$-Gamma functions (see [15] for example) to reduce the case in which $A_n(x)$ is rational to one in which $A_n(x)$ is polynomial, and hence, is also given by (2.12).

THEOREM 2.5. *If $A_0$ and $A_N$ are semisimple with eigenvalues $\theta_1, \ldots, \theta_m$ and $\kappa_1, \ldots, \kappa_m$ respectively, with*

$$(2.14) \qquad\qquad \frac{\lambda_i}{\lambda_j}, \frac{\kappa_i}{\kappa_j} \notin q^{\mathbb{N}^+}, \qquad \forall i, j,$$

*then we have formal solutions*

$$(2.15a) \qquad\qquad Y_0(x) = \widehat{Y}_0(x) \mathrm{diag}\left( e_{q,\lambda_1}(x) \right)$$

$$(2.15b) \qquad\qquad Y_{\infty}(x) = \widehat{Y}_{\infty}(x) \mathrm{diag}\left( \theta_q(x/q)^{-N} e_{q,\lambda_i}(x) \right)$$

*where $\widehat{Y}_0(x)$ and $\widehat{Y}_{\infty}(x)$ are series around $x = 0$ and $x = \infty$ respectively.*

The functions $\theta_q(x)$ and $e_{q,c}(x)$ in this theorem satisfy

$$qx\theta_q(qx) = \theta_q(x), \qquad\qquad e_{q,c}(qx) = ce_{q,c}(x).$$

There is a generalization of this symbolic form in cases in which some of the eigenvalues are 0 in the work of Birkhoff and Guenther [5], and when the (2.14) is not satisfied by Adams [1]. A cleaner and even more general existence theorem based on vector bundles on Riemann surfaces is due to Praagman [37]. The difference analogue of the monodromy matrices is considered to be the (Birkhoff's) connection matrix, which is a invariant under $\Delta_x$ that relates the two formal series solutions [3, 4].

Since in both cases $A_n(x)$ is polynomial, we write

$$\det A_n(x) = \prod_{j=1}^{m} \kappa_j \prod_{i=1}^{mN} (x - a_i),$$

where $a_i \neq 0$. This expression in the $q$-difference case gives us a relation between the $\theta_i$'s, $\kappa_j$'s and the $a_k$'s. Just as the $M_i$ were periodic in the values of $T_{\nu,0}$ the differential case, the connection matrices are periodic or quasi-periodic (i.e., $f(a) = f(qa)$) in the $\theta_i$'s, $\kappa_j$'s and the $a_k$'s. The way in which the discrete Painlevé equations arise is that we associate a shift in a collection of the periodic or quasi-periodic variables with the transformation $n \to n + 1$.

THEOREM 2.6. *Given a system of the form* (1.1a), *a discrete isomonodromic deformation is governed by*

$$(2.16) \qquad Y_{n+1}(x) = R_n(x)Y_n(x).$$

We may compute $R_n(x)$ using (2.13) and (2.15) to give (1.1b). Using (1.1a) and (2.16), we obtain the compatibility when we require the solutions satisfy $\Delta_x \Delta_n Y_n(x) = \Delta_n \Delta_x Y_n(x)$. For the cases (1.2), (1.3) and (1.4) we have

$$(2.17a) \qquad A_{n+1}(x)R_n(x) = R_n(x)A_n(x) + \frac{\mathrm{d}R(x)}{\mathrm{d}x},$$
$$(2.17b) \qquad A_{n+1}(x)R_n(x) = R_n(x+h)A_n(x),$$
$$(2.17c) \qquad A_{n+1}(x)R_n(x) = R_n(qx)A_n(x),$$

which may be solved for $A_{n+1}(x)$ to induce a map of the form (1.5).

We now turn to why we are drawn to (1.6). In the differential setting for isospectral deformations the Hamiltonians were connected to the spectral curve. In the difference setting, it is the invariants are connected to the spectral curve [**55**]. These invariants and Hamiltonians essentially play the same role. We seek a discrete evolution on the spectral curve that is linear on the Jacobian of the curve. In the cases we consider, the characteristic equation defines a biquadratic curve, hence, we consider an action defined by the group law on biquadratics.

If $\Gamma$ defines a fibration of the plane by biquadratics in coordinates $(y, z) \in \mathbb{P}_1^2$, then the QRT map is given by (1.6) where $\Gamma = \hat{\Gamma} = \tilde{\Gamma}$. A fundamental result of Tsuda is that if we embed biquadratic fibres in $\mathbb{P}_2$ as a cubic plane curves, the QRT map admits the description

$$(2.18) \qquad \hat{Q} + P_9 = Q + P_8,$$

where $P_9$ and $P_8$ are the images of $y = \infty$ and $z = \infty$. We have depicted this in Figure 1. These points are two of nine base points [**51**]. In particular, when we identify the elliptic curve with its Jacobian the action of the QRT map is discrete and linear.

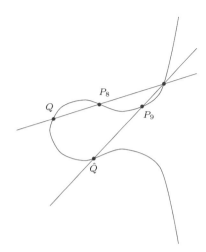

FIGURE 1. The geometric interpretation of the QRT map.

The discrete Painlevé equations are nonautonomous versions of the QRT maps, but they are not directly associated with a fibration of the plane by biquadratic curves, but rather an intermediate fibration. The evolution of the discrete Painlevé equations may be described in terms of moving biquadratic curves [**22**] in the following way; let $P_1, \ldots, P_9$ be 9 points in $\mathbb{P}_2$ and $\Gamma_0$ the unique cubic curve passing through these points. We let $T$ be the birational mapping that fixes $P_1, \ldots, P_7$ and sends $P_8$ and $P_9$ to $\hat{P}_8$ and $\hat{P}_9$ respectively in accordance with

$$P_1 + \ldots + P_7 + P_8 + \hat{P}_9 = 0,$$
$$P_8 + P_9 = \hat{P}_8 + \hat{P}_9,$$

which interpreted in terms of the group law $\Gamma_0$. In this setting, the Painlevé variables, encoded in the point $Q = [y : z : 1]$, is sent to $\hat{Q}$, which is determined by

(2.19) $$\hat{Q} + \hat{P}_9 = \hat{Q} + P_8,$$

on the fibre containing $Q$ in the fibration by cubic plane curves with base points $P_1, \ldots, P_8, \hat{P}_9$. This elegant description of the discrete Painlevé equations is the non-autonomous variant of (2.18). What distinguishes the QRT maps is that $P_1 + \ldots + P_9 = 0$ on $\Gamma_0$. If the autonomous limits of the discrete Painlevé equations as QRT maps were to make any sense, $P_8$ and $P_9$ need to be chosen on the fibre of (2.19) in the same way as the QRT case. This also explains why we seek an intermediate curve, $\tilde{\Gamma}(x)$, since a priori the points $P_1, \ldots, P_8, \hat{P}_9$ are not associated with the parameter values of either $A_n(x)$ or $A_{n+1}(x)$. Naturally, any valid choice of basepoints should give a Schlesinger transformation, which seems a natural geometric setting for the Schlesinger transformations. We do not pursue this here, but this seems linked to the setting of Rains [**42**].

## 3. Examples

We have tried to capture the above theory in a set of examples that demonstrate the principles. We start with something simple to demonstrate the mechanics, then we choose three associated linear problems that are regular in the sense of the theorems provided. We consider the sixth Painlevé equation [**47**], the $q$-analogue of the sixth Painlevé equation [**21**] and the discrete analogue of the sixth Painlevé equations [**2**].

### 3.1. Schlesinger transformations for the second Painlevé equation.
The second Painlevé equation arises an isomonodromic deformation of an irregular system of linear differential equations [**13**]. It is an illustrative example since there is only one parameter involved. This makes it simple to determine the change in parameter required in (1.6).

The second Painlevé equation may be written as

(3.1) $$\frac{\mathrm{d}^2 y}{\mathrm{d}t^2} = 2y^3 + yt + \alpha.$$

The associated linear problem for (3.1) is given by

$$
(3.2) \quad \frac{dY_n(x)}{dx} = \left( \begin{pmatrix} 1 & 0 \\ 0 & -1 \end{pmatrix} x^2 + \begin{pmatrix} 0 & u \\ \frac{2z}{u} & 0 \end{pmatrix} x + \begin{pmatrix} \frac{t}{2} - z & -uy \\ \frac{2\theta + 2yz}{u} & z - \frac{t}{2} \end{pmatrix} \right) Y_n(x),
$$

where $\alpha = \theta - 1/2$. The isomondromic deformation may be written as the compatibility of (3.2) with

$$
\frac{dY_n(x)}{dt} = \frac{1}{2} \left( \begin{pmatrix} 1 & 0 \\ 0 & -1 \end{pmatrix} x + \begin{pmatrix} 0 & u \\ \frac{2z}{u} & 0 \end{pmatrix} \right) Y_n(x).
$$

By computing entries of the compatibility, we obtain

$$
(3.3a) \qquad\qquad\qquad y' = z - y^2 - \frac{t}{2},
$$

$$
(3.3b) \qquad\qquad\qquad z' = 2yz + \theta,
$$

whose equivalence to (3.1) is easily verified.

We consider the characteristic equation for $A_n(x)$,

$$
(3.4) \qquad \Gamma(\lambda, x) = \left( \lambda^2 - \frac{1}{4} \left(t + 2x^2\right)^2 - 2\theta x \right) + \frac{1}{2} \left( \frac{z(t-z)}{2} + y^2 z + \theta y \right),
$$

where we have bracketed terms that depend on the Painlevé variables, $y$ and $z$, and those that do not. Using Theorem 2.1 for systems of linear differential equations with irregular singular points we obtain the following Hamiltonian description of the evolution.

COROLLARY 3.1. *The ismonodromic deformation admits a Hamiltonian description with respect to the following Hamiltonian*

$$
(3.5) \qquad\qquad H_{II} = H_{II}(y, z, \theta) = \frac{z(t-z)}{2} + y^2 z + \theta y.
$$

From the discussion in the previous section, the appearance of the Hamiltonian in the characteristic equation is natural. The other interesting feature of this Hamiltonian is biquadratic in $y$ and $z$. This is an important feature for the evolution we wish to describe, but it was also a feature exploited in numerically integrating biquadratic Hamiltonians to give discrete Painlevé equations [**29**].

There are just two Schlesinger transformations, inducing changes $\theta \to \theta \pm 1$. If there exists a transformation inducing one of these changes, this would clearly be an isomonodromic deformation given (2.6). In the following theorem, we let $y = y_n$, $z = z_n$ and associate the transformation $n \to n+1$ with a change $\theta_{n+1} = \theta_n - 1$.

PROPOSITION 3.2. *The Schlesinger transformation corresponding to the transformation* $\theta_{n+1} = \theta_n - 1$ *is given by*

$$
(3.6a) \qquad\qquad\qquad y_{n+1} + y_n = -\frac{\theta_n}{z_n},
$$

$$
(3.6b) \qquad\qquad\qquad z_{n+1} + z_n = 2y_{n+1}^2 + t.
$$

PROOF. To derive the isomonodromic deformation specified by the change $\theta_{n+1} = \theta_n - 1$, we use the general form of the solution around $\infty$ written as

$$Y_n(x) = \left( I + \frac{Y_1}{x} + \frac{Y_2}{x^2} + \dots \right) e^{T(x)}$$

$$T(x) = \begin{pmatrix} 1 & 0 \\ 0 & -1 \end{pmatrix} \frac{x^3}{3} + \begin{pmatrix} t & 0 \\ 0 & -t \end{pmatrix} \frac{x}{2} + \begin{pmatrix} \theta & 0 \\ 0 & -\theta \end{pmatrix} \log \frac{1}{x}.$$

coupled with the formal solution around $x = 0$. From this expansion, we compute the expansions for $R_n(x) = Y_{n+1}(x) Y_n(x)^{-1}$ around $x = 0$ and $x = \infty$, which tells us $R_n(x) = R_1 + R_0 x + O(x^{-1}) = R_0 + R_1 x + O(x^2) = R_0 + R_1 x$. Furthermore, using the first few terms of $Y_n(x)$ gives us the expression

(3.7) $$R_n(x) = \begin{pmatrix} 0 & -\dfrac{u_{n+1}}{2} \\ -\dfrac{z_n}{u_n} & x - y_{n+1} \end{pmatrix}.$$

The map $(y_n, z_n) \to (y_{n+1}, z_{n+1})$ can be computed using (2.17) which gives (3.6). □

This is essentially the computation given in [20]. The following theorem is our first demonstration the role of (1.6).

PROPOSITION 3.3. *The transformation given by (3.6) may be obtained by (1.6) where $\tilde{\theta}_n = \theta_n$ and $\theta_{n+1} = \theta_n - 1$.*

PROOF. If $\tilde{\theta}_n$ is some intermediate value, then solving (1.6) for $y_{n+1}$ and $z_{n+1}$ gives

$$y_{n+1} + y_n = -\frac{\tilde{\theta}_n}{z_n}, \qquad z_{n+1} + z_n = -t - 2y_{n+1}^2,$$

When we compare the differential equation for $z_{n+1}$ and $z_{n+1}$, we find

$$y'_{n+1} = \frac{\left( \theta_n - \tilde{\theta}_n \right) \tilde{\theta}_n}{\left( t + 2y_{n+1}^2 + z_{n+1} \right)^2} - \frac{t}{2} - y_{n+1}^2 - z_{n+1},$$

$$z'_{n+1} = \frac{4y_{n+1} \tilde{\theta} \left( \tilde{\theta}_n - \theta_n \right)}{\left( t + 2y_{n+1}^2 + z_{n+1} \right)^2} + 2y_{n+1} z_{n+1} + 2\tilde{\theta}_n - \theta_n - 1,$$

which, when $\tilde{\theta}_n = \theta_n$ gives (3.3) for $\theta_{n+1} = \theta_n - 1$, confirming (1.6) coincides with (3.6). □

As mentioned above, because of the explicit appearance of the Hamiltonian in the characteristic equation this is equivalent to the observation made from a numerical algorithms perspective by Murata et al. [29]. This remarkable connection further emphasizes a possible link between (1.6) and a Hamiltonian description. Another observation is that the resulting map is symplectic, which can be shown using only the rules defined for all Poisson brackets.

**3.2. The sixth Painlevé equation.** The sixth Painlevé equation is at the top of the hierarchy for differential Painlevé equations [**30**]. The linear problem for the sixth Painlevé equation holds a special place in the theory of integrable systems as it is perhaps the simplest to understand in terms of the theory. The sixth Painlevé equation is presented as

$$(3.8) \quad \frac{\mathrm{d}^2 y}{\mathrm{d}t^2} = \frac{1}{2}\left(\frac{1}{y} + \frac{1}{y-1} + \frac{1}{y-t}\right)\left(\frac{\mathrm{d}y}{\mathrm{d}t}\right)^2 - \left(\frac{1}{t} + \frac{1}{x-1} + \frac{1}{y-x}\right)\frac{\mathrm{d}y}{\mathrm{d}t}$$
$$+ \frac{y(y-1)(y-t)}{t^2(t-1)^2}\left(\alpha + \frac{\beta t}{y^2} + \frac{\gamma(t-1)}{(y-1)^2} + \frac{\delta t(t-1)}{(y-t)^2}\right).$$

The linear problem for the sixth Painlevé equation is of the form

$$(3.9) \quad \frac{\mathrm{d}Y_n(x)}{\mathrm{d}x} = \left(\frac{A_0}{x} + \frac{A_1}{x-1} + \frac{A_t}{x-t}\right)Y_n(x),$$

where the coefficient matrices are

$$A_i = \begin{pmatrix} z_i + \theta_i & -w_i z_i \\ \dfrac{z_i + \theta_i}{w_i} & -z_i \end{pmatrix},$$

with

$$A_0 + A_1 + A_t + A_\infty = 0, \qquad A_\infty = \begin{pmatrix} \kappa_1 & 0 \\ 0 & \kappa_2 \end{pmatrix},$$

where $\kappa_1 + \kappa_2 + \theta_0 + \theta_1 + \theta_t = 0$ and

$$\kappa_1 - \kappa_2 = \theta_\infty.$$

The correspondence between the $\theta_i$'s and the parameters in (3.8) is

$$\alpha = \frac{(\theta_\infty - 1)^2}{2}, \quad \beta = -\frac{\theta_0^2}{2}, \quad \gamma = \frac{\theta_1^2}{2}, \quad \delta = \frac{1 - \theta_t^2}{2}.$$

If we use notation $A_n(x) = (a_{i,j}(x))$, then we specify the spectral Darboux coordinates, $y$ and $z$, by

$$(3.10) \qquad\qquad a_{1,2}(x) = w(x-y),$$
$$(3.11) \qquad\qquad a_{1,1}(y) = z,$$

where $w$ is a gauge variable and tracelessness determines $a_{2,2}(y)$. These conditions are sufficient to express the entries of $A_0$, $A_1$ and $A_t$ in terms of $w$, $y$ and $z$ alone. These variables parameterize the moduli space, moreover, if we consider stability under gauge invariance, the resulting moduli space is two dimensional.

The isomonodromic deformation in the variable $t$ may be written as the compatibility between (3.9) and

$$\frac{\mathrm{d}Y_n(x)}{\mathrm{d}x} = -\frac{A_t}{x-t}Y_n(x),$$

which is equivalent to the system

(3.12a)
$$y' = \frac{(y-1)\left(t\theta_t + (\kappa_1 + \kappa_2)\,(t-y)\right) + \theta_1(y-t)}{(t-1)t} + \frac{2(y-1)yz(t-y)}{(t-1)t},$$

(3.12b)
$$z' = \frac{z\left(\theta_t + \theta_0(t+1) + \theta_1 t + 2\,(\kappa_1 + \kappa_2)\,y\right)}{(t-1)t} + \frac{\kappa_1 \kappa_2}{(t-1)t} + \frac{z^2\left(t + 3y^2 - 2(t+1)y\right)}{(t-1)t}.$$

The equivalence of (3.12) with (3.8) may be easily verified.

The spectral curve, in terms of the Painlevé variable is

$$(3.13) \quad \Gamma(\lambda) = \lambda^2 - \lambda\left(\frac{\theta_0}{x} + \frac{\theta_1}{x-1} + \frac{\theta_t}{x-t}\right) + \frac{(y-1)yz^2(t-y) + \kappa_1\kappa_2(x-y)}{(x-1)x(x-t)}$$
$$\frac{z\left((y-1)\left(t\theta_t + (\kappa_1 + \kappa_2)\,(t-y)\right) + \theta_1(y-t)\right)}{(x-1)x(x-t)}.$$

To find the Hamiltonian associated with this isomonodromic deformation we appeal to Theorem 2.1.

COROLLARY 3.4. *The Hamiltonian describing the isomonodromic deformation for the sixth Painlevé equation is*

$$(3.14) \qquad H_{VI} = \frac{1}{t(t-1)}\left(\kappa_1\kappa_2 y(y-1)(y-t)z^2 + +\theta_t\left(\theta_0(t-1) + \theta_1 t\right)\right.$$
$$\left. ((t-y)\left(\theta_0(y-1) + \theta_1 y\right) - (y-1)y\theta_t) z\right).$$

While we observed that (1.6) presented a succinct way of expressing a Schlesinger transformation for the second Painlevé equation, a priori, it is not clear why a birational mapping of the form (1.6) should yield a symmetry of the sixth Painlevé equation. Firstly, the relation is defined on the spectral curve, which is a property of the linear system. Secondly, in the case of the second Painlevé equation, the set of translational symmetries may be identified with $\mathbb{Z}$, hence, the change in parameters is canonical. For the sixth Painlevé equation, there is no canonical translational direction per se as we may identify the set of translational symmetries with $\mathbb{Z}^4$. For this reason, we proceed in a different way, which is to show that (1.6) defines a symmetry, and that the symmetry arises as a Schlesinger transformation.

Our first difficulty is the spectral curve appears not to be a biquadratic in the Painlevé variables. It can be made to be a biquadratic over the variables $(y, \zeta)$, where $\zeta = zy$. We use (1.6) to identify a shift $n \to n+1$. We let $y = y_n$, $z = z_n$ and $\zeta = \zeta_n = y_n z_n$ and use the notation $\hat{\theta}_i$ and $\hat{\kappa}_j$ to denote the values of $\theta_i$'s and $\kappa_j$'s shifted in the $n$-direction respectively.

PROPOSITION 3.5. *The discrete evolution equations, (1.6), defines the transformation*

$$(3.15a) \qquad\qquad y_{n+1} = \frac{tz_n(y_n z_n - \theta_0)}{(y_n z_n + \kappa_1)(y_n z_n + \kappa_2)},$$

$$(3.15b) \qquad\qquad z_{n+1} = \frac{\theta_t}{y_{n+1} - t} + \frac{\theta_1 + 1}{y_{n+1} - 1} + \frac{\theta_0 - y_n z_n}{y_{n+1}},$$

*for the shift* $\hat{\theta}_1 = \theta_1 + 1$, $\hat{\theta}_t = \theta_t + 1$, $\hat{\kappa}_1 = \kappa_1 - 1$ *and* $\hat{\kappa}_2 = \kappa_2 - 1$.

PROOF. If we use (1.6) for some set of intermediate values we obtain

$$
(3.16a) \qquad y_{n+1} y_n = \frac{t \zeta_n \left( \zeta_n - \tilde{\theta}_0 \right)}{\left( \tilde{\kappa}_1 + \zeta_n \right) \left( \tilde{\kappa}_2 + \zeta_n \right)},
$$

$$
(3.16b) \qquad \zeta_{n+1} + \zeta_n = \tilde{\theta}_0 + \tilde{\theta}_1 + \tilde{\theta}_t + \frac{\tilde{\theta}_1}{y_{n+1} - 1} + \frac{t \tilde{\theta}_t}{y_{n+1} - t}.
$$

The simplest way to proceed is to compare the derivatives from (3.15) using (3.12) and (3.16a) and (3.16b). Using (3.15a), (3.12) and (3.16a) we find

$$
\begin{aligned}
y'_{n+1} =\ & \frac{\tilde{\kappa}_2 \left( \tilde{\kappa}_2 - \kappa_1 \right) \left( \tilde{\theta}_1 + \tilde{\kappa}_1 + \tilde{\theta}_t \right) \left( \tilde{\theta}_0 + \tilde{\theta}_1 + \tilde{\kappa}_1 + \tilde{\theta}_t + \kappa_2 \right)}{(t-1)\left( \tilde{\kappa}_1 - \tilde{\kappa}_2 \right) \left( \tilde{\kappa}_2 + \zeta_n \right)^2} \\
& + \frac{\left( \theta_0 - \tilde{\theta}_0 \right) y_{n+1}^2 \left( \tilde{\theta}_0 + \tilde{\kappa}_1 \right) \left( \tilde{\theta}_1 + \tilde{\kappa}_1 + \tilde{\theta}_t \right)}{(t-1)t \left( \tilde{\theta}_0 - \zeta_n \right)^2} \\
& + \frac{y_{n+1}^2 \left( \tilde{\theta}_1 + \tilde{\theta}_s + \theta_0 \right) + t \left( -\tilde{\theta}_1 - \tilde{\theta}_t + \theta_0 + \theta_1 + \theta_t \right)}{(t-1)t} \\
& - \frac{y_{n+1} \left( \theta_t + \theta_1 t + \theta_0(t+1) - t + 1 \right)}{(t-1)t} \\
& + \frac{\left( \kappa_1 - \tilde{\kappa}_1 \right) \tilde{\kappa}_1 \left( \tilde{\kappa}_1 - \kappa_2 \right) \left( \tilde{\theta}_0 + \tilde{\kappa}_1 \right)}{(t-1)\left( \tilde{\kappa}_1 - \tilde{\kappa}_2 \right) \left( \tilde{\kappa}_1 + \zeta_n \right)^2} + \frac{2\zeta_n \left( y_{n+1} - 1 \right) \left( t - y_{n+1} \right)}{(t-1)t}.
\end{aligned}
$$

By shifting (3.12) in $n$ and using (3.16b), the resulting expression for $y_{n+1}$ in $y_{n+1}$ and $\zeta_n$ is

$$
\begin{aligned}
y'_{n+1} =\ & \frac{y_{n+1} \left( 2 \left( \tilde{\theta}_t + t \tilde{\theta}_1 \right) - \hat{\theta}_t - t \hat{\theta}_1 \right) + y_{n+1}^2 \left( \tilde{\theta}_1 + \tilde{\theta}_t + \hat{\theta}_1 + \hat{\theta}_t \right)}{(t-1)t} \\
& + \frac{\left( \hat{\theta}_0 - 2\tilde{\theta}_0 \right) \left( y_{n+1} - 1 \right) \left( y_{n+1} - t \right)}{(t-1)t} - \frac{2\zeta_n \left( y_{n+1} - 1 \right) \left( t - y_{n+1} \right)}{(t-1)t}.
\end{aligned}
$$

Similar, albeit longer relations for $z'_{n+1}$ may be used to show

$$
(3.17a) \qquad \tilde{\theta}_0 = \hat{\theta}_0 = \theta_0, \quad \tilde{\theta}_1 = \hat{\theta}_1 = \theta_1 + 1, \quad \tilde{\theta}_t = \theta_t = \hat{\theta}_t - 1,
$$

$$
(3.17b) \qquad \hat{\kappa}_1 = \tilde{\kappa} = \kappa_1 - 1, \quad \hat{\kappa}_2 = \tilde{\kappa}_2 - 1 = \kappa_2 - 1,
$$

which proves (3.15) for the chosen parameters. □

We mention that because of the correspondence between the characteristic equation and the Hamiltonian, this change of variables and computation of (3.15) was also derived by Murata et al. in the context of integrable discretizations of biquadratic Hamiltonian systems [**29**]. The surface of initial conditions for (3.15) coincides with the surface for the sixth Painlevé equation. To show that a transformation of the form (1.6) in this case corresponds to a Schlesinger transformation, we still need to show this transformation arises as a Schlesinger transformation.

PROPOSITION 3.6. *The transformation* (3.15) *arises as the Schlesinger transformation.*

PROOF. The formal solutions are of the form $Y_i(x) = Y(x-a_i)(x-a_i)^{T_i}$ where $T_i = \mathrm{diag}(\theta_i, 0)$, and $Y_\infty = Y(1/x)(1/x)_\infty^T$ where $T_\infty = \mathrm{diag}(\kappa_1, \kappa_2)$. Using the elementary Schlesinger transformations computed by Muğan and Sakka [47], it is a relatively simple task to find the $R_n(x)$ arising as a product of two matrices inducing elementary Schlesinger transformations. Since all elementary transformations commute, the two elementary Schlesinger transformations may be chosen to correspond to the change (3.17). With an $R_n(x)$ determined, it is a simple, yet tedious task to confirm (3.15).                                                                  □

What is telling about the form of the Hamiltonian is that the evolution arises as the product of two elementary Schlesinger transformations (in the sense of [20]). We observed in [32] that this a common feature of many of the Lax pairs for many discrete Painlevé equations [2, 7] and $q$-Painlevé equations [21, 28].

### 3.3. The $q$-analogue of the sixth Painlevé equation.

The $q$-analogue of the sixth Painlevé equation was presented with its Lax pair for the first time by Jimbo and Sakai [21]. While Lax pairs for discrete Painlevé equations as pairs of commuting difference operators was presented in [36], a remarkable consequence of [21] was that the commutation relations are equivalent to preservation of a connection matrix. For this reason, we call this a connection (matrix) preserving deformation.

The $q$-analogue of the sixth Painlevé equation is the system whose evolution is defined by

$$(3.18a) \qquad a_1 a_2 y_{n+1} y_n = \frac{(q\theta_1 z_n - t_n a_1 a_2)(q\theta_2 z_n - t_n a_1 a_2)}{(q\kappa_1 z_n - 1)(q\kappa_2 z_n - 1)},$$

$$(3.18b) \qquad q^2 \kappa_1 \kappa_2 z_{n+1} z_n = \frac{(y_{n+1} - qt_n a_1)(y_{n+1} - qt_n a_2)}{(y_{n+1} - a_3)(y_{n+1} - a_4)},$$

where $t_{n+1} = qt_n$. The parameters are constrained by the equation

$$\theta_1 \theta_2 = \kappa_1 \kappa_2 a_1 a_2 a_3 a_4.$$

The evolution in the spectral variable is a specified by (1.1a) with (1.4) with

$$(3.19) \quad A_n(x) = \begin{pmatrix} \kappa_1((x-y_n)(x-\alpha_n) + z_{1,n}) & \kappa_2 w(x - y_n) \\ \dfrac{\kappa_2}{w_n}(\gamma_n x + \delta_n) & \kappa_1((x-y_n)(x-\beta_n) + z_{2,n}) \end{pmatrix},$$

where $\alpha_n$, $\beta_n$, $\gamma_n$ and $\delta_n$ are functions of $y_n$ and $z_n$ determined by the conditions

$$(3.20) \quad \det A_n(x) = \kappa_1 \kappa_2 (x - a_1 t_n)(x - a_2 t_n)(x - a_3)(x - a_4),$$

$$(3.21) \qquad z_{1,n} = \frac{(y_n - a_1 t_n)(y_n - a_2 t_n)}{q\kappa_1 z_n}, \qquad z_{2,n} = (y_n - a_3)(y_n - a_4)q\kappa_1 z_n,$$

and that $A(0)$ has eigenvalues $\theta_1 t_n$ and $\theta_2 t_n$. This specifies that $\alpha_n$, $\beta_n$, $\gamma_n$ and $\delta_n$ are given by

$$\alpha_n = \frac{1}{\kappa_1 - \kappa_2}\left(\kappa_2(2y_n - a_1 t_n - a_2 t_n - a_3 - a_4) + \frac{t_n\theta_1 + t_n\theta_2 - \kappa_1 z_{1,n} - \kappa_2 z_{2,n}}{y_n}\right),$$

$$\beta_n = \frac{1}{\kappa_1 - \kappa_2}\left(\kappa_1(a_1 t_n + a_2 t_n + a_3 + a_4 - 2y_n) + \frac{\kappa_1 z_{1,n} + \kappa_2 z_{2,n} - t_n\theta_1 - t_n\theta_2}{y_n}\right),$$

$$\gamma_n = \frac{1}{y_n}\left(t_n^2 a_1 a_2 a_3 a_4 - (y_n\alpha_n + z_{1,n})(y_n\beta_n + z_{2,n})\right),$$

$$\delta_n = y_n^2 + 2y_n(\alpha_n + \beta_n) + \alpha_n\beta_n + z_{1,n} + z_{2,n} - a_1 a_2 t_n^2$$
$$- t_n(a_1 + a_2)(a_3 + a_4) + a_3 a_4.$$

The connection matrix preserving deformation is given by the following result.

PROPOSITION 3.7 (Jimbo and Sakai [21]). *The sixth Painlevé arises as the connection preserving deformations described by* $t_{n+1} = qt_n$.

PROOF. Looking at the determinant of the matrix equation, we have that the matrix satisfies a linear $q$-difference equation, which may be solved by $q$-Pochhammer symbols. The ratio of which gives us that

$$\det R_n(x) = \frac{1}{(x - qa_1 t_n)(x - qa_2 t_n)}.$$

Using form of the solutions at $x = 0$ and $x = \infty$, we compute expansions

$$R_n(x) = Y_{n+1}(x)Y_n(x)^{-1},$$

which implies that $R_n(x)$ takes the form

$$R_n(x) = \frac{x(xI + R_0)}{(x - qa_1 t_n)(x - qa_2 t_n)}.$$

There are numerous equivalent ways of calculating $R_0 = (r_{ij})$. One way to compute $R_0$ is to use the first few terms series expansion around $x = 0$ or $x = \infty$. In particular, the value of $r_{12}$ found using the expansion around $x = \infty$ gives

$$r_{12} = \frac{q\kappa_2 w - q\kappa_2 \hat{w}}{\kappa_1 q - \kappa_2}.$$

Taking the residues of the top right entry of the compatibility, (2.17), at the values $x = qa_1 t_n$ and $x = qa_2 t_n$, results in the alternative expression

$$r_{12} = \frac{q\kappa_2 w_{n+1} z_{n+1}}{1 - q\kappa_1 z_{n+1}}.$$

Equating these two expressions is equivalent to

(3.22) $$\frac{w_{n+1}}{w_n} = \frac{q\kappa_1 z_{n+1} - 1}{\kappa_2 z_{n+1} - 1}.$$

Alternatively, taking the residues of $x = a_1 t$ and $x = a_2 t$, gives

$$r_{12} = -\frac{q w_n (a_1 t_n - y_n)(a_2 t_n - y_n)}{(a_1 t_n - y_n)(a_2 t_n - y_n) - z_{2,n}},$$

whose compatibility with previous values of $r_{12}$ gives (3.18b). With these values, comparing corresponding values for $r_{11}$ gives (3.18). The combination of (3.18) and (3.22) solve (2.17). □

We highlight that the steps in this discrete isomonodromic deformation are the
same as the continuous cases that we have treated. The matrix, $R(x)$, governing the
isomonodromic deformation may be evaluated directly from the fundamental solutions. We remark that while we have given one connection preserving deformation
in a system of commuting transformations that make the full lattice of connection
preserving deformations [32]. One can decompose these transformations into analogous elementary Schlesinger transformations. The step used to derive the form of
the $R_n(x)$ matrix also works for irregular solutions at $x = 0$ and $x = \infty$ using the
so-called Birkhoff-Guenther form of the solutions, which we explored in [32].

What we wish to show is that this isomonodromic deformation arises naturally
from (1.6). We express the spectral curve in terms of the matrix coefficients as

$$(3.23) \quad \Gamma(\lambda) = \lambda^2 + \kappa_1\kappa_2(x - a_1t_n)(x - a_2t_n)(x - a_3)(x - a_4)$$
$$- \lambda \left( \kappa_1((x - y_n)(x - \alpha_n) + z_{1,n}) + \kappa_2((x - y_n)(x - \beta_n) + z_{2,n}) \right).$$

When we expand this in terms of $y$ and $z$, we find that the spectral curve takes the
form

$$(3.24) \quad \Gamma(\lambda) = \lambda^2 + \kappa_1\kappa_2(x - a_1t_n)(x - a_2t_n)(x - a_3)(x - a_4) +$$
$$\lambda \left( \frac{\kappa_1\kappa_2 q x z_n (a_3 - y_n)(y_n - a_4)}{y_n} - \frac{x(a_1t_n - y_n)(a_2t_n - y_n)}{qyz_n} \right.$$
$$\left. + \frac{(x - y_n)((\theta_1 + \theta_2)t_n - (\kappa_1 + \kappa_2)xy_n)}{y_n} \right).$$

The variable $t_n$ is a somewhat artificial in the context of connection preserving
deformations, or more generally the full affine Weyl group of Bäcklund transformations of type $D_5^{(1)}$ [45]. Both the connection preserving deformation and the
Schlesinger transformations arise in the same manner, hence, to determine whether
the (1.6) determines (3.18), we are required to consider the evolution of (3.18) as
a deformation in which $t_n = 1$ and $\hat{a}_1/a_1 = \hat{a}_2/a_2 = \hat{\theta}_1/\theta_1 = \hat{\theta}_2/\theta_2 = q$, which
defines a direction $n$.

THEOREM 3.8. *The evolution equations* (3.18) *admit the representation* (1.6).

PROOF. The intermediate change of variables is the change shifts $a_1$ and $a_2$ by
$q$, where the resulting application of (1.6) to (3.24) gives (3.18a) and (3.18b).  □

For linear systems of $q$-difference equations, we have tested this relation on all
the Lax pairs featured in the work by Murata [28], and found that (1.6) is a succinct
way of describing the evolution of each discrete isomonodromic deformation listed,
including the $q$-Painlevé equation known as $q$-P($A_2^{(1)}$) [28, 44].

**3.4. The discrete analogue of the sixth Painlevé equation.** The last
example we wish to give is the discrete version of the sixth Painlevé equation ($d$-
P$_{VI}$) [2]. We call it the discrete version of the sixth Painlevé equation because
it possesses a continuum limit to the sixth Painlevé equation. There are two Lax
pairs for $d$-P$_{VI}$, a difference-difference Lax pair of the form (1.3) and (1.1b) [2],
and a recent differential difference Lax pair of the form (1.2) and (1.1b) [12]. We
recently found a reduction from the lattice potential Korteweg-de Vries equation
to $d$-P$_{VI}$ using a parameterization of the Lax pair from [2].

The form of discrete version of the sixth Painlevé equation we chose may be written

(3.25a)
$$(z_{n+1} + y_n)(y_{n+1} + z_{n+1}) = \frac{(y_{n+1} - a_3)(y_{n+1} - a_4)(y_{n+1} - a_5)(y_{n+1} - a_6)}{(y_{n+1} - a_1 + t_n)(y_{n+1} - a_2 + t_n)},$$

(3.25b) $\quad (y_{n+1} + z_n)(y_n + z_n) = \dfrac{(z_n + a_3)(z_n + a_4)(z_n + a_5)(z_n + a_6)}{(z_n + a_7 + t_n)(z_n + a_8 + t_n)},$

where $t_{n+1} = t_n + h$ and

$$a_1 + a_2 + a_3 + a_4 + a_5 + a_6 + a_7 + a_8 = h.$$

We start with a Lax pair of the form

(3.26)

$$A_n(x) = x^3 I +$$
$$\begin{pmatrix} (\kappa_1 + t_n)\left((x - \alpha_n)(x - y_n) + z_{1,n}\right) & (\kappa_2 + t_n)\, w_n(x - y_n) \\ (\kappa_1 + t_n)\dfrac{(\gamma_n x + \delta_n)}{w_n} & (\kappa_2 + t)\left((x - \beta_n)(x - y_n) + z_{2,n}\right) \end{pmatrix},$$

where the function $w$ is related to the gauge freedom. The functions, $\alpha$, $\beta$, $\gamma$ and $\delta$ are determined by

(3.27) $\quad \det A(x) = (x - a_1 + t_n)(x - a_2 + t_n)(x - a_3)(x - a_4)(x - a_5)(x - a_6).$

There is also a relation between $z_{1,n}$ and $z_{2,n}$, which means that $z_{1,n}$ and $z_{2,n}$ may be written in terms of a single variable, $z_n$, chosen later to simplify the evolution equations.

As we did in a previous study [33], we give expressions for these functions in terms of the coefficients of the determinant;

$$\sum_{k=0}^{6} \mu_i x^k = \det A_n(x).$$

The functions $\alpha_n$, $\beta_n$, $\gamma_n$ and $\delta_n$ are given, in terms of these

(3.28a)
$$\alpha_n = \frac{t_n^2}{\kappa_1 - \kappa_2} + \frac{\mu_3 + (\kappa_2 - \kappa_1)\left(y_n^2 - z_{2,n}\right) + \mu_4 y_n}{(\kappa_1 - \kappa_2)(\kappa_1 + t_n)} + \frac{t_n\left(\kappa_1 + \kappa_2 - y_n\right)}{\kappa_1 - \kappa_2}$$
$$- \frac{\mu_4 - 2y_n^2 + \kappa_1 y_n + z_{1,n} + z_{2,n} - \kappa_1 \kappa_2}{\kappa_1 - \kappa_2},$$

(3.28b)
$$\beta_n = \frac{t_n^2}{\kappa_2 - \kappa_1} - \frac{\mu_3 + (\kappa_1 - \kappa_2)\left(y_n^2 - z_{1,n}\right) + \mu_4 y_n}{(\kappa_1 - \kappa_2)(\kappa_2 + t_n)} - \frac{t_n\left(\kappa_1 + \kappa_2 - y_n\right)}{\kappa_1 - \kappa_2}$$
$$+ \frac{\mu_4 - 2y_n^2 + \kappa_2 y_n + z_{1,n} + z_{2,n} - \kappa_1 \kappa_2}{\kappa_1 - \kappa_2},$$

$$\gamma_n = \alpha_n \beta_n + \frac{\mu_0 + \mu_1 y_n}{y_n^2\,(\kappa_1 + t_n)(\kappa_2 + t_n)} - \frac{z_{1,n} z_{2,n}}{y_n^2} + y_n(\alpha_n + \beta_n) + z_{1,n} + z_{2,n},$$

(3.28c) $\quad \delta_n = \dfrac{\mu_0}{y\,(\kappa_1 + t_n)(\kappa_2 + t_n)} - \dfrac{(\alpha y_n + z_{1,n})(\beta y_n + z_{2,n})}{y_n}.$

We just need to parameterize this moduli space in

$$(3.29a) \qquad y_n^3 + z_{1,n} \left( \kappa_1 + t_n \right) = \frac{(y_n - a_3)(y_n - a_4)(y_n - a_5)(y_n - a_6)}{z_n + y_n},$$

$$(3.29b) \qquad y_n^3 + z_{2,n} \left( \kappa_2 + t_n \right) = (z_n + y_n)(y_n - a_1 + t_n)(y_n - a_2 + t_n).$$

We may think of the moduli space as being parameterized by $y_n$, $z_n$ and $w_n$.

THEOREM 3.9. *The connection preserving deformation is given by* (3.25).

PROOF. There is a matrix, $R_n(x)$, relating the systems by (1.1b). The determinant of the solution is the solution of a scalar difference equation, and can be solved explicitly, giving that

$$\det R_n(x) = \frac{1}{(x - a_1 + t_{n+1})(x - a_1 + t_{n+1})}.$$

Reading off the formal solutions, (2.13) above, and solving for $\rho_1$ and $\rho_2$ gives a formal solution of the form

$$(3.30) \quad Y_{\pm\infty}(x) = x^{3x/h} e^{-3x/h} \left( I + \frac{Y_1}{x} + \frac{Y_2}{x^2} + \dots \right) \mathrm{diag} \left( x^{\frac{\kappa_1 + t_n}{h} - \frac{3}{2}}, x^{\frac{\kappa_1 + t_n}{h} - \frac{3}{2}} \right).$$

Computing $R_n(x) = \hat{Y}_{\pm\infty}(x) Y_{\pm\infty}(x)^{-1}$, gives us a rational matrix of the form

$$R_n(x) = \frac{x(xI + R_0)}{(x - a_1 + t_n)(x - a_2 + t_n)}.$$

Comparing the residue of (2.17) at $x = a_1 + t_n$ and $x = a_1 + t_n$ gives a value for the top right entry, which we compare against the value obtained by considering leading asymptotics of the top right entry for (2.17) to obtain

$$(3.31) \qquad \frac{w_{n+1}}{w_n} = \frac{(\kappa_2 + t_n)(a_3 + a_4 + a_5 + a_6 - h + \kappa_2 + z_n + t_n)}{(\kappa_2 + t_n)(a_3 + a_4 + a_5 + a_6 + \kappa_1 + z_n + t_{n+1})}.$$

Using this expression in the value for the top right entry obtained by considering residues at $x = a_1 + t - h$ and $x = a_2 + t - h$ gives (3.25a), where

$$a_7 = -\kappa_1 - a_1 - a_2, \qquad a_8 = a_3 + a_4 + a_5 + a_6 + \kappa_1.$$

Comparing the top left entry using (3.31) and (3.25a) readily gives (3.25b). Furthermore, the compatibility under these values is an identity.  □

We now turn to the expression for the isomonodromic deformations using the spectral curve. We first write the characteristic equation as

$$\Gamma(\lambda, x) = \lambda^2 + (x - a_1 + t_n)(x - a_2 + t_n)(x - a_3)(x - a_4)(x - a_5)(x - a_6)$$
$$+ \lambda \left( x^2 \left( \kappa_1 + \kappa_2 + 2t_n \right) - x \left( \alpha_n \left( \kappa_1 + t_n \right) + \beta_n \left( \kappa_2 + t_n \right) + y_n \left( \kappa_1 + \kappa_2 + 2t_n \right) \right) \right)$$
$$- y_n \left( \alpha_n \left( \kappa_1 + t_n \right) + \beta \left( \kappa_2 + t_n \right) \right) - z_{1,n} \left( \kappa_1 + t_n \right) - z_{2,n} \left( \kappa_2 + t_n \right) - 2x^3 \right),$$

which may be expressed in terms of $y$ and $z$. The discrete isomonodromic deformations are described by the following theorem.

THEOREM 3.10. *The evolution equations* (3.25) *admit the representation* (1.6).

PROOF. We simply note that the above approach works where the intermediate change is moves $a_1$ and $a_2$ but not $a_7$ and $a_8$, in which (1.6a) is equivalent to (3.25a) while (1.6b) is equivalent to (3.25b). Demanding that the resulting characteristic equation is of the same form ensures the particular change in parameters is uniquely defined.  □

## 4. Discussion

We have shown that a certain class of discrete isomomondromic deformations admit an incredibly simple formulation in terms of the characteristic equation of the associated linear problem. The form of the evolution defining the discrete isomonodromic deformation seems to be the same regardless whether we have a differential-difference, $q$-difference-difference or difference-difference Lax pair. It would be an interesting task to show that deformations of the type described here are Frobenius integrable in some sense.

## 5. Acknowledgements

We would like to acknowledge helpful discussions with Prof. Eric Rains and Prof. Anton Dzhamay, and we would like to acknowledge Prof. Basil Grammaticos for alerting us to some relevant literature.

## References

[1] C. R. Adams, *On the linear ordinary q-difference equation*, Ann. of Math. (2) **30** (1928/29), no. 1-4, 195–205, DOI 10.2307/1968274. MR1502876

[2] D. Arinkin and A. Borodin, *Moduli spaces of d-connections and difference Painlevé equations*, Duke Math. J. **134** (2006), no. 3, 515–556, DOI 10.1215/S0012-7094-06-13433-6. MR2254625 (2007h:14047)

[3] G. D. Birkhoff, *General theory of linear difference equations*, Trans. Amer. Math. Soc. **12** (1911), no. 2, 243–284, DOI 10.2307/1988577. MR1500888

[4] G. D. Birkhoff, The generalized Riemann problem for linear differential equations and the allied problems for linear difference equations. *Amer. Acad. Proc.*, 49, 521–568, 1914.

[5] G. D. Birkhoff and P. E. Guenther, *Note on a canonical form for the linear q-difference system*, Proc. Nat. Acad. Sci. U. S. A. **27** (1941), 218–222. MR0004047 (2,310d)

[6] P. Boalch, *Quivers and difference Painlevé equations*, Groups and symmetries, CRM Proc. Lecture Notes, vol. 47, Amer. Math. Soc., Providence, RI, 2009, pp. 25–51. MR2500553 (2011g:39028)

[7] A. Borodin, *Isomonodromy transformations of linear systems of difference equations*, Ann. of Math. (2) **160** (2004), no. 3, 1141–1182, DOI 10.4007/annals.2004.160.1141. MR2144976 (2007b:37149)

[8] R. D. Carmichael, *The General Theory of Linear q-Difference Equations*, Amer. J. Math. **34** (1912), no. 2, 147–168, DOI 10.2307/2369887. MR1506145

[9] L. Di Vizio, J.-P. Ramis, J. Sauloy, and C. Zhang, *Équations aux q-différences* (French), Gaz. Math. **96** (2003), 20–49. MR1988639 (2004e:39023)

[10] J. J. Duistermaat, *Discrete integrable systems*, Springer Monographs in Mathematics, Springer, New York, 2010. QRT maps and elliptic surfaces. MR2683025 (2012g:37178)

[11] A. Dzhamay, *On the Lagrangian structure of the discrete isospectral and isomonodromic transformations*, Int. Math. Res. Not. IMRN, posted on 2008, Art. ID rnn 102, 22, DOI 10.1093/imrn/rnn102. MR2439540 (2010b:37152)

[12] A. Dzhamay, H. Sakai and T. Takenawa, Discrete Hamiltonian structure of Schlesinger transformations. *arXiv:1302.2972*, 2013

[13] H. Flaschka and A. C. Newell, *Monodromy- and spectrum-preserving deformations. I*, Comm. Math. Phys. **76** (1980), no. 1, 65–116. MR588248 (82g:35103)

[14] P. J. Forrester, C. M. Ormerod, and N. S. Witte, *Connection preserving deformations and q-semi-classical orthogonal polynomials*, Nonlinearity **24** (2011), no. 9, 2405–2434, DOI 10.1088/0951-7715/24/9/002. MR2819929 (2012e:39013)

[15] G. Gasper and M. Rahman, *Basic hypergeometric series*, Encyclopedia of Mathematics and its Applications, vol. 35, Cambridge University Press, Cambridge, 1990. With a foreword by Richard Askey. MR1052153 (91d:33034)

[16] J. Harnad and M. A. Wisse, *Loop algebra moment maps and Hamiltonian models for the Painlevé transcendants*, Mechanics day (Waterloo, ON, 1992), Fields Inst. Commun., vol. 7, Amer. Math. Soc., Providence, RI, 1996, pp. 155–169. MR1365775 (97d:58076)

[17] N. Hitchin, *Stable bundles and integrable systems*, Duke Math. J. **54** (1987), no. 1, 91–114, DOI 10.1215/S0012-7094-87-05408-1. MR885778 (88i:58068)

[18] M. Jimbo, T. Miwa, Y. Môri, and M. Sato, *Density matrix of an impenetrable Bose gas and the fifth Painlevé transcendent*, Phys. D **1** (1980), no. 1, 80–158, DOI 10.1016/0167-2789(80)90006-8. MR573370 (84k:82037)

[19] M. Jimbo, T. Miwa, and K. Ueno, *Monodromy preserving deformation of linear ordinary differential equations with rational coefficients. I. General theory and $\tau$-function*, Phys. D **2** (1981), no. 2, 306–352, DOI 10.1016/0167-2789(81)90013-0. MR630674 (83k:34010a)

[20] M. Jimbo and T. Miwa, *Monodromy preserving deformation of linear ordinary differential equations with rational coefficients. II*, Phys. D **2** (1981), no. 3, 407–448, DOI 10.1016/0167-2789(81)90021-X. MR625446 (83k:34010b)

[21] M. Jimbo and H. Sakai, *A q-analog of the sixth Painlevé equation*, Lett. Math. Phys. **38** (1996), no. 2, 145–154, DOI 10.1007/BF00398316. MR1403067 (97m:39002)

[22] K. Kajiwara, T. Masuda, M. Noumi, Y. Ohta, and Y. Yamada, $_{10}E_9$ *solution to the elliptic Painlevé equation*, J. Phys. A **36** (2003), no. 17, L263–L272, DOI 10.1088/0305-4470/36/17/102. MR1984002 (2004g:39033)

[23] I. Krichever, *Algebraic versus Liouville integrability of the soliton systems*, XIVth International Congress on Mathematical Physics, World Sci. Publ., Hackensack, NJ, 2005, pp. 50–67. MR2227821 (2007b:14078)

[24] I. M. Krichever, *Analytic theory of difference equations with rational and elliptic coefficients and the Riemann-Hilbert problem* (Russian, with Russian summary), Uspekhi Mat. Nauk **59** (2004), no. 6 (360), 111–150, DOI 10.1070/RM2004v059n06ABEH000798; English transl., Russian Math. Surveys **59** (2004), no. 6, 1117–1154. MR2138470 (2005m:39036)

[25] I. M. Krichever and D. H. Phong, *Symplectic forms in the theory of solitons*, Surveys in differential geometry: integral systems [integrable systems], Surv. Differ. Geom., IV, Int. Press, Boston, MA, 1998, pp. 239–313, DOI 10.4310/SDG.1998.v4.n1.a6. MR1726930 (2001k:37114)

[26] I. Krichever, *Vector bundles and Lax equations on algebraic curves*, Comm. Math. Phys. **229** (2002), no. 2, 229–269, DOI 10.1007/s002200200659. MR1923174 (2003m:14054)

[27] A. P. Magnus, *Painlevé-type differential equations for the recurrence coefficients of semi-classical orthogonal polynomials*, Proceedings of the Fourth International Symposium on Orthogonal Polynomials and their Applications (Evian-Les-Bains, 1992), J. Comput. Appl. Math. **57** (1995), no. 1-2, 215–237, DOI 10.1016/0377-0427(93)E0247-J. MR1340938 (96f:42027)

[28] M. Murata, *Lax forms of the q-Painlevé equations*, J. Phys. A **42** (2009), no. 11, 115201, 17, DOI 10.1088/1751-8113/42/11/115201. MR2485835 (2010f:33040)

[29] M. Murata, J. Satsuma, A. Ramani, and B. Grammaticos, *Discretising systematically the Painlevé equations*, Phys. D **240** (2011), no. 3, 305–309, DOI 10.1016/j.physd.2010.09.007. MR2748736 (2011m:34266)

[30] K. Okamoto, *Isomonodromic deformation and Painlevé equations, and the Garnier system*, J. Fac. Sci. Univ. Tokyo Sect. IA Math. **33** (1986), no. 3, 575–618. MR866050 (88e:58042)

[31] C. M. Ormerod. A study of the associated linear problem for $q$-$P_V$. 2009.

[32] C. M. Ormerod, *The lattice structure of connection preserving deformations for q-Painlevé equations I*, SIGMA Symmetry Integrability Geom. Methods Appl. **7** (2011), Paper 045, 22, DOI 10.3842/SIGMA.2011.045. MR2804591 (2012d:34228)

[33] C. M. Ormerod, *Symmetries and special solutions of reductions of the lattice potential KdV equation*, SIGMA Symmetry Integrability Geom. Methods Appl. **10** (2014), Paper 002, 19. MR3210633

[34] C. M. Ormerod, P. H. van der Kamp, and G. R. W. Quispel, *Discrete Painlevé equations and their Lax pairs as reductions of integrable lattice equations*, J. Phys. A **46** (2013), no. 9, 095204, 22, DOI 10.1088/1751-8113/46/9/095204. MR3030178

[35] C. M. Ormerod, P. H. van der Kamp, J. Hietarinta, and G.R.W. Quispel. Twisted reductions of integrable lattice equations, and their Lax representations. *arXiv preprint arXiv:1307.5208*, 2013.

[36] V. G. Papageorgiou, F. W. Nijhoff, B. Grammaticos, and A. Ramani, *Isomonodromic deformation problems for discrete analogues of Painlevé equations*, Phys. Lett. A **164** (1992), no. 1, 57–64, DOI 10.1016/0375-9601(92)90905-2. MR1162062 (93c:58092)

[37] C. Praagman, *Fundamental solutions for meromorphic linear difference equations in the complex plane, and related problems*, J. Reine Angew. Math. **369** (1986), 101–109, DOI 10.1515/crll.1986.369.101. MR850630 (88b:39004)

[38] G. R. W. Quispel, J. A. G. Roberts, and C. J. Thompson, *Integrable mappings and soliton equations*, Phys. Lett. A **126** (1988), no. 7, 419–421, DOI 10.1016/0375-9601(88)90803-1. MR924318 (88m:58084)

[39] G. R. W. Quispel, J. A. G. Roberts, and C. J. Thompson, *Integrable mappings and soliton equations. II*, Phys. D **34** (1989), no. 1-2, 183–192, DOI 10.1016/0167-2789(89)90233-9. MR982386 (90e:58066)

[40] A. Ramani, B. Grammaticos, and J. Hietarinta, *Discrete versions of the Painlevé equations*, Phys. Rev. Lett. **67** (1991), no. 14, 1829–1832, DOI 10.1103/PhysRevLett.67.1829. MR1125951 (92j:39011)

[41] E. M. Rains, *An isomonodromy interpretation of the elliptic Painlevé equation. I*, arXiv.0807.0258

[42] E. M. Rains, *Generalized Hitchin systems on rational surfaces* arXiv:1307.4033

[43] H. Sakai, *Hypergeometric solution of q-Schlesinger system of rank two*, Lett. Math. Phys. **73** (2005), no. 3, 237–247, DOI 10.1007/s11005-005-0020-z. MR2188296 (2006j:39029)

[44] H. Sakai, *Lax form of the q-Painlevé equation associated with the $A_2^{(1)}$ surface*, J. Phys. A **39** (2006), no. 39, 12203–12210, DOI 10.1088/0305-4470/39/39/S13. MR2266221 (2008b:33034)

[45] H. Sakai, *Rational surfaces associated with affine root systems and geometry of the Painlevé equations*, Comm. Math. Phys. **220** (2001), no. 1, 165–229, DOI 10.1007/s002200100446. MR1882403 (2003c:14030)

[46] H. Sakai, *A q-analog of the Garnier system*, Funkcial. Ekvac. **48** (2005), no. 2, 273–297, DOI 10.1619/fesi.48.273. MR2177121 (2006e:33033)

[47] U. Muğan and A. Sakka, *Schlesinger transformations for Painlevé VI equation*, J. Math. Phys. **36** (1995), no. 3, 1284–1298, DOI 10.1063/1.531121. MR1317441 (95m:34015)

[48] J. Sauloy, *Galois theory of Fuchsian q-difference equations* (English, with English and French summaries), Ann. Sci. École Norm. Sup. (4) **36** (2003), no. 6, 925–968 (2004), DOI 10.1016/j.ansens.2002.10.001. MR2032530 (2004j:39027)

[49] L Schlesinger, Über eine Klasse von Differentialsystemen beliebiger Ordnung mit festen kritischen Punkten, *J. Reine Angew. Math.* **141**(912), 96–145.

[50] T. Shioda and K. Takano, *On some Hamiltonian structures of Painlevé systems. I*, Funkcial. Ekvac. **40** (1997), no. 2, 271–291. MR1480279 (99c:34007)

[51] T. Tsuda, *Integrable mappings via rational elliptic surfaces*, J. Phys. A **37** (2004), no. 7, 2721–2730, DOI 10.1088/0305-4470/37/7/014. MR2047557 (2004m:14078)

[52] M. van der Put and M. Reversat, *Galois theory of q-difference equations* (English, with English and French summaries), Ann. Fac. Sci. Toulouse Math. (6) **16** (2007), no. 3, 665–718. MR2379057 (2009d:39033)

[53] M. van der Put and M. F. Singer, *Galois theory of difference equations*, Lecture Notes in Mathematics, vol. 1666, Springer-Verlag, Berlin, 1997. MR1480919 (2000e:39008)

[54] B. Riemann and F. Klein, *Oeuvres mathématiques de Riemann*, Gauthier-Villars et fils, 1898

[55] P. H. van der Kamp and G. R. W. Quispel, *The staircase method: integrals for periodic reductions of integrable lattice equations*, J. Phys. A **43** (2010), no. 46, 465207, 34, DOI 10.1088/1751-8113/43/46/465207. MR2735224 (2012b:37180)

[56] Y. Yamada, *Lax formalism for q-Painlevé equations with affine Weyl group symmetry of type $E_n^{(1)}$*, Int. Math. Res. Not. IMRN **17** (2011), 3823–3838. MR2836394

[57] Y. Yamada, *A Lax formalism for the elliptic difference Painlevé equation*, SIGMA Symmetry Integrability Geom. Methods Appl. **5** (2009), Paper 042, 15, DOI 10.3842/SIGMA.2009.042. MR2506170 (2010h:37164)

DEPARTMENT OF MATHEMATICS, CALIFORNIA INSTITUTE OF TECHNOLOGY, 1200 EAST CALIFORNIA BLVD., PASADENA, CALIFORNIA 91125

*E-mail address*: `cormerod@caltech.edu`

Contemporary Mathematics
Volume **651**, 2015
http://dx.doi.org/10.1090/conm/651/13044

# Geometric Analysis of Reductions from Schlesinger Transformations to Difference Painlevé Equations

Anton Dzhamay and Tomoyuki Takenawa

ABSTRACT. We present two examples of reductions from the evolution equations describing discrete Schlesinger transformations of Fuchsian systems to difference Painlevé equations: difference Painlevé equation d-$P\left(A_2^{(1)*}\right)$ with the symmetry group $E_6^{(1)}$ and difference Painlevé equation d-$P\left(A_1^{(1)*}\right)$ with the symmetry group $E_7^{(1)}$. In both cases we describe in detail how to compute the Okamoto space of the initial conditions for these equations and emphasize the role played by geometry in helping us to understand the structure of the reduction, a choice of a good coordinate system describing the equation, and how to compare it with other instances of equations of the same type.

## 1. Introduction

It is well-known that many special functions, e.g. the Airy, Bessel, or Legendre functions, originate as series solutions of *linear* ordinary differential equations (ODEs). Such functions play a crucial role in describing a wide range of important physical and mathematical phenomena. An essential point here is that for linear ordinary differential equations singularities of solutions can only occur at the points where the coefficients of the equation itself become singular. This makes it possible to talk about global properties of solutions (and hence, of the corresponding special functions) and the asymptotic behavior of solutions near those fixed singular points.

For nonlinear equations the situation is very different. Although the Cauchy existence theorem guarantees *local* existence of the solution to a given initial value problem at an ordinary (regular) point, in general the domain in which this solution is defined depends not just on the equation itself, but on the initial conditions as well — solutions acquire *movable* (i.e., dependent on the initial values or, equivalently, on the constants of integration) singularities. Such singularities are called *critical* if a corresponding solution looses its single-valued character in a neighborhood of the singularity (e.g., when a singular point is a *branch point*). An ODE is said to satisfy the *Painlevé property* if its general solution is free of *movable critical* singular points. Otherwise, the Riemann surface of a solution becomes dependent on the constants of integration, which prevents global analysis. Thus, equivalently,

---

2010 *Mathematics Subject Classification*. Primary 34M55, 34M56, 14E07.

*Key words and phrases*. Integrable systems, Painlevé equations, difference equations, isomonodromic transformations, birational transformations.

the Painlevé property of an ODE is the uniformizability of its general solution; see
[**Con99a**] (as well as the other excellent articles in the volume [**Con99b**]) for a
careful overview of these ideas.

It is clear that linear equations satisfy the Painlevé property. The importance
of the Painlevé property for nonlinear equations is that, similar to the linear case,
equations that satisfy the Painlevé property give rise to new transcendental functions. In that sense, according to M. Kruskal as quoted in [**GR14**], nonlinear
equations satisfying the Painlevé property are on the border between trivially integrable linear equations and nonlinear equations that are not integrable, and so the
Painlevé property is essentially equivalent to (and is a criterion of) integrability.

The search for new transcendental functions was the original motivation in the
work of P. Painlevé who, together with his student B. Gambier, had classified all
of the rational second-order differential equations that have the Painlevé property,
[**Pai02**], [**Pai73**], [**Gam10**]. Among 50 classes of equations that they found, only
*six* can not be reduced to linear equations or integrated by quadratures. These
equations are now known as *Painlevé equations* $P_I - P_{VI}$. Solutions to these equations, the so-called *Painlevé transcendents*, are playing an increasingly important
role in describing a wide range of nonlinear phenomena in mathematics and physics
[**IKSY91**].

Almost simultaneously with the work of P. Painlevé and B. Gambier, the most
general Painlevé VI equation was obtained by R. Fuchs [**Fuc05**] in the theory
of *isomonodromic deformations* of Fuchsian systems. This theory, developed in
the works of R. Fuchs [**Fuc07**], L. Schlesinger [**Sch12**], R. Garnier [**Gar26**], and
then extended to the non-Fuchsian case by M. Jimbo, T. Miwa, and K. Ueno
[**JMU81, JM81, JM82**] and also by H. Flaschka and A. Newell [**FN80**], as well
as the related Riemann-Hilbert approach [**IN86**], [**FIKN06**], are now among the
most powerful methods for studying the structure of the Painlevé transcendents.

Over the last thirty years a significant effort has been directed towards understanding and generalizing results and methods of the classical theory of completely
integrable systems to the *discrete* case. This is true for the theory of Painlevé
equations as well. Discrete Painlevé equations were originally defined as second
order nonlinear difference equations that have usual Painlevé equations as continuous limits [**BK90**], [**GM90**]. A systematic study of discrete Painlevé equations
was started by B. Grammaticos, J. Hietarinta, F. Nijhoff, V. Papageorgiou and
A. Ramani, [**NP91**], [**RGH91**], [**GRP91**], and many different examples of such
equations were obtained in a series of papers by Grammaticos, Ramani, and their
collaborators by a systematic application of the singularity confinement criterion,
see reviews [**GR04**], [**GR14**], and many references therein. Discrete Painlevé equations also appear in a broad spectrum of important nonlinear problems in mathematics and physics, among which are the theory of orthogonal polynomials, quantum gravity, determinantal random point processes, reductions of integrable lattice
equations, and, notably, as Bäcklund transformations of differential Painlevé equations. Some of these problems are discrete analogues or direct discretizations of the
corresponding continuous problems, and some describe purely discrete phenomena.

It turned out that classifying discrete Painlevé equations by their continuous
limits is not a very good approach, since such correspondence is far from being bijec-
tive. It is both possible for the same discrete equation to have different differential
Painlevé equations as continuous limits under different limiting procedures, and for

different discrete equations to have the same continuous limit. In the seminal paper [**Sak01**], H. Sakai showed that an effective way to understand and classify discrete Painlevé equations is through algebraic geometry.

In this approach, we first rewrite our second-order discrete Painlevé equation as a two-dimensional first-order nonlinear system. This system defines a rational map $\varphi : \mathbb{C}^2 \dashrightarrow \mathbb{C}^2$ that we then extend to the $\mathbb{P}^2$ (or $\mathbb{P}^1 \times \mathbb{P}^1$) compactification of $\mathbb{C}^2$. The resulting map $\varphi : \mathbb{P}^2 \dashrightarrow \mathbb{P}^2$ is birational, and so there exist some points where either the map $\varphi$ or its inverse is undefined. These indeterminacies are then resolved with the help of a blow-up construction. In the Painlevé case, according to Sakai's theory, the complete resolution of indeterminacies is obtained by blowing-up nine, possibly infinitely close, points on $\mathbb{P}^2$ (or eight points on the birationally equivalent $\mathbb{P}^1 \times \mathbb{P}^1$ compactification of $\mathbb{C}^2$). Using automorphisms of $\mathbb{P}^2$ these points can be put into some standard configurations, where we can fix some of their coordinates and the remaining coordinates become parameters $\mathbf{b} = \{b_i\}$ of our dynamic. Here different configurations correspond to different types of discrete Painlevé equations. As a result of this procedure we obtain a compact rational surface $\mathcal{X}_{\mathbf{b}}$, where the subscript $\mathbf{b}$ indicates that this surface is equipped with some additional data given by the configuration and the coordinates of the blow-up points. This surface $\mathcal{X}_{\mathbf{b}}$ is called the *Okamoto space of initial conditions* of the equation. In fact, in this way we actually obtain a whole family of surfaces whose base is given by the parameters $\mathbf{b} = \{b_i\}$. Our dynamic then lifts to an *isomorphism* $\varphi : \mathcal{X}_{\mathbf{b}} \to \mathcal{X}_{\bar{\mathbf{b}}}$, where the parameter evolution $\mathbf{b} \mapsto \bar{\mathbf{b}}$ can be of *additive* (for difference Painlevé equations), *multiplicative* (for $q$-difference Painlevé equations), or *elliptic* (for elliptic Painlevé equation) type. This evolution of parameters reflects the fact that Painlevé equations are *non-autonomous*.

Since all surfaces $\mathcal{X}_{\mathbf{b}}$ are of the same type, we sometimes omit the subscript $\mathbf{b}$ and talk about the *Okamoto surface* $\mathcal{X}$ of our discrete Painlevé equation. This surface has a special property that it admits a unique anti-canonical divisor $D \in |-K_{\mathcal{X}}|$ of *canonical type*. In [**Sak01**] such surfaces are called *generalized Halphen surfaces*. The orthogonal complement of $-K_{\mathcal{X}}$ in the Picard lattice $\mathrm{Pic}(\mathcal{X}) \simeq H^2(\mathcal{X}; \mathbb{Z})$ is described by the affine Dynkin diagram $E_8^{(1)}$ that has two intersecting root subsystems of affine type: $R$, that is generated by classes of irreducible components $D_i$ of the anti-canonical divisor $D$, and its orthogonal complement $R^\perp$ whose corresponding root lattice $Q(R^\perp)$ is called the *symmetry sub-lattice*. Then the *type* of the discrete Painlevé equation is the same as the *type of its surface* $\mathcal{X}_{\mathbf{b}}$, which is just the type of an affine Dynkin diagram describing the root system $R$ of irreducible components $D_i$ of $D$ (essentially, the configuration, including degeneration structure, of the positions of the blow-up points). Moreover, nonlinear Painlevé dynamic now becomes a translation in the symmetry sub-lattice $Q(R^\perp)$; see [**Sak01**] for details. This is somewhat similar to the algebro-geometric integrability of classical integrable systems and soliton equations, where nonlinear dynamic is mapped to commuting linear flows on the Jacobian of the spectral curve of the associated linear problem via the Abel-Jacobi map.

One important observation from Sakai's geometric approach is that in addition to *additive* (difference) and *multiplicative* ($q$-difference) *discretizations* of continuous Painlevé equations, there are some purely discrete Painlevé equations. It also led to the discovery of the master *elliptic* discrete Painlevé equation such that all of the other Painlevé equations can be obtained from it through degenerations (which corresponds to more and more special configurations of the blow-up points). This degenerations can be described by the following scheme, where letters stand for the symmetry type of the equation (which is the type of the affine Dynkin diagram of the root subsystem $R^\perp$), and the subscripts $e$, $q$, $\delta$, and $c$ stand for elliptic, multiplicative, additive and differential Painlevé equations respectively, see Figure 1.

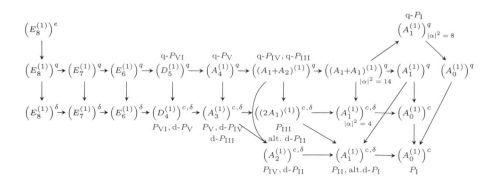

FIGURE 1. Inclusion scheme for the symmetry sub-lattices and corresponding Painlevé equations

The following question then becomes natural and important: *how to represent these new purely discrete equations in the isomonodromic framework*. This question was posed by Sakai in [**Sak07**] (**Problem A** for the difference case and **Problems B,C** for the $q$-difference case).

More precisely, in both the continuous and the discrete difference case we start with some Fuchsian system and consider its isomonodromic deformations. In the continuous case, deformation parameters are locations of singular points of the system. The resulting isomonodromic flows on the space of coefficients of the Fuchsian system are given by a system of nonlinear differential equations called the *Schlesinger equations* (also referred to as the *Schlesinger system* in some literature). In particular, for a $2 \times 2$ Fuchsian system with four poles, Schlesinger equations reduce to the most general $P_{\mathrm{VI}}$ equation. In the discrete difference case, deformation parameters are the characteristic indices of the system and since the isomonodromy condition requires that the indices change by integral shifts, the resulting dynamic is indeed *discrete*. It is expressed in the form of *difference* equations called Schlesinger *transformations*. It is also possible to get the isomonodromic description of difference and $q$-difference Painlevé equations by studying deformations of linear difference and $q$-difference analogues of Fuchsian systems; see [**JS96**] and [**Bor04**] for details.

In [**DST13**] we studied in detail a particular class of Schlesinger transformations that are called *elementary*. These transformations change only two of the characteristic indices of our Fuchsian system (and any other Schlesinger transformation not involving characteristic indices with multiplicity can be represented as a composition of the elementary ones). In particular, we obtained explicit evolution equations governing the resulting discrete dynamic. Our objective for the present paper is to carefully and very explicitly describe reductions of these *discrete Schlesinger evolution equations* to *difference* Painlevé equations.

Since Painlevé equations are of the second order, we focus on Fuchsian systems that have two-dimensional moduli spaces (coordinates on such moduli space are known as accessory parameters). It turns out that, modulo two natural transformations called *Katz's addition* and *middle convolution* [**Kat96**] that preserve the deformation equations [**HF07**], there are only four such systems, [**Kos01**], that have the *spectral type* $(11, 11, 11, 11)$, $(111, 111, 111)$, $(22, 1111, 1111)$, and $(33, 222, 111111)$ (the spectral type of a Fuchsian system encodes the degeneracies of the characteristic indices, or eigenvalues of residue matrices at singular points, and it is carefully defined in the next section).

Isomonodromic deformations of a $(11, 11, 11, 11)$ spectral type Fuchsian system are well known — continuous deformations reduce to the Painlevé VI equation and Schlesinger transformations reduce to the difference Painlevé d-$P\left(D_4^{(1)}\right)$ equation, also known as d-$P_V$, and in [**DST13**] we showed that in this case our *discrete Schlesinger evolution equations* indeed can be reduced to the standard form of d-$P\left(D_4^{(1)}\right)$.

In [**Boa09**] P. Boalch showed that for Fuchsian systems with the spectral types $(111, 111, 111)$, $(22, 1111, 1111)$, and $(33, 222, 111111)$ the Schlesinger transformations reduce to difference Painlevé equations with the required symmetry groups $E_6^{(1)}$, $E_7^{(1)}$, and $E_8^{(1)}$ respectively, thus giving the answer to Sakai's **Problem A**, but without explicit coordinate computations.

Our goal for the present paper is to make this statement very concrete via explicit direct computation of reductions of discrete Schlesinger evolution to difference Painlevé equations with symmetry groups $E_6^{(1)}$ and $E_7^{(1)}$ (we plan to consider deformations of a Fuchsian system of the spectral type $(33, 222, 111111)$ with the symmetry group $E_8^{(1)}$ elsewhere). In addition to verifying that the resulting difference Painlevé equations have the required types d-$P\left(A_2^{(1)*}\right)$ and d-$P\left(A_1^{(1)*}\right)$, we explicitly compare the resulting equations with the previously known instances of equations of the same type. We do so by first choosing a particularly simple instance of a discrete Painlevé equation of a given type that we refer to as a *model* (or *standard*) *example*. We then compute its Okamoto space of initial conditions $\mathcal{X}_\mathbf{b}$, where $\mathbf{b} = \{b_i\}$ are some generic parameters in that model example. Next, we find coordinates (that we denote by $f$ and $g$) on the space of Fuchsian systems of the required spectral type that give an explicit isomorphism between $\mathcal{X}_\mathbf{b}$ and the Okamoto space of initial conditions $\mathcal{X}_\theta$ for the Schlesinger transformation dynamic. This isomorphism also makes it possible to match the generic parameters $\{b_i\}$ of the model example and the characteristic indices $\{\theta_i^j\}$ of our Fuchsian system. Thus, using these coordinates $f$ and $g$ gives us an actual correspondence between the spaces of initial conditions $\mathcal{X}_\mathbf{b} = \mathcal{X}_\theta$, and we can compute and compare the translation directions for the model example dynamic and the elementary Schlesinger transformation dynamic with respect to the same root basis. Further, the matching of parameters $\{b_i\}$ and $\{\theta_i^j\}$ allows us to see the action of the model example dynamic on the Riemann Scheme of our Fuchsian system, i.e., in this way we obtain the Schlesinger transformation that corresponds to the model example, and then we can represent it as a composition of the elementary Schlesinger transformations, which we then verify by an explicit coordinate computation. Of particular interest here is the $E_7^{(1)}$ case which has two characteristic indices of multiplicity 2. We show that in that case the standard form of the equation can not be represented as a composition of elementary Schlesinger transformations of rank one. Thus we first generalize our discrete Schlesinger evolution equations from [**DST13**] to elementary Schlesinger transformations of rank two, and then show how to represent the standard dynamic as a composition of two such rank two transformations. We also provide many details on how to compute the Okamoto space of initial conditions for our equations and how to find explicit isomorphisms between two different instances of such spaces, hoping that this will be helpful for other researchers who are interested in the geometric approach to discrete Painlevé equations. Please note that these examples were also studied in [**AB06**] using the related framework of $d$-connections and modifications.

The paper is organized as follows. In Section 2 we briefly describe our parameterization of a Fuchsian system by its spectral and eigensystem data, define elementary Schlesinger transformations, present evolution equations for elementary Schlesinger transformations as a dynamic on the space of coefficient of our Fuchsian system, and then show how to split them to get the dynamic on the space of eigenvectors of the coefficient matrices (this is a brief overview of our paper [**DST13**]). Next we show how to generalize this to elementary Schlesinger transformations of rank two, which is a new result. In Section 3 we consider two examples of reductions of the elementary Schlesinger transformation dynamic. The first example of a difference Painlevé equation of type d-$P\left(A_2^{(1)*}\right)$ with the symmetry group $E_6^{(1)}$ was also briefly presented in [**DST13**], here we go into considerably more detail and show how the choice of good coordinates, which was essentially guessed in [**DST13**], is really forced on us by the geometric considerations. The next example of a $4 \times 4$ Fuchsian system of the *spectral type* $(22, 1111, 1111)$ (i.e., with three poles and two double eigenvalues at one pole) is the first example that we have which has degenerate eigenvalues, and this is a completely new result. We show that it reduces to difference Painlevé equation of type d-$P\left(A_1^{(1)*}\right)$ with the symmetry group $E_7^{(1)}$. Finally, we give a brief summary in Section 4.

**Acknowledgements.** Part of this work was done when A.D. was visiting T.T. at the Tokyo University of Marine Science and Technology and Nalini Joshi at the University of Sydney, and A.D. would like to thank both Universities for the stimulating working environment and, together with the University of Northern Colorado, for the generous travel support. We are also very grateful to the referee for useful suggestions.

## 2. Preliminaries

The goal of this section is to write down evolution equations for elementary Schlesinger transformations, as well as to introduce the necessary notation. Our presentation here is very brief and we refer the interested reader to [**DST13**] for details. The main new and important result of this section is the generalization of equations governing elementary Schlesinger transformation dynamic on the decomposition space from rank-one to rank-two Schlesinger transformations.

**2.1. Fuchsian Systems.** Consider a generic *Fuchsian system* (or a *Fuchsian equation*) on the Riemann sphere $\mathbb{P}^1$ written in the *Schlesinger normal form*:

$$(2.1) \qquad \frac{d\mathbf{Y}}{dz} = \mathbf{A}(z)\mathbf{Y} = \left( \sum_{i=1}^n \frac{\mathbf{A}_i}{z - z_i} \right) \mathbf{Y}, \qquad z_i \neq z_j \text{ for } i \neq j,$$

where $\mathbf{A}_i = \operatorname{res}_{z_i} \mathbf{A}(z)\, dz$ are constant $m \times m$ matrices. In addition to simple poles at finite distinct points $z_1, \ldots, z_n$, this system also has another simple pole at $z_\infty = \infty \in \mathbb{P}^1$ if $\mathbf{A}_\infty = \operatorname{res}_\infty \mathbf{A}(z)\, dz = -\sum_{i=1}^{n} \mathbf{A}_i \neq \mathbf{0}$. The *spectral data* of system (2.1) consists of locations of the simple poles $z_1, \ldots, z_n$ and the eigenvalues (also called the *characteristic indices*) $\theta_i^j$ of $\mathbf{A}_i$ and their multiplicities. These multiplicities are encoded by the *spectral type* of the system,

$$\mathfrak{m} = m_1^1 \cdots m_1^{l_1}, m_2^1 \cdots m_2^{l_2}, \cdots, m_n^1 \cdots m_n^{l_n}, m_\infty^1 \cdots m_\infty^{l_\infty},$$

where partitions $m = m_i^1 + \cdots + m_i^{l_i}$, $m_i^1 \geq \cdots \geq m_i^{l_i} \geq 1$ describe the multiplicities of the eigenvalues of $\mathbf{A}_i$. Spectral type classifies Fuchsian systems up to isomorphisms and the operations of addition and middle convolution.

**2.2. Schlesinger Transformations.** Schlesinger transformations are discrete analogues of the usual Schlesinger differential equations describing isomonodromic deformations of the underlying Fuchsian system. They are rational transformations $(\mathbf{Y}(z), \mathbf{A}(z)) \mapsto (\bar{\mathbf{Y}}(z), \bar{\mathbf{A}}(z))$ preserving the singularity structure and the monodromy data of the system (2.1), except for the integral shifts in the characteristic indices $\theta_i^j$, and so the coefficient matrix now depends on $\theta_i^j$, $\mathbf{A} = \mathbf{A}(z; \boldsymbol{\Theta})$. These transformations were introduced by L. Schlesinger [**Sch12**] in the Fuchsian case and studied extensively, including the extension to the case of irregular singular points, by M. Jimbo, T. Miwa, and K. Ueno in [**JM81**]. Schlesinger transformations are given by the following *differential–difference Lax Pair*:

$$\begin{cases} \dfrac{d\mathbf{Y}}{dz} = \mathbf{A}(z; \boldsymbol{\Theta})\mathbf{Y} = \left( \sum_{i=1}^{n} \dfrac{\mathbf{A}_i(\boldsymbol{\Theta})}{z - z_i} \right) \mathbf{Y}, \\ \bar{\mathbf{Y}}(z) = \mathbf{R}(z)\mathbf{Y}(z). \end{cases}$$

where $\mathbf{R}(z)$ is a specially chosen rational matrix function called the *multiplier* of the transformation. The compatibility condition of this Lax Pair is

$$(2.2) \qquad \bar{\mathbf{A}}(z; \boldsymbol{\Theta})\mathbf{R}(z) = \mathbf{R}(z)\mathbf{A}(z; \boldsymbol{\Theta}) + \frac{d\mathbf{R}(z)}{dz}.$$

In [**DST13**] we considered a special class of Schlesinger transformations for which the multiplier matrix has the form

$$(2.3) \qquad \mathbf{R}(z) = \mathbf{I} + \frac{z_0 - \zeta_0}{z - z_0}\mathbf{P}, \quad \text{where } \mathbf{P} = \mathbf{P}^2 \text{ is a } projector.$$

It turns out that in this case it is possible to solve equation (2.2) explicitly to obtain a discrete dynamic on the space of coefficient matrices. First, let $\mathbf{Q} = \mathbf{I} - \mathbf{P}$ be the complementary projector, $\mathbf{Q}^2 = \mathbf{Q}$, $\mathbf{PQ} = \mathbf{QP} = \mathbf{0}$. Then, after substituting $\mathbf{R}(z)$ of the form (2.3) in (2.2) (and its inverse) we immediately see that $z_0, \zeta_0 \in \{z_i\}_{i=1}^{n}$, and if we put $z_0 = z_\alpha$ and $\zeta_0 = z_\beta$, we get the following equations on the coefficient matrices:

$$\bar{\mathbf{A}}_i = \mathbf{R}(z_i)\mathbf{A}_i\mathbf{R}^{-1}(z_i) \quad \text{for} \quad i \neq \alpha, \beta \qquad (\text{and therefore } \bar{\Theta}_i = \Theta_i),$$

$$(2.4) \quad \mathbf{Q}\bar{\mathbf{A}}_\alpha = \mathbf{A}_\alpha \mathbf{Q}, \qquad \bar{\mathbf{A}}_\beta \mathbf{Q} = \mathbf{Q}\mathbf{A}_\beta,$$

$$(2.5) \quad \bar{\mathbf{A}}_\alpha \mathbf{P} = \mathbf{P}\mathbf{A}_\alpha - \mathbf{P}, \qquad \mathbf{P}\bar{\mathbf{A}}_\beta = \mathbf{A}_\beta \mathbf{P} + \mathbf{P},$$

$$\bar{\mathbf{A}}_\alpha = \mathbf{A}_\alpha + \sum_{i \neq \alpha} \left( \frac{z_\beta - z_\alpha}{z_i - z_\alpha} \right)(\mathbf{P}\mathbf{A}_i - \bar{\mathbf{A}}_i \mathbf{P}),$$

$$\bar{\mathbf{A}}_\beta = \mathbf{A}_\beta + \sum_{i \neq \beta} \left( \frac{z_\alpha - z_\beta}{z_i - z_\beta} \right)(\mathbf{A}_i \mathbf{P} - \mathbf{P}\bar{\mathbf{A}}_i).$$

Then either (2.4) or (2.5) imposes important constraints on the projector $\mathbf{P}$:

$$(2.6) \quad \mathbf{P}\mathbf{A}_\alpha \mathbf{Q} = 0 \quad (\text{or } \mathbf{P}\mathbf{A}_\alpha \mathbf{P} = \mathbf{P}\mathbf{A}_\alpha), \qquad \mathbf{Q}\mathbf{A}_\beta \mathbf{P} = 0 \quad (\text{or } \mathbf{P}\mathbf{A}_\beta \mathbf{P} = \mathbf{A}_\beta \mathbf{P}),$$

and if this condition is satisfied, we get the following dynamic on the space of coefficient matrices:

$$(2.7) \qquad \bar{\mathbf{A}}_i = \mathbf{R}(z_i)\mathbf{A}_i\mathbf{R}^{-1}(z_i), \qquad i \neq \alpha, \beta,$$

$$(2.8) \qquad \bar{\mathbf{A}}_\alpha = \mathbf{A}_\alpha - \mathbf{Q}\mathbf{A}_\alpha \mathbf{P} - \mathbf{P} + \sum_{i \neq \alpha} \left( \frac{z_\beta - z_\alpha}{z_i - z_\alpha} \right)\mathbf{P}\mathbf{A}_i\mathbf{Q},$$

$$(2.9) \qquad \bar{\mathbf{A}}_\beta = \mathbf{A}_\beta - \mathbf{P}\mathbf{A}_\beta \mathbf{Q} + \mathbf{P} + \sum_{i \neq \beta} \left( \frac{z_\alpha - z_\beta}{z_i - z_\beta} \right)\mathbf{Q}\mathbf{A}_i\mathbf{P}.$$

Indeed,

$$\bar{\mathbf{A}}_\alpha = \bar{\mathbf{A}}_\alpha \mathbf{P} + \bar{\mathbf{A}}_\alpha \mathbf{Q} = \mathbf{P}\mathbf{A}_\alpha - \mathbf{P} + \mathbf{A}_\alpha(\mathbf{I} - \mathbf{P}) + \sum_{i \neq \alpha} \left( \frac{z_\alpha - z_\beta}{z_\alpha - z_i} \right)\mathbf{P}\mathbf{A}_i\mathbf{Q}$$

$$= \mathbf{A}_\alpha - \mathbf{Q}\mathbf{A}_\alpha \mathbf{P} - \mathbf{P} + \sum_{i \neq \alpha} \left( \frac{z_\beta - z_\alpha}{z_i - z_\alpha} \right)\mathbf{P}\mathbf{A}_i\mathbf{Q},$$

and the equation for $\bar{\mathbf{A}}_\beta$ is obtained in a similar fashion. We call equations (2.7–2.9) *Discrete Schlesinger Evolution Equations*.

**2.3. The Decomposition Space.** It is sometimes more convenient to extend the dynamic from the space of coefficients of the Fuchsian system to the space of eigenvectors of the coefficient matrices, we call this space the *decomposition space*. In particular, this is the space on which both the continuous ([**JMMS80**]) and the discrete ([**DST13**]) Hamiltonian equations for Schlesinger deformations can be written. Before defining this space it is convenient to reduce the number of parameters in our system by using scalar gauge transformations of the form $\tilde{\mathbf{Y}}(z) = w(z)^{-1}\mathbf{Y}(z)$, where $w(z)$ is a solution of the *scalar* equation

$$\frac{dw}{dz} = \sum_{i=1}^n \frac{\theta_i^j}{z - z_i}w.$$

Such transformations change the residue matrices by $\tilde{\mathbf{A}}_i = \mathbf{A}_i - \theta_i^j \mathbf{I}$ (and consequently change the residue matrix at infinity by $\tilde{\mathbf{A}}_\infty = \mathbf{A}_\infty + \theta_i^j \mathbf{I}$). Hence we can always make one of the eigenvalues $\theta_i^j = 0$ by choosing a good representative with respect to the action by the group of local *scalar gauge transformations*. So we make the following assumption.

ASSUMPTION 2.1. *We always assume that at all but one of the points $z_i$ (including $z_\infty$) the eigenvalue $\theta_i^1$ of the highest multiplicity $m_i^1$ is zero.*

We also need the following important *semi-simplicity* assumption.

ASSUMPTION 2.2. *We assume that the coefficient matrices $\mathbf{A}_i$ are diagonalizable (even when we have multiple eigenvalues).*

In view of these assumptions, coefficient matrices $\mathbf{A}_i$ are similar to diagonal matrices $\mathrm{diag}\{\theta_i^1, \ldots, \theta_i^{r_i}, 0, \ldots, 0\}$, where $r_i = \mathrm{rank}(\mathbf{A}_i)$. Omitting the zero eigenvalues, we put

$$(2.10) \qquad \mathbf{\Theta}_i = \mathrm{diag}\{\theta_i^1, \ldots, \theta_i^{r_i}\}.$$

Further, in view of Assumption (2.2) there exist full sets of *right* eigenvectors $\mathbf{b}_{i,j}$, $\mathbf{A}_i \mathbf{b}_{i,j} = \theta_i^j \mathbf{b}_{i,j}$, and *left* eigenvectors $\mathbf{c}_i^{j\dagger}$, $\mathbf{c}_i^{j\dagger} \mathbf{A}_i = \theta_i^j \mathbf{c}_i^{j\dagger}$ (we use the $\dagger$ symbol to indicate a *row*-vector or a matrix of row vectors). In the matrix form, omitting vectors with indices $j > r_i$ that are in the kernel of $\mathbf{A}_i$, we can write

$$\mathbf{B}_i = \begin{bmatrix} \mathbf{b}_{i,1} \cdots \mathbf{b}_{i,r_i} \end{bmatrix}, \quad \mathbf{A}_i \mathbf{B}_i = \mathbf{B}_i \mathbf{\Theta}_i, \quad \mathbf{C}_i^\dagger = \begin{bmatrix} \mathbf{c}_i^{1\dagger} \\ \vdots \\ \mathbf{c}_i^{r_i\dagger} \end{bmatrix}, \quad \mathbf{C}_i^\dagger \mathbf{A}_i = \mathbf{\Theta}_i \mathbf{C}_i^\dagger,$$

with $\mathbf{\Theta}_i$ defined by (2.10). Then we have a decomposition $\mathbf{A}_i = \mathbf{B}_i \mathbf{C}_i^\dagger$, provided that $\mathbf{C}_i^\dagger \mathbf{B}_i = \mathbf{\Theta}_i$. We call this last condition the *orthogonality condition* (since $\mathbf{\Theta}_i$ is diagonal) and we assume that it holds even when we have repeating eigenvalues. This condition is related to the normalization ambiguity of the eigenvectors. Thus, given $\mathbf{A}_i$, we can construct (in a non-unique way) a corresponding decomposition pair $(\mathbf{B}_i, \mathbf{C}_i^\dagger)$. The space of all such pairs for all finite indices $1 \le i \le n$, without any additional conditions, is our *decomposition space*. We denote it as

$$\mathcal{B} \times \mathcal{C} = (\mathbb{C}^{r_1} \times \cdots \times \mathbb{C}^{r_n}) \times ((\mathbb{C}^{r_1})^\dagger \times \cdots \times (\mathbb{C}^{r_n})^\dagger)$$
$$\simeq (\mathbb{C}^{r_1} \times (\mathbb{C}^{r_1})^\dagger) \times \cdots \times (\mathbb{C}^{r_n} \times (\mathbb{C}^{r_n})^\dagger)$$

and write an element $(\mathbf{B}, \mathbf{C}^\dagger)$ of this space as a list of $n$ pairs $(\mathbf{B}_1, \mathbf{C}_1^\dagger; \cdots; \mathbf{B}_n, \mathbf{C}_n^\dagger)$. Then, given a Riemann Scheme of a Fuchsian system (equivalently, a collection $\mathbf{\Theta} = \{\theta_i^j\}$ of the characteristic indices having the correct multiplicities and satisfying the Fuchs relation), we denote by
$$(2.11)$$

$$(\mathcal{B} \times \mathcal{C})_{\mathbf{\Theta}} = \{(\mathbf{B}_1, \mathbf{C}_1^\dagger; \cdots; \mathbf{B}_n, \mathbf{C}_n^\dagger) \in \mathcal{B} \times \mathcal{C} \mid \mathbf{C}_i^\dagger \mathbf{B}_i = \mathbf{\Theta}_i, \sum_{i=1}^{n} \mathbf{B}_i \mathbf{C}_i^\dagger = \mathbf{A}_\infty \sim \mathbf{\Theta}_\infty\}$$

the corresponding fiber in the decomposition space (since for Schlesinger transformations locations of the poles are just fixed parameters of the dynamic, we occasionally omit them, as in the above notation).

REMARK 2.3. There are two natural actions on the decomposition space $\mathcal{B} \times \mathcal{C}$. First, the group $\mathbb{GL}_m$ of global gauge transformations of the Fuchsian system induces the following action. Given $\mathbf{P} \in \mathbb{GL}_m$, we have the action $\mathbf{A}_i \mapsto \mathbf{P}\mathbf{A}_i \mathbf{P}^{-1}$ which translates into the action $(\mathbf{B}_i, \mathbf{C}_i^\dagger) \mapsto (\mathbf{P}\mathbf{B}_i, \mathbf{C}_i^\dagger \mathbf{P}^{-1})$. We refer to such transformations as *similarity transformations*. It is often necessary to restrict this action to the subgroup $G_{\mathbf{A}_\infty}$ preserving the form of $\mathbf{A}_\infty$. Second, for any pair $(\mathbf{B}_i, \mathbf{C}_i^\dagger)$

the pair $(\mathbf{B}_i\mathbf{Q}_i, \mathbf{Q}_i^{-1}\mathbf{C}_i^{\dagger})$ determines the same matrix $\mathbf{A}_i$ for $\mathbf{Q}_i \in \mathbb{GL}_{r_i}$. The condition $\mathbf{Q}_i^{-1}\mathbf{C}_i^{\dagger}\mathbf{B}_i\mathbf{Q}_i = \mathbf{Q}_i^{-1}\mathbf{\Theta}_i\mathbf{Q}_i = \mathbf{\Theta}_i$ restricts $\mathbf{Q}_i$ to the stabilizer subgroup $G_{\mathbf{\Theta}_i}$ of $\mathbb{GL}_{r_i}$. In particular, when all $\theta_i^j$ are distinct, $\mathbf{Q}_i$ must be a diagonal matrix. We refer to such transformations as *trivial transformations*. These two actions obviously commute with each other. The phase space for the Schlesinger dynamic is the quotient space of $(\mathcal{B} \times \mathcal{C})_{\mathbf{\Theta}}$ by this action.

**2.4. Schlesinger Dynamic on the Decomposition Space.** In this section we explain how to lift the Schlesinger evolution equations to the decomposition space.

2.4.1. *Rank One.* In [**DST13**] we focused on the *elementary* Schlesinger transformations $\left\{ \begin{smallmatrix} \alpha & \beta \\ \mu & \nu \end{smallmatrix} \right\}$ that only change two of the characteristic indices by unit shifts, i.e., $\bar{\theta}_\alpha^\mu = \theta_\alpha^\mu - 1$ and $\bar{\theta}_\beta^\nu = \theta_\beta^\nu + 1$, $\alpha \neq \beta$. For such transformations the projector matrix $\mathbf{P}$ has rank one and the multiplier matrix has the form (2.3) with

$$(2.12) \quad \mathbf{R}(z) = \mathbf{I} + \frac{z_\alpha - z_\beta}{z - z_\alpha}\mathbf{P}, \quad \text{where } \mathbf{P} = \frac{\mathbf{b}_{\beta,\nu}\mathbf{c}_\alpha^{\mu\dagger}}{\mathbf{c}_\alpha^{\mu\dagger}\mathbf{b}_{\beta,\nu}}, \quad \text{and we put } \mathbf{Q} = \mathbf{I} - \mathbf{P}.$$

In this case, under the semi-simplicity Assumption (2.2), it is possible to decompose equations (2.7–2.9) to get the dynamic on the space $(\mathcal{B} \times \mathcal{C})_{\mathbf{\Theta}}$.

THEOREM 2.4 ([**DST13**]). *An elementary Schlesinger transformation $\left\{ \begin{smallmatrix} \alpha & \beta \\ \mu & \nu \end{smallmatrix} \right\}$ defines the map*

$$(\mathcal{B} \times \mathcal{C})_{\mathbf{\Theta}} \to (\bar{\mathcal{B}} \times \bar{\mathcal{C}})_{\bar{\mathbf{\Theta}}}$$

*given by the following evolution equations (grouped for convenience), where $c_i^j$ are arbitrary non-zero constants.*

*(i) Transformation vectors:*

$$(2.13) \qquad \bar{\mathbf{b}}_{\alpha,\mu} = \frac{1}{c_\alpha^\mu}\mathbf{b}_{\beta,\nu}, \qquad \bar{\mathbf{c}}_\beta^{\nu\dagger} = c_\beta^\nu\mathbf{c}_\alpha^{\mu\dagger}.$$

*(ii) Generic indices:*

$$(2.14) \qquad \bar{\mathbf{b}}_{i,j} = \frac{1}{c_i^j}\mathbf{R}(z_i)\mathbf{b}_{i,j}, \ (i \neq \alpha \text{ and if } i = \beta, j \neq \nu);$$

$$(2.15) \qquad \bar{\mathbf{c}}_i^{j\dagger} = c_i^j\mathbf{c}_i^{j\dagger}\mathbf{R}^{-1}(z_i), \ (i \neq \beta \text{ and if } i = \alpha, j \neq \mu).$$

*(iii) Special indices:*

$$(2.16) \qquad \bar{\mathbf{b}}_{\alpha,j} = \frac{1}{c_\alpha^j} \left( \mathbf{I} - \frac{\mathbf{P}}{\theta_\alpha^\mu - \theta_\alpha^j - 1} \left( \sum_{i \neq \alpha} \frac{z_\beta - z_\alpha}{z_i - z_\alpha} \mathbf{A}_i \right) \right) \mathbf{b}_{\alpha,j}, \qquad j \neq \mu;$$

$$(2.17) \qquad \bar{\mathbf{c}}_\beta^{j\dagger} = c_\beta^j \mathbf{c}_\beta^{j\dagger} \left( \mathbf{I} - \left( \sum_{i \neq \beta} \frac{z_\alpha - z_\beta}{z_i - z_\beta} \mathbf{A}_i \right) \frac{\mathbf{P}}{\theta_\beta^\nu - \theta_\beta^j + 1} \right), \qquad j \neq \nu;$$

$$\bar{\mathbf{b}}_{\beta,\nu} = \frac{1}{c_\beta^\nu} \left( (\theta_\beta^\nu + 1)\mathbf{I} + \right.$$

$$(2.18) \qquad \left. \mathbf{Q} \left( \mathbf{I} + \sum_{j \neq \nu} \frac{\mathbf{b}_{\beta,j} \mathbf{c}_\beta^{j\dagger}}{\theta_\beta^\nu - \theta_\beta^j + 1} \right) \left( \sum_{i \neq \beta} \frac{z_\alpha - z_\beta}{z_i - z_\beta} \mathbf{A}_i \right) \right) \frac{\mathbf{b}_{\beta,\nu}}{\mathbf{c}_\alpha^{\mu\dagger} \mathbf{b}_{\beta,\nu}};$$

$$\bar{\mathbf{c}}_\alpha^{\mu\dagger} = c_\alpha^\mu \frac{\mathbf{c}_\alpha^{\mu\dagger}}{\mathbf{c}_\alpha^{\mu\dagger} \mathbf{b}_{\beta,\nu}} \left( (\theta_\alpha^\mu - 1)\mathbf{I} + \right.$$

$$(2.19) \qquad \left. \left( \sum_{i \neq \alpha} \frac{z_\beta - z_\alpha}{z_i - z_\alpha} \mathbf{A}_i \right) \left( \mathbf{I} + \sum_{j \neq \mu} \frac{\mathbf{b}_{\alpha,j} \mathbf{c}_\alpha^{j\dagger}}{\theta_\alpha^\mu - \theta_\alpha^j - 1} \right) \mathbf{Q} \right).$$

2.4.2. *Rank Two.* For the difference Painlevé equation d-$P\left(A_1^{(1)*}\right)$ we need to study Schlesinger transformations of a Fuchsian system that has the spectral type $(22, 1111, 1111)$, and so we need to consider Schlesinger transformations that change not one but two eigenvalues at each point $z_\alpha$ and $z_\beta$. In this section we show how to obtain the corresponding dynamic on the decomposition space. The resulting equations suggest what happens in the general case of a projector $\mathbf{P}$ of arbitrary rank, but since the focus of the present paper is on examples, we plan to consider the general case elsewhere.

Naively, we want to consider Schlesinger transformations of the form

$$(2.20) \qquad \left\{ \begin{smallmatrix} \alpha & \beta \\ \mu_1 & \nu_1 \\ \mu_2 & \nu_2 \end{smallmatrix} \right\} = \left\{ \begin{smallmatrix} \alpha & \beta \\ \mu_1 & \nu_1 \end{smallmatrix} \right\} \circ \left\{ \begin{smallmatrix} \alpha & \beta \\ \mu_2 & \nu_2 \end{smallmatrix} \right\} = \left\{ \begin{smallmatrix} \alpha & \beta \\ \mu_2 & \nu_2 \end{smallmatrix} \right\} \circ \left\{ \begin{smallmatrix} \alpha & \beta \\ \mu_1 & \nu_1 \end{smallmatrix} \right\}.$$

However, if one of the characteristic indices (say $\alpha$) has multiplicity, applying a rank-one elementary Schlesinger transformation will change the spectral type of the equation (e.g., in our example, a rank-one transformation $\left\{ \begin{smallmatrix} 1 & 2 \\ 1 & 1 \end{smallmatrix} \right\}$ maps the moduli space of Fuchsian equations of spectral type $(22, 1111, 1111)$ to a smaller moduli space $(112, 1111, 1111)$, and in fact our formulas in this case do not work, since some of the expressions become undefined). Thus, we need to develop the rank-two version of the elementary Schlesinger transformation separately. We start with a composition of two rank-one maps to get an insight on the structure of the multiplier matrix in the rank-two case, but then proceed to derive the dynamic equations independently. The resulting equations are then defined on moduli spaces of Fuchsian systems that have multiplicity in the spectral type (e.g., in our example, the map $\left\{ \begin{smallmatrix} 1 & 2 \\ 1 & 1 \\ 2 & 2 \end{smallmatrix} \right\}$ is defined on both moduli spaces of Fuchsian systems of spectral type $(1111, 1111, 1111)$ and $(22, 1111, 1111)$). So we start with the multiplier matrix that is a product (and for simplicity we put $\mu_i = i$ and $\nu_j = j$ for this derivation)

$$\mathbf{R}(z) = \bar{\mathbf{R}}_1(z)\mathbf{R}_2(z) = \left(\mathbf{I} + \frac{z_\alpha - z_\beta}{z - z_\alpha}\bar{\mathbf{P}}_1\right)\left(\mathbf{I} + \frac{z_\alpha - z_\beta}{z - z_\alpha}\mathbf{P}_2\right),$$

where, in view of (2.12) and (2.14–2.15),

$$\mathbf{P}_i = \frac{\mathbf{b}_{\beta,i}\mathbf{c}_\alpha^{i\dagger}}{\mathbf{c}_\alpha^{i\dagger}\mathbf{b}_{\beta,i}}, \quad \text{and} \quad \bar{\mathbf{P}}_1 = \frac{\bar{\mathbf{b}}_{\beta,1}\bar{\mathbf{c}}_\alpha^{1\dagger}}{\bar{\mathbf{c}}_\alpha^{1\dagger}\bar{\mathbf{b}}_{\beta,1}} = \frac{\mathbf{Q}_2\mathbf{b}_{\beta,1}\mathbf{c}_\alpha^{1\dagger}\mathbf{Q}_2}{\mathbf{c}_\alpha^{1\dagger}\mathbf{Q}_2\mathbf{b}_{\beta,1}} = \frac{\mathbf{Q}_2\mathbf{P}_1\mathbf{Q}_2}{\mathrm{Tr}(\mathbf{Q}_2\mathbf{P}_1)}.$$

Here $\mathbf{Q}_i = \mathbf{I} - \mathbf{P}_i$ is, as usual, the complementary projector. Then, since clearly $\bar{\mathbf{P}}_1\mathbf{P}_2 = 0$,

$$\mathbf{R}(z) = \mathbf{I} + \frac{z_\alpha - z_\beta}{z - z_\alpha}\mathcal{P},$$

where $\mathcal{P} = \bar{\mathbf{P}}_1 + \mathbf{P}_2$ is also a projector. Let us now rewrite $\mathcal{P}$ in a more symmetric form. First note that, since $\mathbf{P}_i$ are rank-one projectors,

$$\mathrm{Tr}(\mathbf{Q}_2\mathbf{P}_1) = \mathrm{Tr}(\mathbf{P}_1 - \mathbf{P}_2\mathbf{P}_1) = 1 - \mathrm{Tr}(\mathbf{P}_1\mathbf{P}_2) = \mathrm{Tr}(\mathbf{Q}_1\mathbf{P}_2).$$

Also, note that for any rank-one projector $\mathbf{S}$ and for any matrix $\mathbf{M}$ we have an identity $\mathbf{SMS} = \mathrm{Tr}(\mathbf{MS})\mathbf{S}$. Therefore,

$$\mathcal{P} = \frac{\mathbf{Q}_2\mathbf{P}_1\mathbf{Q}_2 + \mathbf{P}_2\mathbf{Q}_1\mathbf{P}_2}{\mathrm{Tr}(\mathbf{Q}_2\mathbf{P}_1)} = \frac{\mathbf{Q}_2\mathbf{P}_1}{\mathrm{Tr}(\mathbf{Q}_2\mathbf{P}_1)} + \frac{\mathbf{Q}_1\mathbf{P}_2}{\mathrm{Tr}(\mathbf{Q}_1\mathbf{P}_2)} = \mathcal{P}_1 + \mathcal{P}_2,$$

where

$$(2.21) \qquad \mathcal{P}_1 = \frac{\mathbf{Q}_2\mathbf{P}_1}{\mathrm{Tr}(\mathbf{Q}_2\mathbf{P}_1)} = \frac{\mathbf{Q}_2\mathbf{b}_{\beta,1}\mathbf{c}_\alpha^{1\dagger}}{\mathbf{c}_\alpha^{1\dagger}\mathbf{Q}_2\mathbf{b}_{\beta,1}}, \qquad \mathcal{P}_2 = \frac{\mathbf{Q}_1\mathbf{P}_2}{\mathrm{Tr}(\mathbf{Q}_1\mathbf{P}_2)} = \frac{\mathbf{Q}_1\mathbf{b}_{\beta,2}\mathbf{c}_\alpha^{2\dagger}}{\mathbf{c}_\alpha^{2\dagger}\mathbf{Q}_1\mathbf{b}_{\beta,2}}$$

are two mutually orthogonal rank-one projectors, $\mathcal{P}_i^2 = \mathcal{P}_i$, $\mathcal{P}_1\mathcal{P}_2 = \mathcal{P}_2\mathcal{P}_1 = 0$. At the same time, since obviously $\mathbf{Q}_2\mathbf{P}_1 + \mathbf{Q}_1\mathbf{P}_2 = \mathbf{P}_1\mathbf{Q}_2 + \mathbf{P}_2\mathbf{Q}_1$, $\mathcal{P} = \tilde{\mathcal{P}}_1 + \tilde{\mathcal{P}}_2$, where

$$(2.22) \qquad \tilde{\mathcal{P}}_1 = \frac{\mathbf{P}_1\mathbf{Q}_2}{\mathrm{Tr}(\mathbf{Q}_2\mathbf{P}_1)} = \frac{\mathbf{b}_{\beta,1}\mathbf{c}_\alpha^{1\dagger}\mathbf{Q}_2}{\mathbf{c}_\alpha^{1\dagger}\mathbf{Q}_2\mathbf{b}_{\beta,1}}, \qquad \tilde{\mathcal{P}}_2 = \frac{\mathbf{P}_2\mathbf{Q}_1}{\mathrm{Tr}(\mathbf{Q}_1\mathbf{P}_2)} = \frac{\mathbf{b}_{\beta,2}\mathbf{c}_\alpha^{2\dagger}\mathbf{Q}_1}{\mathbf{c}_\alpha^{2\dagger}\mathbf{Q}_1\mathbf{b}_{\beta,2}}.$$

We also put $\mathcal{Q}_i = \mathbf{I} - \mathcal{P}_i$, $\tilde{\mathcal{Q}}_i = \mathbf{I} - \tilde{\mathcal{P}}_i$, and $\mathcal{Q} = \mathbf{I} - \mathcal{P}$. In view of the orthogonality conditions $\mathbf{C}_i^\dagger\mathbf{B}_i = \mathbf{\Theta}_i$, it is easy to describe the eigenvectors for each of those projectors. We do this just for $\mathcal{P}$s since for $\mathcal{Q}$s eigenvectors are the same but eigenvalues swap between 0 and 1. To describe the spectrum, we use the notation $(\theta; \mathbf{w}^\dagger, \mathbf{v})$, where $\theta$ is an eigenvalue (which is either 0 or 1 for projectors), $\mathbf{w}^\dagger$ is a row (or left) eigenvector and $\mathbf{v}$ is a column (or right) eigenvector. Then we get

$$\mathrm{Eigen}(\mathbf{P}_i) = \{(1; \mathbf{c}_\alpha^{i\dagger}, \mathbf{b}_{\beta,i}), (0; \mathbf{c}_\beta^{j\dagger}, \mathbf{b}_{\alpha,j}) \text{ for } j \neq i\}, \quad i = 1, 2;$$

$$\mathrm{Eigen}(\mathcal{P}_1) = \{(1; \mathbf{c}_\alpha^{1\dagger}, \mathbf{Q}_2\mathbf{b}_{\beta,1}), (0; \mathbf{c}_\alpha^{2\dagger}, \mathbf{b}_{\alpha,2}), (0; \mathbf{c}_\beta^{j\dagger}, \mathbf{b}_{\alpha,j}) \text{ for } j > 2\};$$

$$\mathrm{Eigen}(\mathcal{P}_2) = \{(1; \mathbf{c}_\alpha^{2\dagger}, \mathbf{Q}_1\mathbf{b}_{\beta,2}), (0; \mathbf{c}_\alpha^{1\dagger}, \mathbf{b}_{\alpha,1}), (0; \mathbf{c}_\beta^{j\dagger}, \mathbf{b}_{\alpha,j}) \text{ for } j > 2\};$$

$$\mathrm{Eigen}(\tilde{\mathcal{P}}_1) = \{(1; \mathbf{c}_\alpha^{1\dagger}\mathbf{Q}_2, \mathbf{b}_{\beta,1}), (0; \mathbf{c}_\beta^{2\dagger}, \mathbf{b}_{\beta,2}), (0; \mathbf{c}_\beta^{j\dagger}, \mathbf{b}_{\alpha,j}) \text{ for } j > 2\};$$

$$\mathrm{Eigen}(\tilde{\mathcal{P}}_1) = \{(1; \mathbf{c}_\alpha^{2\dagger}\mathbf{Q}_1, \mathbf{b}_{\beta,2}), (0; \mathbf{c}_\beta^{1\dagger}, \mathbf{b}_{\beta,1}), (0; \mathbf{c}_\beta^{j\dagger}, \mathbf{b}_{\alpha,j}) \text{ for } j > 2\};$$

$$\mathrm{Eigen}(\mathcal{P}) = \{(1; \mathbf{c}_\alpha^{1\dagger}, \mathbf{Q}_2\mathbf{b}_{\beta,1}), (1; \mathbf{c}_\alpha^{2\dagger}, \mathbf{Q}_1\mathbf{b}_{\beta,2}), (0; \mathbf{c}_\beta^{j\dagger}, \mathbf{b}_{\alpha,j}) \text{ for } j > 2\}$$
$$= \{(1; \mathbf{c}_\alpha^{1\dagger}\mathbf{Q}_2, \mathbf{b}_{\beta,1}), (1; \mathbf{c}_\alpha^{2\dagger}\mathbf{Q}_1, \mathbf{b}_{\beta,2}), (0; \mathbf{c}_\beta^{j\dagger}, \mathbf{b}_{\alpha,j}) \text{ for } j > 2\}.$$

REMARK 2.5. Note that the sum of two rank-one projectors is not a projector. Here $\mathcal{P}$ is the "correct" way to add $\mathbf{P}_1$ and $\mathbf{P}_2$ so that the result is a rank-two projector that is a sum of two orthogonal rank-one projectors that has the same row and column spaces as $\mathbf{P}_1 + \mathbf{P}_2$. Also, note that there are many ways to choose bases in the row and column ranges of $\mathcal{P}$, the choices above reflect the splittings $\mathcal{P} = \mathcal{P}_1 + \mathcal{P}_2 = \tilde{\mathcal{P}}_1 + \tilde{\mathcal{P}}_2$.

We can now use these projectors to split the discrete Schlesinger evolution equations to define dynamic on eigenvectors. The proof is very similar in spirit to the rank-one case proof in [**DST13**].

THEOREM 2.6. *Consider a multiplier matrix in the form*

$$(2.23) \qquad \mathbf{R}(z) = \mathbf{I} + \frac{z_\alpha - z_\beta}{z - z_\alpha}\mathcal{P}, \quad \text{where } \mathcal{P} = \mathcal{P}_1 + \mathcal{P}_2 = \tilde{\mathcal{P}}_1 + \tilde{\mathcal{P}}_2,$$

*and $\mathcal{P}_i$, $\tilde{\mathcal{P}}_i$ are given by (2.21–2.22). Then $\mathcal{P}$ satisfies the constraints (2.6) and so defines a Schlesinger transformation. This Schlesinger transformation has the type $\left\{\begin{smallmatrix} \alpha & \beta \\ 1 & 1 \\ 2 & 2 \end{smallmatrix}\right\}$ and the corresponding map*

$$(\mathcal{B} \times \mathcal{C})_{\Theta} \to (\bar{\mathcal{B}} \times \bar{\mathcal{C}})_{\bar{\Theta}}$$

*is given by the following evolution equations, where $c_i^j$ are again arbitrary non-zero constants.*

*(i) Transformation vectors:*

$$(2.24) \quad \bar{\mathbf{b}}_{\alpha,1} = \frac{1}{c_\alpha^1}\mathbf{Q}_2\mathbf{b}_{\beta,1}, \; \bar{\mathbf{b}}_{\alpha,2} = \frac{1}{c_\alpha^2}\mathbf{Q}_1\mathbf{b}_{\beta,2}, \quad \bar{\mathbf{c}}_\beta^{1\dagger} = c_\beta^1 c_\alpha^{1\dagger}\mathbf{Q}_2, \; \bar{\mathbf{c}}_\beta^{2\dagger} = c_\beta^2 c_\alpha^{2\dagger}\mathbf{Q}_1.$$

*(ii) Generic indices:*

$$(2.25) \qquad \bar{\mathbf{b}}_{i,j} = \frac{1}{c_i^j}\mathbf{R}(z_i)\mathbf{b}_{i,j}, \; (i \neq \alpha \text{ and if } i = \beta, j > 2);$$

$$(2.26) \qquad \bar{\mathbf{c}}_i^{j\dagger} = c_i^j \mathbf{c}_i^{j\dagger}\mathbf{R}^{-1}(z_i), \; (i \neq \beta \text{ and if } i = \alpha, j > 2).$$

*(iii) Special indices (here $k = 1, 2$, $k' = 3 - k$, and $j > 2$):*

$$(2.27) \quad \bar{\mathbf{b}}_{\alpha,j} = \frac{1}{c_\alpha^j}\left(\mathbf{I} - \left(\frac{\mathcal{P}_1}{\theta_\alpha^1 - \theta_\alpha^j - 1} + \frac{\mathcal{P}_2}{\theta_\alpha^2 - \theta_\alpha^j - 1}\right)\left(\sum_{i \neq \alpha}\frac{z_\beta - z_\alpha}{z_i - z_\alpha}\mathbf{A}_i\right)\right)\mathbf{b}_{\alpha,j};$$

$$(2.28) \quad \bar{\mathbf{c}}_\beta^{j\dagger} = c_\beta^j \mathbf{c}_\beta^{j\dagger}\left(\mathbf{I} - \left(\sum_{i \neq \beta}\frac{z_\alpha - z_\beta}{z_i - z_\beta}\mathbf{A}_i\right)\left(\frac{\tilde{\mathcal{P}}_1}{\theta_\beta^1 - \theta_\beta^j + 1} + \frac{\tilde{\mathcal{P}}_2}{\theta_\beta^2 - \theta_\beta^j + 1}\right)\right);$$

$$(2.29) \quad \bar{\mathbf{b}}_{\beta,k} = \frac{1}{c_\beta^k} \left( (\theta_\beta^k + 1)\mathbf{I} + \right.$$

$$\mathcal{Q}\left(\mathbf{I} + \sum_{j>2} \frac{\mathbf{b}_{\beta,j}\mathbf{c}_\beta^{j\dagger}}{\theta_\beta^1 - \theta_\beta^j + 1}\right)\left(\sum_{i\neq\beta} \frac{z_\alpha - z_\beta}{z_i - z_\beta}\mathbf{A}_i\right)\right) \frac{\mathbf{b}_{\beta,k}}{\mathbf{c}_\alpha^{k\dagger}\mathbf{Q}_{k'}\mathbf{b}_{\beta,k}};$$

$$(2.30) \quad \bar{\mathbf{c}}_\alpha^{k\dagger} = c_\alpha^k \frac{\mathbf{c}_\alpha^{k\dagger}}{\mathbf{c}_\alpha^{k\dagger}\mathbf{Q}_{k'}\mathbf{b}_{\beta,k}}\left( (\theta_\alpha^k - 1)\mathbf{I} + \right.$$

$$\left(\sum_{i\neq\alpha} \frac{z_\beta - z_\alpha}{z_i - z_\alpha}\mathbf{A}_i\right)\left(\mathbf{I} + \sum_{j>2} \frac{\mathbf{b}_{\alpha,j}\mathbf{c}_\alpha^{j\dagger}}{\theta_\alpha^k - \theta_\alpha^j - 1}\right)\mathcal{Q}\right).$$

PROOF. Of course the statement that $\mathcal{P}$ defines an elementary Schlesinger transformation of the type $\left\{\begin{smallmatrix} \alpha & \beta \\ 1 & 1 \\ 2 & 2 \end{smallmatrix}\right\}$ follows from how we derived it, but it can also be seen directly. E.g., conditions (2.21–2.22) follow immediately from

$$(2.31) \qquad \mathcal{P}\mathbf{A}_\alpha = (\theta_\alpha^1\mathcal{P}_1 + \theta_\alpha^2\mathcal{P}_2), \qquad \mathbf{A}_\beta\mathcal{P} = \theta_\beta^1\tilde{\mathcal{P}}_1 + \theta_\beta^2\tilde{\mathcal{P}}_2,$$

and the fact that

$$\bar{\theta}_\alpha^i = \theta_\alpha^i - 1, \quad \bar{\theta}_\beta^i = \theta_\beta^i + 1 \quad \text{for } i = 1, 2 \quad \text{and} \quad \bar{\theta}_i^j = \theta_i^j \quad \text{otherwise},$$

can be seen, in particular, from our derivation of the evolution equations below.

To establish (i), we use (2.5):

$$\bar{\mathbf{A}}_\alpha\mathcal{P} = \mathcal{P}\mathbf{A}_\alpha - \mathcal{P} = (\theta_\alpha^1 - 1)\mathcal{P}_1 + (\theta_\alpha^2 - 1)\mathcal{P}_2.$$

Since $\mathcal{P}_1\mathbf{Q}_2\mathbf{b}_{\beta,1} = \mathbf{Q}_2\mathbf{b}_{\beta,1}$ and $\mathcal{P}_2\mathbf{Q}_2\mathbf{b}_{\beta,1} = \mathbf{0}$, we see that $\bar{\theta}_\alpha^1 = \theta_\alpha^1 - 1$ and $\bar{\mathbf{b}}_{\alpha,1} \sim \mathbf{Q}_2\mathbf{b}_{\beta,1}$, where $\sim$ stands for 'proportional'. Then $\bar{\mathbf{b}}_{\alpha,1} = \mathbf{Q}_2\mathbf{b}_{\beta,1}/c_\alpha^1$, where $c_\alpha^1$ is some non-zero proportionality constant. The other equations in this part are proved similarly. Note that the consequence of (i) is that we can write

$$(2.32) \quad \mathcal{P}_i = \frac{\bar{\mathbf{b}}_{\alpha,i}\mathbf{c}_\alpha^{i\dagger}}{\mathbf{c}_\alpha^{i\dagger}\bar{\mathbf{b}}_{\alpha,i}}, \quad \tilde{\mathcal{P}}_i = \frac{\mathbf{b}_{\beta,i}\bar{\mathbf{c}}_\beta^{i\dagger}}{\bar{\mathbf{c}}_\beta^{i\dagger}\mathbf{b}_{\beta,i}}, \quad \mathcal{P} = \frac{\bar{\mathbf{b}}_{\alpha,1}\mathbf{c}_\alpha^{1\dagger}}{\mathbf{c}_\alpha^{1\dagger}\bar{\mathbf{b}}_{\alpha,1}} + \frac{\bar{\mathbf{b}}_{\alpha,2}\mathbf{c}_\alpha^{2\dagger}}{\mathbf{c}_\alpha^{2\dagger}\bar{\mathbf{b}}_{\alpha,2}} = \frac{\mathbf{b}_{\beta,1}\bar{\mathbf{c}}_\beta^{1\dagger}}{\bar{\mathbf{c}}_\beta^{1\dagger}\mathbf{b}_{\beta,1}} + \frac{\mathbf{b}_{\beta,2}\bar{\mathbf{c}}_\beta^{2\dagger}}{\bar{\mathbf{c}}_\beta^{2\dagger}\mathbf{b}_{\beta,2}}.$$

For the generic case $i \neq \alpha, \beta$ in (ii) the proof is identical to the rank-one case. (Since it is also short, we opted to include it to make the paper more self-contained). From (2.7) we see that

$$\bar{\mathbf{A}}_i\mathbf{R}(z_i)\mathbf{B}_i = \mathbf{R}(z_i)\mathbf{A}_i\mathbf{B}_i = \mathbf{R}(z_i)\mathbf{B}_i\mathbf{\Theta}_i,$$

and so $\bar{\mathbf{\Theta}}_i = \mathbf{\Theta}_i$ and $\bar{\mathbf{B}}_i\bar{\mathbf{D}}_i = \mathbf{R}(z_i)\mathbf{B}_i$, where $\bar{\mathbf{D}}_i = \text{diag}\{c_i^j\}$ is a diagonal matrix of non-zero proportionality constants. Similarly, $\bar{\mathbf{\Delta}}_i\bar{\mathbf{C}}_i^\dagger = \mathbf{C}_i^\dagger\mathbf{R}^{-1}(z_i)$. The orthogonality condition $\bar{\mathbf{C}}_i^\dagger\bar{\mathbf{B}}_i = \mathbf{\Theta}_i$ implies that $\bar{\mathbf{\Delta}}_i\bar{\mathbf{D}}_i = \mathbf{I}$, which gives (ii) for generic indices. For $i = \alpha$, from (2.4) we see that $\mathbf{C}_\alpha^\dagger\mathcal{Q}\bar{\mathbf{A}}_\alpha = \mathbf{\Theta}_\alpha\mathbf{C}_\alpha^\dagger\mathcal{Q}$, and so again $\bar{\mathbf{\Delta}}_\alpha\bar{\mathbf{C}}_\alpha^\dagger = \mathbf{C}_\alpha^\dagger\mathcal{Q}$. However, since $\mathbf{c}_\alpha^{1\dagger}\mathcal{Q} = \mathbf{c}_\alpha^{2\dagger}\mathcal{Q} = 0$, $(\bar{\mathbf{\Delta}}_\alpha)_1^1 = (\bar{\mathbf{\Delta}}_\alpha)_2^2 = 0$ and we can not recover $\bar{\mathbf{c}}_\alpha^{1\dagger}$ and $\bar{\mathbf{c}}_\alpha^{2\dagger}$. The case $i = \beta$ is similar.

Finally, let us consider special indices. To find $\bar{\mathbf{b}}_{\alpha,j}$ for $j > 2$ start with (2.8) and (2.26):

$$\bar{\mathbf{A}}_\alpha = \bar{\mathbf{b}}_{\alpha,1}\bar{\mathbf{c}}_\alpha^1 + \bar{\mathbf{b}}_{\alpha,2}\bar{\mathbf{c}}_\alpha^2 + \sum_{j>2} \bar{\mathbf{b}}_{\alpha,j}(c_\alpha^j\mathcal{Q}\mathbf{c}_\alpha^j) = \mathbf{A}_\alpha - \mathcal{Q}\mathbf{A}_\alpha\mathcal{P} + \sum_{i\neq\alpha}\left(\frac{z_\beta - z_\alpha}{z_i - z_\alpha}\right)\mathcal{P}\mathbf{A}_i\mathcal{Q}.$$

Multiplying on the right by $\mathbf{b}_{\alpha,j}$, using the orthogonality conditions and $\mathcal{P}\mathbf{b}_{\alpha,j} = \mathbf{0}$, $\mathcal{Q}\mathbf{b}_{\alpha,j} = \mathbf{b}_{\alpha,j}$, we get
(2.33)
$$\bar{\mathbf{b}}_{\alpha,1}(\bar{\mathbf{c}}_{\alpha}^{1\dagger}\mathbf{b}_{\alpha,j}) + \bar{\mathbf{b}}_{\alpha,2}(\bar{\mathbf{c}}_{\alpha}^{2\dagger}\mathbf{b}_{\alpha,j}) + c_{\alpha}^{j}\theta_{\alpha}^{j}\bar{\mathbf{b}}_{\alpha,j} = \theta_{\alpha}^{j}\mathbf{b}_{\alpha,j} + \sum_{i\neq\alpha}\left(\frac{z_{\beta}-z_{\alpha}}{z_{i}-z_{\alpha}}\right)\mathcal{P}\mathbf{A}_{i}\mathbf{b}_{\alpha,j}.$$

Now left-multiply by $\bar{\mathbf{c}}_{\alpha}^{1\dagger}$ and use expression (2.32) for $\mathcal{P}$ and orthogonality conditions again to get
$$\bar{\theta}_{\alpha}^{1}(\bar{\mathbf{c}}_{\alpha}^{1\dagger}\mathbf{b}_{\alpha,j}) = \theta_{\alpha}^{j}(\bar{\mathbf{c}}_{\alpha}^{1\dagger}\mathbf{b}_{\alpha,j}) + \bar{\theta}_{\alpha}^{1}\frac{\mathbf{c}_{\alpha}^{1\dagger}}{\mathbf{c}_{\alpha}^{1\dagger}\bar{\mathbf{b}}_{\alpha,1}}\sum_{i\neq\alpha}\left(\frac{z_{\beta}-z_{\alpha}}{z_{i}-z_{\alpha}}\right)\mathbf{A}_{i}\mathbf{b}_{\alpha,j}.$$

This gives
$$(\bar{\mathbf{c}}_{\alpha}^{1\dagger}\mathbf{b}_{\alpha,j}) = \frac{\bar{\theta}_{\alpha}^{1}}{\bar{\theta}_{\alpha}^{1}-\theta_{\alpha}^{j}}\frac{\mathbf{c}_{\alpha}^{1\dagger}}{\mathbf{c}_{\alpha}^{1\dagger}\bar{\mathbf{b}}_{\alpha,1}}\sum_{i\neq\alpha}\left(\frac{z_{\beta}-z_{\alpha}}{z_{i}-z_{\alpha}}\right)\mathbf{A}_{i}\mathbf{b}_{\alpha,j},$$
$$\bar{\mathbf{b}}_{\alpha,1}(\bar{\mathbf{c}}_{\alpha}^{1\dagger}\mathbf{b}_{\alpha,j}) = \frac{\bar{\theta}_{\alpha}^{1}}{\bar{\theta}_{\alpha}^{1}-\theta_{\alpha}^{j}}\mathcal{P}_{1}\sum_{i\neq\alpha}\left(\frac{z_{\beta}-z_{\alpha}}{z_{i}-z_{\alpha}}\right)\mathbf{A}_{i}\mathbf{b}_{\alpha,j}.$$

Repeating the same steps for $\bar{\mathbf{b}}_{\alpha,2}(\bar{\mathbf{c}}_{\alpha}^{2\dagger}\mathbf{b}_{\alpha,j})$, substituting the result into (2.33), solving for $\bar{\mathbf{b}}_{\alpha,j}$ and simplifying gives (2.27); (2.28) is proved in a similar fashion.

Finally, to get expressions for $\bar{\mathbf{b}}_{\beta,1}$ and $\bar{\mathbf{b}}_{\beta,2}$, use all of the previously obtained expressions to write
$$\bar{\mathbf{A}}_{\beta} = \bar{\mathbf{b}}_{\beta,1}\bar{\mathbf{c}}_{\beta}^{1\dagger} + \bar{\mathbf{b}}_{\beta,2}\bar{\mathbf{c}}_{\beta}^{2\dagger} + \sum_{j>2}\bar{\mathbf{b}}_{\beta,j}\bar{\mathbf{c}}_{\beta}^{j\dagger}$$
$$= c_{\beta}^{1}\bar{\mathbf{b}}_{\beta,1}\mathbf{c}_{\alpha}^{1\dagger}\mathbf{Q}_{2} + c_{\beta}^{2}\bar{\mathbf{b}}_{\beta,2}\mathbf{c}_{\alpha}^{2\dagger}\mathbf{Q}_{1}$$
$$+ \sum_{j>2}\mathcal{Q}\mathbf{b}_{\beta,j}\mathbf{c}_{\beta}^{j\dagger}\left(\mathbf{I} - \left(\sum_{i\neq\beta}\frac{z_{\alpha}-z_{\beta}}{z_{i}-z_{\beta}}\mathbf{A}_{i}\right)\left(\frac{\tilde{\mathcal{P}}_{1}}{\theta_{\beta}^{1}-\theta_{\beta}^{j}+1} + \frac{\tilde{\mathcal{P}}_{2}}{\theta_{\beta}^{2}-\theta_{\beta}^{j}+1}\right)\right)$$

which, in view of (2.9), also can be written as
$$\bar{\mathbf{A}}_{\beta} = \mathbf{A}_{\beta} - \mathcal{P}\mathbf{A}_{\beta}\mathcal{Q} + \mathcal{P} + \sum_{i\neq\beta}\left(\frac{z_{\alpha}-z_{\beta}}{z_{i}-z_{\beta}}\right)\mathcal{Q}\mathbf{A}_{i}\mathcal{P}.$$

Multiplying on the right by $\mathbf{b}_{\beta,1}$ we get
$$\bar{\mathbf{A}}_{\beta}\mathbf{b}_{\beta,1} = c_{\beta}^{1}(\mathbf{c}_{\alpha}^{1\dagger}\mathbf{Q}_{2}\mathbf{b}_{\beta,1})\bar{\mathbf{b}}_{\beta,1} + \sum_{j>2}\mathcal{Q}\mathbf{b}_{\beta,j}\mathbf{c}_{\beta}^{j\dagger}\left(\mathbf{I} - \sum_{i\neq\beta}\frac{z_{\alpha}-z_{\beta}}{z_{i}-z_{\beta}}\frac{\mathbf{A}_{i}}{\theta_{\beta}^{1}-\theta_{\beta}^{j}+1}\right)\mathbf{b}_{\beta,1}$$
$$= c_{\beta}^{1}(\mathbf{c}_{\alpha}^{1\dagger}\mathbf{Q}_{2}\mathbf{b}_{\beta,1})\bar{\mathbf{b}}_{\beta,1} - \mathcal{Q}\left(\sum_{i\neq\beta}\frac{z_{\alpha}-z_{\beta}}{z_{i}-z_{\beta}}\mathbf{A}_{i}\right)\left(\sum_{j>2}\frac{\mathbf{b}_{\beta,j}\mathbf{c}_{\beta}^{j\dagger}}{\theta_{\beta}^{1}-\theta_{\beta}^{j}+1}\right)\mathbf{b}_{\beta,1}$$
$$= \theta_{\beta}^{1}\mathbf{b}_{\beta,1} + \mathbf{b}_{\beta,1} + \mathcal{Q}\left(\sum_{i\neq\beta}\frac{z_{\alpha}-z_{\beta}}{z_{i}-z_{\beta}}\mathbf{A}_{i}\right)\mathbf{b}_{\beta,1}.$$

Solving for $\bar{\mathbf{b}}_{\beta,1}$ gives (2.29) for $k=1$, and the expression for $\mathbf{b}_{\beta,2}$ is obtained by right-multiplying by $\mathbf{b}_{\beta,2}$ instead. Equations (2.30) are obtained along the same lines. $\qquad\square$

## 3. Reductions from Schlesinger Transformations to Difference Painlevé Equations

In this section, which is the central section of the paper, we consider two examples of reductions from the Schlesinger dynamic on the decomposition space to difference Painlevé equations. First we consider Schlesinger transformations of a Fuchsian system of spectral type $(111, 111, 111)$. Resulting difference Painlevé equation is of type d-$P\left(A_2^{(1)*}\right)$ and has the symmetry group $E_6^{(1)}$. We have previously considered this example in [DST13], but the exposition there was very brief and relied on a nontrivial observation on how to choose good coordinates parameterizing our Fuchsian system. Here we not only provide more details but also show how geometric considerations *lead us* to the appropriate coordinate choice. In the second example we consider Schlesinger transformations of a Fuchsian system of spectral type $(22, 1111, 1111)$, which gives difference Painlevé equation of type d-$P\left(A_1^{(1)*}\right)$ with the symmetry group $E_7^{(1)}$. This example is completely new and here, in addition to elementary Schlesinger transformations of rank one, we also, for the first time, consider elementary Schlesinger transformations of rank two — we need such transformations to represent the standard example of a difference Painlevé equation of type d-$P\left(A_1^{(1)*}\right)$, as written in [GRO03], [Sak07], as a composition of elementary Schlesinger transformations.

### 3.1. Reduction to difference Painlevé equation of type d-$P\left(A_2^{(1)*}\right)$ with the symmetry group $E_6^{(1)}$.

3.1.1. *Model Example.* For our model example of type d-$P\left(A_2^{(1)*}\right)$ we take the equation that was first written by Grammaticos, Ramani, and Ohta, [GRO03], see also Murata [Mur04] and Sakai [Sak07]. Following Sakai's geometric approach, we view this equation as a birational map $\varphi : \mathbb{P}^1 \times \mathbb{P}^1 \dashrightarrow \mathbb{P}^1 \times \mathbb{P}^1$ with parameters $b_1, \ldots, b_8$

$$(3.1) \qquad \begin{pmatrix} b_1 & b_2 & b_3 & b_4 \\ b_5 & b_6 & b_7 & b_8 \end{pmatrix}; f, g \mapsto \begin{pmatrix} \bar{b}_1 & \bar{b}_2 & \bar{b}_3 & \bar{b}_4 \\ \bar{b}_5 & \bar{b}_6 & \bar{b}_7 & \bar{b}_8 \end{pmatrix}; \bar{f}, \bar{g} \end{pmatrix},$$

where $\bar{b}_1 = b_1, \bar{b}_2 = b_2, \bar{b}_3 = b_3, \bar{b}_4 = b_4, \bar{b}_5 = b_5 + \delta, \bar{b}_6 = b_6 + \delta, \bar{b}_7 = b_7 - \delta,$ $\bar{b}_8 = b_8 - \delta, \delta = b_1 + \cdots + b_8$, and $\bar{f}$ and $\bar{g}$ are given by the equation

$$(3.2) \qquad \begin{cases} (f+g)(\bar{f}+g) = \dfrac{(g+b_1)(g+b_2)(g+b_3)(g+b_4)}{(g-b_5)(g-b_6)} \\ (\bar{f}+g)(\bar{f}+\bar{g}) = \dfrac{(\bar{f}-b_1)(\bar{f}-b_2)(\bar{f}-b_3)(\bar{f}-b_4)}{(\bar{f}+b_7-\delta)(\bar{f}+b_8-\delta)} \end{cases}.$$

This map has the following eight indeterminate points:

$$p_1(b_1, -b_1), \quad p_3(b_3, -b_3), \quad p_5(\infty, b_5), \quad p_7(-b_7, \infty),$$
$$p_2(b_2, -b_2), \quad p_4(b_4, -b_4), \quad p_6(\infty, b_6), \quad p_8(-b_8, \infty),$$

resolving which by the blow-up procedure then gives us a rational surface $\mathcal{X}_\mathbf{b}$, known as the *Okamoto space of initial conditions* for this difference Painlevé equation, that is described by the blow-up diagram on Figure 2.

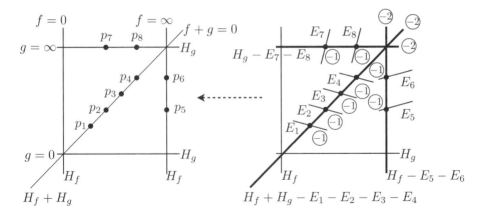

FIGURE 2. Okamoto surface $\mathcal{X}_{\mathbf{b}}$ for the model form of d-$P\left(A_2^{(1)*}\right)$.

The Picard lattice of $\mathcal{X}_{\mathbf{b}}$ is generated by the total transforms $H_f$ and $H_g$ of the coordinate lines and the classes of the exceptional divisors $E_i$,

$$\mathrm{Pic}(\mathcal{X}) = \mathbb{Z}H_f \oplus \mathbb{Z}H_g \oplus \bigoplus_{i=1}^{8} \mathbb{Z}E_i.$$

The anti-canonical divisor $-K_{\mathcal{X}} = 2H_f + 2H_g - \sum_{i=1}^{8} E_i$ uniquely decomposes as a positive linear combination of $-2$-curves $D_i$, $-K_{\mathcal{X}} = D_0 + D_1 + D_2$, where the irreducible components $D_i$, in bold on Figure 2, are given by

$$D_0 = H_f + H_g - E_1 - E_2 - E_3 - E_4, \quad D_1 = H_f - E_5 - E_6, \quad D_2 = H_g - E_7 - E_8.$$

The configuration of components $D_i$ is described by the Dynkin diagram of type $A_2^{(1)}$ (with nodes corresponding to classes of self-intersection $-2$ and edges connecting classes of intersection index 1). To this diagram correspond two different types of surfaces, the generic one corresponds to the multiplicative system of type $A_2^{(1)}$, and the degenerate configuration, where all three components $D_i$ intersect at one point, corresponds to the additive system denoted by $A_2^{(1)*}$, which is clearly our case, see Figure 3.

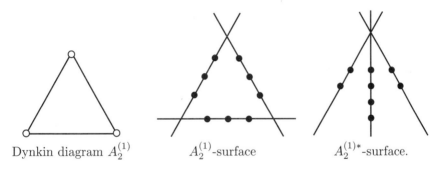

Dynkin diagram $A_2^{(1)}$          $A_2^{(1)}$-surface          $A_2^{(1)*}$-surface.

FIGURE 3. Configurations of type $A_2^{(1)}$

Components $D_i$ of $-K_{\mathcal{X}}$ span the sub-lattice $R = \text{Span}_{\mathbb{Z}}\{D_0, D_1, D_2\}$, and its orthogonal complement $R^{\perp}$ is called the *symmetry sub-lattice*. In our case, it is easy to see that $R^{\perp} = \text{Span}_{\mathbb{Z}}\{\alpha_0, \dots, \alpha_6\}$ is of type $E_6^{(1)}$, see Figure 4.

$$
\begin{aligned}
\alpha_0 &= E_3 - E_4, \quad & \alpha_1 &= E_2 - E_3, \\
\alpha_2 &= E_1 - E_2, \quad & \alpha_3 &= H_f - E_1 - E_7, \\
\alpha_4 &= E_7 - E_8, \quad & \alpha_5 &= H_g - E_1 - E_5, \\
\alpha_6 &= E_5 - E_6
\end{aligned}
$$

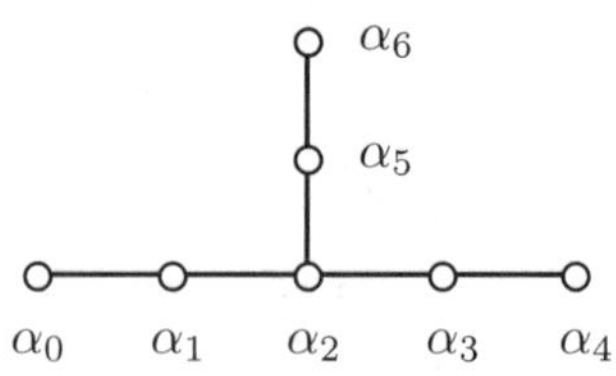

FIGURE 4. Symmetry sub-lattice for d-$P(A_2^{(1)*})$

Finally, we compute the action of $\varphi_*$ on $\text{Pic}(\mathcal{X})$ to be

$$
\begin{aligned}
(3.3) \quad H_f &\mapsto 6H_f + 3H_g - 2E_1 - 2E_2 - 2E_3 - 2E_4 - E_5 - E_6 - 3E_7 - 3E_8, \\
H_g &\mapsto 3H_f + H_g - E_1 - E_2 - E_3 - E_4 - E_7 - E_8, \\
E_1 &\mapsto 2H_f + H_g - E_2 - E_3 - E_4 - E_7 - E_8, \\
E_2 &\mapsto 2H_f + H_g - E_1 - E_3 - E_4 - E_7 - E_8, \\
E_3 &\mapsto 2H_f + H_g - E_1 - E_2 - E_4 - E_7 - E_8, \\
E_4 &\mapsto 2H_f + H_g - E_1 - E_2 - E_3 - E_7 - E_8, \\
E_5 &\mapsto 3H_f + H_g - E_1 - E_2 - E_3 - E_4 - E_6 - E_7 - E_8, \\
E_6 &\mapsto 3H_f + H_g - E_1 - E_2 - E_3 - E_4 - E_5 - E_7 - E_8, \\
E_7 &\mapsto H_f - E_8, \\
E_8 &\mapsto H_f - E_7,
\end{aligned}
$$

and so the induced action $\varphi_*$ on the sub-lattice $R^{\perp}$ is given by the following *translation*:

$$
(\alpha_0, \alpha_1, \alpha_2, \alpha_3, \alpha_4, \alpha_5, \alpha_6) \mapsto (\alpha_0, \alpha_1, \alpha_2, \alpha_3, \alpha_4, \alpha_5, \alpha_6) + (0, 0, 0, 1, 0, -1, 0)(-K_{\mathcal{X}}),
$$

as well as the permutation $(D_0 D_1 D_2)$ of the irreducible components of $-K_{\mathcal{X}}$. We now want to compare this standard picture with the one that is obtained from Schlesinger transformations.

3.1.2. *Schelsinger Transformations.* Consider a $3 \times 3$ Fuchsian system of the spectral type $111, 111, 111$. This system has three poles and it is convenient to assume that one of them is at $z_3 = \infty$, since our elementary Schlesinger transformations preserve $\mathbf{A}_{\infty}$. Also, in view of scalar gauge transformations we can assume that $\text{rank}(\mathbf{A}_i) = 2$ at finite poles (and using Möbius transformation preserving $z = \infty$, we can in principle map those poles to $z_1 = 0$ and $z_2 = 1$). Thus,

$$
\mathbf{A}_i = \mathbf{B}_i \mathbf{C}_i^{\dagger} = \begin{bmatrix} \mathbf{b}_{i,1} & \mathbf{b}_{i,2} \end{bmatrix} \begin{bmatrix} \mathbf{c}_i^{1\dagger} \\ \mathbf{c}_i^{2\dagger} \end{bmatrix}, \qquad i = 1, 2.
$$

So the *Riemann scheme* and the *Fuchs relation* for our system are

$$
\left\{ \begin{array}{ccc} z_1 & z_2 & z_3 \\ \theta_1^1 & \theta_2^1 & \theta_3^1 \\ \theta_1^2 & \theta_2^2 & \theta_3^2 \\ 0 & 0 & \theta_3^3 \end{array} \right\}, \qquad \theta_1^1 + \theta_1^2 + \theta_2^1 + \theta_2^2 + \sum_{j=1}^{3} \theta_3^j = 0.
$$

This example does not have any continuous deformation parameters but it admits non-trivial Schlesinger transformation. Consider an elementary Schlesinger transformation $\{\begin{smallmatrix} 1 & 2 \\ 1 & 1 \end{smallmatrix}\}$ that changes $\bar{\theta}_1^1 = \theta_1^1 - 1$, $\bar{\theta}_1^2 = \theta_1^2 + 1$, and fixes the remaining characteristic indices. The projector matrices for this map are

$$\mathbf{P} = \frac{\mathbf{b}_{2,1}\mathbf{c}_1^{1\dagger}}{\mathbf{c}_1^{1\dagger}\mathbf{b}_{2,1}}, \qquad \mathbf{Q} = \mathbf{I} - \mathbf{P},$$

and the evolution equations (2.13–2.19) take the form

$$\bar{\mathbf{b}}_{1,1} = \frac{1}{c_1^1}\mathbf{b}_{2,1}, \qquad \bar{\mathbf{b}}_{1,2} = \frac{1}{c_1^2}\left(\mathbf{I} - \frac{\mathbf{P}\mathbf{A}_2}{\theta_1^1 - \theta_1^2 - 1}\right)\mathbf{b}_{1,2},$$

$$\bar{\mathbf{b}}_{2,2} = \frac{1}{c_2^2}\mathbf{Q}\mathbf{b}_{2,2}, \qquad \bar{\mathbf{b}}_{2,1} = \frac{1}{c_2^1}\left((\theta_2^1 + 1)\mathbf{I} + \mathbf{Q}\left(\mathbf{I} + \frac{\mathbf{b}_{2,2}\mathbf{c}_2^{2\dagger}}{\theta_2^1 - \theta_2^2 + 1}\right)\mathbf{A}_1\right)\frac{\mathbf{b}_{2,1}}{\mathbf{c}_1^{1\dagger}\mathbf{b}_{2,1}},$$

$$\bar{\mathbf{c}}_1^{2\dagger} = c_1^2\mathbf{c}_1^{2\dagger}\mathbf{Q}, \qquad \bar{\mathbf{c}}_1^{1\dagger} = c_1^1\frac{\mathbf{c}_1^{1\dagger}}{\mathbf{c}_1^{1\dagger}\mathbf{b}_{2,1}}\left((\theta_1^1 - 1)\mathbf{I} + \mathbf{A}_2\left(\mathbf{I} + \frac{\mathbf{b}_{1,2}\mathbf{c}_1^{2\dagger}}{\theta_1^1 - \theta_1^2 - 1}\right)\mathbf{Q}\right),$$

$$\mathbf{c}_2^{1\dagger} = c_2^1\mathbf{c}_1^{1\dagger}, \qquad \mathbf{c}_2^{2\dagger} = c_2^2\mathbf{c}_2^{2\dagger}\left(\mathbf{I} - \frac{\mathbf{A}_1\mathbf{P}}{\theta_2^1 - \theta_2^2 + 1}\right),$$

where $c_i^j$ are arbitrary non-zero constants (corresponding to trivial gauge transformations).

We now explicitly show that the space of accessory parameters for Fuchsian systems of this type is two-dimensional by using various gauge transformations to put vectors $\mathbf{b}_{i,j}$ and $\mathbf{c}_i^{j\dagger}$ in some normal form, and then introduce a coordinate system on this phase space. First, assuming that we are in a generic situation, we use a global similarity transformation to map the vectors $\mathbf{b}_{1,1}$, $\mathbf{b}_{1,2}$, and $\mathbf{b}_{2,1}$ to the standard basis, and then use trivial gauge transformations (i.e., choose appropriate constants $c_i^j$) to make all components of $\mathbf{b}_{2,2}$ equal to 1. Then the orthogonality conditions $\mathbf{C}_i^\dagger\mathbf{B}_i = \mathbf{\Theta}_i$ give us the following parameterization:

$$\mathbf{B}_1 = \begin{bmatrix} 1 & 0 \\ 0 & 1 \\ 0 & 0 \end{bmatrix}, \quad \mathbf{C}_1^\dagger = \begin{bmatrix} \theta_1^1 & 0 & \alpha \\ 0 & \theta_1^2 & \beta \end{bmatrix}, \quad \mathbf{B}_2 = \begin{bmatrix} 0 & 1 \\ 0 & 1 \\ 1 & 1 \end{bmatrix}, \quad \mathbf{C}_2^\dagger = \begin{bmatrix} -x - \theta_2^1 & x & \theta_2^1 \\ \theta_2^2 - y & y & 0 \end{bmatrix}.$$

Here we choose $x$ and $y$ as our coordinates, and we can express $\alpha = \alpha(x, y)$ and $\beta = \beta(x, y)$ from the condition that the eigenvalues of $\mathbf{A}_\infty = -\mathbf{B}_1\mathbf{C}_1^\dagger - \mathbf{B}_2\mathbf{C}_2^\dagger$ are $\kappa_1$, $\kappa_2$, and $\kappa_3$ (the resulting expressions, although easy to obtain, are quite large and we omit them). We then get the following dynamic in the coordinates $(x, y)$:

$$\begin{cases} \bar{x} = \dfrac{\alpha - \beta}{\alpha(\theta_1^2 - \theta_1^1 + 1)}\left(\alpha(x + y) + \theta_1^1 y\right) \\ \bar{y} = \dfrac{\alpha - \beta}{\alpha(\theta_1^2 - \theta_1^1 + 1)}\left(\dfrac{\alpha(\alpha(x + y) + y(\theta_1^2 + 1))(\theta_1^1 - \theta_2^2 + 1)}{\alpha(\theta_2^1 + 1) - (\alpha - \beta)y} - \alpha(x + y) - \theta_1^1 y\right) \end{cases},$$

where we still need to substitute $\alpha = \alpha(x, y)$ and $\beta = \beta(x, y)$. So this map is quite complicated and it reflects the fact that our choice of the coordinates was rather arbitrary. To better understand the map we return again to geometry.

The indeterminate points of the map $\psi : (x, y) \to (\bar{x}, \bar{y})$ are

$$p_1 \left( \frac{(\theta_1^1 + \theta_2^1 + \theta_3^1)(\theta_1^2 + \theta_3^1)}{\theta_1^1 - \theta_1^2}, \; -\frac{(\theta_1^1 + \theta_2^1 + \theta_3^1)(\theta_1^2 + \theta_3^1)}{\theta_1^1 - \theta_1^2} \right), \quad p_4(0,0),$$

$$p_2 \left( \frac{(\theta_1^1 + \theta_2^1 + \theta_3^2)(\theta_1^2 + \theta_3^2)}{\theta_1^1 - \theta_1^2}, \; -\frac{(\theta_1^1 + \theta_2^1 + \theta_3^2)(\theta_1^2 + \theta_1^2)}{\theta_1^1 - \theta_1^2} \right), \quad p_5(-\theta_2^1, \theta_2^2),$$

$$p_3 \left( \frac{(\theta_1^1 + \theta_2^1 + \theta_3^3)(\theta_1^2 + \theta_3^2)}{\theta_1^1 - \theta_1^2}, \; -\frac{(\theta_1^1 + \theta_2^1 + \theta_3^3)(\theta_1^2 + \theta_3^3)}{\theta_1^1 - \theta_2^2} \right),$$

as well as the sequence of infinitely close points

$$p_6 \left( \frac{1}{x} = 0, \frac{1}{y} = 0 \right) \longleftarrow p_7 \left( \frac{1}{x} = 0, \frac{x}{y} = -1 \right)$$

$$\longleftarrow p_8 \left( \frac{1}{x} = 0, \frac{x}{y} = -1, \frac{x(x+y)}{y} = \frac{(\theta_1^2 + 1)(\theta_2^1 - \theta_2^2)}{\theta_1^2 - \theta_1^1} \right).$$

Note also that the points $p_1, \dots, p_6$ (and, after blowing up, the point $p_7$ as well) all lie on a $(2,2)$-curve $Q$ given by the equation

$$(3.4) \qquad (\theta_1^1 - \theta_1^2)(x+y)(x+y+\theta_2^1 - \theta_2^2) + (\theta_2^1 - \theta_2^2)(\theta_2^2 x + \theta_2^1 y) = 0.$$

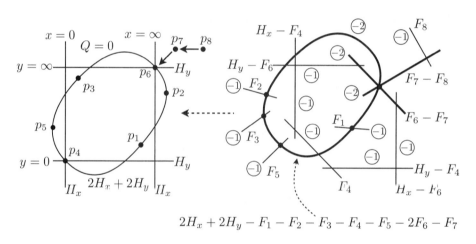

FIGURE 5. Okamoto surface $\mathcal{X}_\theta$ for the Schlesinger transformations reduction to d-$P\left(A_2^{(1)*}\right)$ using $\mathbb{P}^1 \times \mathbb{P}^1$ compactification of $\mathbb{C}^2$.

Resolving indeterminate points of this map using blow-ups gives us the Okamoto surface $\mathcal{X}_\theta$ pictured on Figure 5. We can immediately see that in this case

$$-K_{\mathcal{X}_\theta} = (2H_x + 2H_y - F_1 - F_2 - F_3 - F_4 - F_5 - 2F_6 - F_7) + (F_6 - F_7) + (F_7 - F_8),$$

where $F_i$ stand for classes of exceptional divisors, and, since all of the $-2$-curves intersect at one point, $\mathcal{X}$ indeed has the type $A_2^{(1)*}$. Unfortunately, two of the three irreducible components of $-K_{\mathcal{X}}$ are now completely in the blow-up region. This makes identification with the standard example more difficult since we have to go through a sequence of coordinate charts to do the computation. A better approach is to use $\mathbb{P}^2$ compactification of $\mathbb{C}^2$ (recall that in this case, according to general theory, we expect to have nine blow-up points instead of eight).

In this compactification we still have the same finite points that, in the homogeneous coordinates, are

$$p_1\left(\frac{(\theta_1^1+\theta_2^1+\theta_3^1)(\theta_1^2+\theta_3^1)}{\theta_1^1-\theta_1^2}:-\frac{(\theta_1^1+\theta_2^2+\theta_3^1)(\theta_1^2+\theta_1^1)}{\theta_1^1-\theta_1^2}:1\right),\quad p_4(0:0:1),$$

$$p_4\left(\frac{(\theta_1^1+\theta_2^1+\theta_3^2)(\theta_1^2+\theta_3^2)}{\theta_1^1-\theta_1^2}:-\frac{(\theta_1^1+\theta_2^2+\theta_3^2)(\theta_1^2+\theta_1^2)}{\theta_1^1-\theta_1^2}:1\right),\quad p_5(-\theta_2^1:\theta_2^2:1),$$

$$p_3\left(\frac{(\theta_1^1+\theta_2^1+\theta_3^3)(\theta_1^2+\theta_3^3)}{\theta_1^1-\theta_1^2}:-\frac{(\theta_1^1+\theta_2^1+\theta_3^3)(\theta_1^2+\theta_1^3)}{\theta_1^1-\theta_1^2}:1\right).$$

There are also three more points on the line at infinity, and one infinitely close point $p_9$:

$$p_6(1:-1:0)\leftarrow p_9\left(0,\frac{\theta_1^1-\theta_1^2}{(\theta_2^1-\theta_2^2)(\theta_1^2+1)}\right),\quad p_7(0:1:0),\quad p_8(1:0:0),$$

where coordinates of $p_9$ are w.r.t the coordinate system $u=\frac{X+Y}{X}$, $v=\frac{Z}{X+Y}$ in the chart $X\neq 0$. Points $p_1,\dots,p_6$ lie on the projectivization of the $(2,2)$-curve $Q$ whose homogeneous equation in $\mathbb{P}^2$ is

$$(3.5)\quad (\theta_1^1-\theta_1^2)(X+Y)(X+Y+(\theta_2^1-\theta_2^2)Z)+(\theta_2^1-\theta_2^2)(\theta_2^2 X+\theta_2^1 Y)Z=0.$$

The resulting blow-up diagram is depicted on Figure 6.

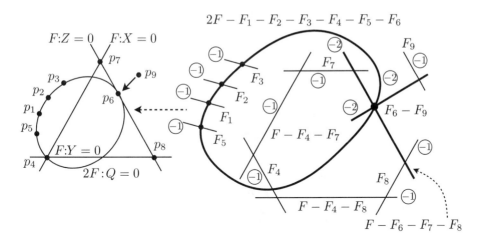

FIGURE 6. Okamoto surface $\mathcal{X}_\theta$ for the Schlesinger transformations reduction to d-$P\left(A_2^{(1)*}\right)$ using $\mathbb{P}^2$ compactification of $\mathbb{C}^2$.

As before, we see that the anti-canonical divisor $-K_\mathcal{X}$ uniquely decomposes as a positive linear combination of $-2$-curves $D_i$,

$$-K_\mathcal{X}=3F-\sum_{i=1}^{9}F_i=D_0+D_1+D_2,$$

where

$$D_0=2F-F_1-F_2-F_3-F_4-F_5-F_6,\quad D_1=F-F_6-F_7-F_8,\quad D_2=F_6-F_9.$$

The configuration of components $D_i$ is again described by the Dynkin diagram of type $A_2^{(1)}$, and since all three $-2$-curves intersect at one point, this is a surface of type $A_2^{(1)*}$. To compare this dynamic with the model example considered earlier we need to find an explicit isomorphism between the corresponding Okamoto surfaces, choose the same bases in the Picard lattice, and then compute the translation directions in the symmetry sub-lattice. This is what we do next.

3.1.3. *Reduction to the standard form.* To match the surface $\mathcal{X}_\theta$ described by the blow-up diagram on Figure 6 with the surface $\mathcal{X}_\mathbf{b}$ described by diagram on Figure 2, we look for the blow-down structure describing $\mathcal{X}_\mathbf{b}$ in $\mathrm{Pic}(\mathcal{X}_\theta)$, i.e., we look for rational classes $\mathcal{H}_f, \mathcal{H}_g, \mathcal{E}_1, \ldots, \mathcal{E}_8$ in $\mathrm{Pic}(\mathcal{X}_\theta)$ such that

$$\mathcal{H}_f \bullet \mathcal{H}_g = 1, \ \mathcal{E}_i^2 = -1, \ \mathcal{H}_f^2 = \mathcal{H}_g^2 = \mathcal{H}_f \bullet \mathcal{E}_i = \mathcal{H}_g \bullet \mathcal{E}_i = \mathcal{E}_i \bullet \mathcal{E}_j = 0, \ 1 \le i \ne j \le 8,$$

and the resulting configuration matches diagram on Figure 2. By the (virtual) genus formula $g(C) = (C^2 + K_\mathcal{X} \bullet C)/2 + 1$, we see that we should look for classes of rational curves of self-intersection zero among $F - F_i$ and for classes of rational curves of self-intersection $-1$ among $F_i$ or $F - F_i - F_j$.

Comparing the $-2$-curves on both diagrams,

$$D_0 = 2F - F_1 - F_2 - F_3 - F_4 - F_5 - F_6 = \mathcal{H}_f + \mathcal{H}_g - \mathcal{E}_1 - \mathcal{E}_2 - \mathcal{E}_3 - \mathcal{E}_4,$$
$$D_1 = F - F_6 - F_7 - F_8 = \mathcal{H}_f - \mathcal{E}_5 - \mathcal{E}_6,$$
$$D_2 = F_6 - F_9 = \mathcal{H}_g - \mathcal{E}_7 - \mathcal{E}_8,$$

we see that it makes sense to choose $\mathcal{E}_i = F_i$ for $i = 1, \ldots, 4$. Then $\mathcal{H}_f + \mathcal{H}_g = F - F_5 - F_6$, and looking at $D_1$ we put $\mathcal{H}_f = F - F_6$, $\mathcal{E}_5 = F_7$, $\mathcal{E}_6 = F_8$. This then requires that $\mathcal{H}_g = F - F_5$, and looking at $D_2$ we get $\mathcal{E}_7 + \mathcal{E}_8 = F - F_5 - F_6 + F_9$. We put $\mathcal{E}_7 = F - F_5 - F_6$ (to ensure that $\mathcal{H}_f \bullet \mathcal{E}_7 = \mathcal{H}_g \bullet \mathcal{E}_7 = 0$), and then $\mathcal{E}_8 = F_9$. To summarize, we get the following identification, which clearly satisfies all of the required conditions

$$\mathcal{H}_f = F - F_6, \quad \mathcal{E}_1 = F_1, \quad \mathcal{E}_3 = F_3, \quad \mathcal{E}_5 = F_7, \quad \mathcal{E}_7 = F - F_5 - F_6,$$
$$\mathcal{H}_g = F - F_5, \quad \mathcal{E}_2 = F_2, \quad \mathcal{E}_4 = F_4, \quad \mathcal{E}_6 = F_8, \quad \mathcal{E}_8 = F_9.$$

To complete the correspondence it remains to define the base coordinates $f$ and $g$ of the linear systems $|\mathcal{H}_f|$ and $|\mathcal{H}_g|$ that will map the exceptional fibers of the divisors $\mathcal{E}_i$ to the points $\pi_i$ such that $\pi_5$ and $\pi_6$ are on the line $f = \infty$, $\pi_7$ and $\pi_8$ are on $g = \infty$, and $\pi_1, \ldots \pi_4$ are on the line $f + g = 0$. Since the pencil $|\mathcal{H}_f|$ consists of all curves on $\mathbb{P}^2$ passing through $p_6(1 : -1 : 0)$,

$$|\mathcal{H}_f| = |F - F_6| = \{aX + bY + cZ = 0 \mid a - b = 0\} = \{a(X + Y) + cZ = 0\},$$

we can define the projective base coordinate as $f_1 = [X + Y : Z]$. Similarly,

$$|\mathcal{H}_g| = |F - F_5| = \{aX + bY + cZ = 0 \mid -\theta_2^1 a + \theta_2^2 b + c = 0\}$$
$$= \{a(X + \theta_2^1 Z) + b(Y - \theta_2^2 Z) = 0\},$$

and $g_1 = [X + \theta_2^1 Z : Y - \theta_2^2 Z]$. Then

$$f_1(\pi_5) = f_1(p_7) = [-1 : 0], \qquad g_1(\pi_7) = g_1(p_6) = [1 : -1],$$
$$f_1(\pi_6) = f_1(p_8) = [1 : 0], \qquad g_1(\pi_8) = g_1(p_6) = [1 : -1].$$

In order to have $g_1(\pi_7) = g_1(\pi_8) = \infty$ we first make an affine change of coordinates $\tilde{g}_1 = [(X + \theta_2^1 Z) + (Y - \theta_2^2 Z) : Y - \theta_2^2 Z)]$ to get $\tilde{g}_1(\pi_7) = \tilde{g}_1(\pi_8) = [0 : -1]$ and then put

$$f_2 = \frac{X+Y}{Z}, \qquad g_2 = \frac{Y - \theta_2^2 Z}{(X+Y) + (\theta_2^1 - \theta_2^2)Z}.$$

Equation (3.5) of the curve $Q$ in these coordinates becomes

$$(3.6) \qquad Z^2(f + \theta_2^1 - \theta_2^2)((\theta_1^1 - \theta_1^2)f_2 + (\theta_2^1 - \theta_2^2)((\theta_2^1 - \theta_2^2)g_2 + \theta_2^2)) = 0,$$

and the points $\pi_1, \ldots, \pi_4$ lie on the line $(\theta_1^1 - \theta_1^2)f_2 + (\theta_2^1 - \theta_2^2)((\theta_2^1 - \theta_2^2)g_2 + \theta_2^2)) = 0$.
Thus, if we finally put

$$(3.7) \qquad f = \frac{(\theta_1^1 - \theta_1^2)}{(\theta_2^1 - \theta_2^2)}f_2 = \frac{(\theta_1^1 - \theta_1^2)(X+Y)}{(\theta_2^1 - \theta_2^2)Z} = \frac{(\theta_1^1 - \theta_1^2)(x+y)}{(\theta_2^1 - \theta_2^2)},$$

$$(3.8) \qquad g = (\theta_2^1 - \theta_2^2)g_2 + \theta_2^2 = \frac{\theta_2^2 X + \theta_2^1 Y}{(X+Y) + (\theta_2^1 - \theta_2^2)Z} = \frac{\theta_2^2 x + \theta_2^1 y}{(x+y) + (\theta_2^1 - \theta_2^2)},$$

points $\pi_1, \ldots, \pi_4$ will be on the line $f + g = 0$, points $\pi_5$ and $\pi_6$ will be on the line $f = \infty$, and points $\pi_7$ and $\pi_8$ will be on the line $g = \infty$, as requires. Specifically, we get

$$\pi_1(\theta_1^2 + \theta_3^1, -\theta_1^2 - \theta_3^1), \quad \pi_3(\theta_1^2 + \theta_3^3, -\theta_1^2 - \theta_3^3), \quad \pi_5(\infty, \theta_2^1), \quad \pi_7(\theta_1^2 - \theta_1^1, \infty),$$
$$\pi_2(\theta_1^2 + \theta_3^3, -\theta_1^2 - \theta_3^3), \quad \pi_4(0,0), \qquad\qquad \pi_6(\infty, \theta_2^2), \quad \pi_8(\theta_1^2 + 1, \infty).$$

Thus, we immediately get the identification between the parameters in the Riemann scheme of our Fuchsian system and the parameters $b_i$ in the model equation:

$$(3.9) \qquad\begin{aligned} b_1 &= \theta_1^2 + \theta_3^1, & b_3 &= \theta_1^2 + \theta_3^3, & b_5 &= \theta_2^1, & b_7 &= \theta_1^1 - \theta_1^2, \\ b_2 &= \theta_1^2 + \theta_3^3, & b_4 &= 0, & b_6 &= \theta_2^2, & b_8 &= -\theta_1^2 - 1. \end{aligned}$$

This, in turn, allows us to see the effect of the standard Painlevé dynamic on the Riemann scheme. Indeed, $\delta = b_1 + \cdots + b_8 = -1$, and, for example, $\bar{k}_1 = \bar{b}_1 + \bar{b}_8 + 1 = b_1 + b_8 - \delta + 1 = k_1 + 1$, and so on. So for the model equation we get

$$\left\{\begin{array}{ccc} z_1 & z_2 & z_3 \\ \theta_1^1 & \theta_2^1 & \theta_3^1 \\ \theta_1^2 & \theta_2^2 & \theta_3^2 \\ 0 & 0 & \theta_3^3 \end{array}\right\} \overset{\text{d-}P(A_2^{(1)*})}{\longmapsto} \left\{\begin{array}{ccc} z_1 & z_2 & z_3 \\ \theta_1^1 & \theta_2^1 - 1 & \theta_3^1 + 1 \\ \theta_1^2 - 1 & \theta_2^2 - 1 & \theta_3^2 + 1 \\ 0 & 0 & \theta_3^3 + 1 \end{array}\right\},$$

whereas our elementary Schlesinger transformation acts as

$$\left\{\begin{array}{ccc} z_1 & z_2 & z_3 \\ \theta_1^1 & \theta_2^1 & \theta_3^1 \\ \theta_1^2 & \theta_2^2 & \theta_3^2 \\ 0 & 0 & \theta_3^3 \end{array}\right\} \overset{\{\begin{smallmatrix}1\,2\\1\,1\end{smallmatrix}\}}{\longmapsto} \left\{\begin{array}{ccc} z_1 & z_2 & z_3 \\ \theta_1^1 - 1 & \theta_2^1 + 1 & \theta_3^1 \\ \theta_1^2 & \theta_2^2 & \theta_3^2 \\ 0 & 0 & \theta_3^3 \end{array}\right\}.$$

Thus, these two transformations correspond to the different translation directions in the symmetry root sub-lattice of the surface $\tilde{X}$ and so are not equivalent. Indeed, we compute the action of $\psi_*$ of an elementary Schlesinger transformation $\{\begin{smallmatrix}1\,2\\1\,1\end{smallmatrix}\}$ on the classes $\mathcal{H}_f$, $\mathcal{H}_g$, and $\mathcal{E}_i$ to be

$$
\begin{aligned}
\mathcal{H}_f &\mapsto 2\mathcal{H}_f + 3\mathcal{H}_g - \mathcal{E}_1 - \mathcal{E}_2 - \mathcal{E}_3 - \mathcal{E}_4 - 2\mathcal{E}_5 - 2\mathcal{E}_8, \\
\mathcal{H}_g &\mapsto 3\mathcal{H}_f + 5\mathcal{H}_g - 2\mathcal{E}_1 - 2\mathcal{E}_2 - 2\mathcal{E}_3 - 2\mathcal{E}_4 - 3\mathcal{E}_5 - \mathcal{E}_6 - 2\mathcal{E}_8, \\
\mathcal{E}_1 &\mapsto \mathcal{H}_f + 2\mathcal{H}_g - \mathcal{E}_2 - \mathcal{E}_3 - \mathcal{E}_4 - \mathcal{E}_5 - \mathcal{E}_8, \\
\mathcal{E}_2 &\mapsto \mathcal{H}_f + 2\mathcal{H}_g - \mathcal{E}_1 - \mathcal{E}_3 - \mathcal{E}_4 - \mathcal{E}_5 - \mathcal{E}_8, \\
\mathcal{E}_3 &\mapsto \mathcal{H}_f + 2\mathcal{H}_g - \mathcal{E}_1 - \mathcal{E}_2 - \mathcal{E}_4 - \mathcal{E}_5 - \mathcal{E}_8, \\
\mathcal{E}_4 &\mapsto \mathcal{H}_f + 2\mathcal{H}_g - \mathcal{E}_1 - \mathcal{E}_2 - \mathcal{E}_3 - \mathcal{E}_5 - \mathcal{E}_8, \\
\mathcal{E}_5 &\mapsto \mathcal{E}_7, \\
\mathcal{E}_6 &\mapsto 2\mathcal{H}_f + 2\mathcal{H}_g - \mathcal{E}_1 - \mathcal{E}_2 - \mathcal{E}_3 - \mathcal{E}_4 - 2\mathcal{E}_5 - \mathcal{E}_8, \\
\mathcal{E}_7 &\mapsto 2\mathcal{H}_f + 3\mathcal{H}_g - \mathcal{E}_1 - \mathcal{E}_2 - \mathcal{E}_3 - \mathcal{E}_4 - 2\mathcal{E}_5 - \mathcal{E}_6 - 2\mathcal{E}_8, \\
\mathcal{E}_8 &\mapsto \mathcal{H}_g - \mathcal{E}_5,
\end{aligned}
$$

and compare with the standard dynamic $\varphi_*$ given by (3.3) to see this explicitly:

$$
\begin{aligned}
\psi_* : (\alpha_0, \alpha_1, \alpha_2, \alpha_3, \alpha_4, \alpha_5, \alpha_6) &\mapsto (\alpha_0, \alpha_1, \alpha_2, \alpha_3, \alpha_4, \alpha_5, \alpha_6) + \\
&\quad (0, 0, 0, -1, 1, 1, -1)\,(-K_{\mathcal{X}}), \\
\varphi_* : (\alpha_0, \alpha_1, \alpha_2, \alpha_3, \alpha_4, \alpha_5, \alpha_6) &\mapsto (\alpha_0, \alpha_1, \alpha_2, \alpha_3, \alpha_4, \alpha_5, \alpha_6) + \\
&\quad (0, 0, 0, 1, 0, -1, 0)\,(-K_{\mathcal{X}}).
\end{aligned}
$$

It is possible to represent the standard Painlevé dynamic as a composition of two elementary Schlesinger transformations, combined with some automorphisms of our Fuchsian system. We first demonstrate this by looking at a sequence of actions on the Riemann scheme:

$$
\left\{
\begin{array}{ccc}
z_1 & z_2 & z_3 \\
\theta_1^1 & \theta_2^1 & \theta_3^1 \\
\theta_1^2 & \theta_2^2 & \theta_3^2 \\
0 & 0 & \theta_3^3
\end{array}
\right\}
\xrightarrow{\left\{\begin{smallmatrix} 2\ 1 \\ 1\ 1 \end{smallmatrix}\right\}}
\left\{
\begin{array}{ccc}
z_1 & z_2 & z_3 \\
\theta_1^1 + 1 & \theta_2^1 - 1 & \theta_3^1 \\
\theta_1^2 & \theta_2^2 & \theta_3^2 \\
0 & 0 & \theta_3^3
\end{array}
\right\}
$$

$$
\xrightarrow{\sigma_1(1,3)}
\left\{
\begin{array}{ccc}
z_1 & z_2 & z_3 \\
0 & \theta_2^1 - 1 & \theta_3^1 \\
\theta_1^2 & \theta_2^2 & \theta_3^2 \\
\theta_1^1 + 1 & 0 & \theta_3^3
\end{array}
\right\}
\xrightarrow{\rho_1(-\theta_1^1-1)}
\left\{
\begin{array}{ccc}
z_1 & z_2 & z_3 \\
-\theta_1^1 - 1 & \theta_2^1 - 1 & \theta_3^1 + \theta_1^1 + 1 \\
\theta_1^2 - \theta_1^1 - 1 & \theta_2^2 & \theta_3^2 + \theta_1^1 + 1 \\
0 & 0 & \theta_3^3 + \theta_1^1 + 1
\end{array}
\right\}
$$

$$
\xrightarrow{\left\{\begin{smallmatrix} 2\ 1 \\ 2\ 1 \end{smallmatrix}\right\}}
\left\{
\begin{array}{ccc}
z_1 & z_2 & z_3 \\
-\theta_1^1 & \theta_2^1 - 1 & \theta_3^1 + \theta_1^1 + 1 \\
\theta_1^2 - \theta_1^1 - 1 & \theta_2^2 - 1 & \theta_3^2 + \theta_2^1 + 1 \\
0 & 0 & \theta_3^3 + \theta_1^1 + 1
\end{array}
\right\}
$$

$$
\xrightarrow{\sigma_1(1,3)}
\left\{
\begin{array}{ccc}
z_1 & z_2 & z_3 \\
0 & \theta_1^1 - 1 & \theta_3^1 + \theta_1^1 + 1 \\
\theta_1^2 - \theta_1^1 - 1 & \theta_2^2 - 1 & \theta_3^2 + \theta_1^1 + 1 \\
\theta_1^1 & 0 & \theta_3^3 + \theta_1^1 + 1
\end{array}
\right\}
\xrightarrow{\rho_1(\theta_1^1)}
\left\{
\begin{array}{ccc}
z_1 & z_2 & z_3 \\
\theta_1^1 & \theta_2^1 - 1 & \theta_3^1 + 1 \\
\theta_1^2 - 1 & \theta_2^2 - 1 & \theta_3^2 + 1 \\
0 & 0 & \theta_3^3 + 1
\end{array}
\right\}.
$$

Here $\left\{\begin{smallmatrix} 2\ 1 \\ 1\ 1 \end{smallmatrix}\right\}$ and $\left\{\begin{smallmatrix} 2\ 1 \\ 2\ 1 \end{smallmatrix}\right\}$ are the usual elementary Schlesinger transformations, the map $\rho_i(s) : \mathbf{A}(z) \mapsto (z - z_i)^s \mathbf{A}(z)$ is a scalar gauge transformation, and $\sigma_i(j, k)$ is a map that exchanges the $j$-th and the $k$-th eigenvectors (and eigenvalues) of $\mathbf{A}_i$. Note that, if the eigenvalues $\theta_i^j$ and $\theta_i^k$ are non-zero, this map is just a permutation on the decomposition space $\mathcal{B} \times \mathcal{C}$. And even though for the map $\sigma_1(1, 3)$ that we use above one of the eigenvectors has the eigenvalue zero, this map is still well-defines as a map on $\mathcal{B} \times \mathcal{C}$, since $\mathbf{b}_1^3 \in \mathrm{Ker}(\mathbf{C}_1^\dagger)$ and $\mathbf{c}_1^{3\dagger} \in \mathrm{Ker}(\mathbf{B}_1)$. In fact, if we combine $\sigma_1(1,3)$ with $\rho_1(-\theta_1^1)$ to define a transformation $\Sigma_1(1,3) = \rho_1(-\theta_1^1) \circ \sigma_1(1,3)$,

$$\Sigma_1(1,3): \left\{\begin{array}{ccc} z_1 & z_2 & z_3 \\ \theta_1^1 & \theta_2^1 & \theta_3^1 \\ \theta_1^2 & \theta_2^2 & \theta_3^2 \\ 0 & 0 & \theta_3^3 \end{array}\right\} \longmapsto \left\{\begin{array}{ccc} z_1 & z_2 & z_3 \\ -\theta_1^1 & \theta_2^1 & \theta_3^1 + \theta_1^1 \\ \theta_1^2 - \theta_1^1 & \theta_2^2 & \theta_3^2 + \theta_1^1 \\ 0 & 0 & \theta_3^3 + \theta_1^1 \end{array}\right\},$$

the action of $\Sigma_1(1,3)$ on the decomposition space is explicitly given by

$$\Sigma_1(1,3): \left(\mathbf{b}_1^1, \mathbf{b}_1^2; \mathbf{c}_1^{1\dagger}, \mathbf{c}_1^{2\dagger}; \mathbf{b}_2^1, \mathbf{b}_2^2; \mathbf{c}_2^{1\dagger}, \mathbf{c}_2^{2\dagger}\right) \longmapsto$$

$$\left((\mathbf{c}_1^{1\dagger} \times \mathbf{c}_1^{2\dagger})^t, \mathbf{b}_1^2; \frac{-\theta_1^1(\mathbf{b}_1^1 \times \mathbf{b}_1^2)^t}{(\mathbf{c}_1^{1\dagger} \times \mathbf{c}_1^{2\dagger})(\mathbf{b}_1^1 \times \mathbf{b}_1^2)}, \mathbf{c}_1^{2\dagger}; \mathbf{b}_2^1, \mathbf{b}_2^2; \mathbf{c}_2^{1\dagger}, \mathbf{c}_2^{2\dagger}\right),$$

where $\times$ is the usual cross-product and $t$ denotes transposition. We also had to use the normalization condition $\mathbf{C}_1^\dagger \mathbf{B}_1 = \mathbf{\Theta}_1$.

It is also possible to show, by a direct computation, that

$$\text{d-}P\left(A_2^{(1)*}\right) = \Sigma_1(1,3) \circ \left\{\begin{smallmatrix}2\,1\\2\,1\end{smallmatrix}\right\} \circ \Sigma_1(1,3) \circ \left\{\begin{smallmatrix}2\,1\\1\,1\end{smallmatrix}\right\}$$

holds on the level of equations as well. Let us now summarize the results of this section.

THEOREM 3.1. *Consider a Fuchsian system of the spectral type* $(111, 111, 111)$ *of the form*

$$\frac{d\mathbf{Y}}{dz} = \left(\frac{\mathbf{A}_1}{z} + \frac{\mathbf{A}_2}{z-1}\right)\mathbf{Y},$$

*where the matrices* $\mathbf{A}_i$ *are normalized by gauge and similarity transformations to be of the form*

$$\mathbf{A}_1 = \begin{bmatrix} 1 & 0 \\ 0 & 1 \\ 0 & 0 \end{bmatrix}\begin{bmatrix} \theta_1^1 & 0 & \alpha(x,y) \\ 0 & \theta_1^2 & \beta(x,y) \end{bmatrix}, \qquad \mathbf{A}_1 = \begin{bmatrix} 0 & 1 \\ 0 & 1 \\ 1 & 1 \end{bmatrix}\begin{bmatrix} -x - \theta_2^1 & x & \theta_2^1 \\ \theta_2^2 - y & y & 0 \end{bmatrix},$$

*and do the change of variables inverse to (3.7–3.8):*

$$(3.10) \qquad x = \frac{(\theta_2^1 - g)f + (\theta_1^2 - \theta_1^1)g}{\theta_1^1 - \theta_1^2}, \qquad y = -\frac{(\theta_2^2 - g)f + (\theta_1^2 - \theta_1^1)g}{\theta_1^1 - \theta_1^2}.$$

*Then in the coordinates* $(f, g)$ *the Okamoto surface for the Schlesinger transformation dynamic* $\mathcal{X}_\theta$ *coincides with the Okamoto surface for the model example dynamic* $\mathcal{X}_\mathbf{b}$ *under the parameter identification (3.9), and the model* $\text{d-}P\left(A_2^{(1)*}\right)$ *dynamic can be represented as a composition of elementary Schlesinger transformations and phase space automorphisms,*

$$\text{d-}P\left(A_2^{(1)*}\right) = \Sigma_1(1,3) \circ \left\{\begin{smallmatrix}2\,1\\2\,1\end{smallmatrix}\right\} \circ \Sigma_1(1,3) \circ \left\{\begin{smallmatrix}2\,1\\1\,1\end{smallmatrix}\right\}.$$

**Remark:** After the paper had already been accepted, we were able to show that in this case the standard dynamic and the Schlesinger transformation dynamic are conjugated via a very non-trivial change of variables. We plan to explain this result in a separate publication.

### 3.2. Reductions to difference Painlevé equation of type d-$P\left(A_1^{(1)*}\right)$ with the symmetry group $E_7^{(1)}$.

3.2.1. *Model Example.* For our model example of d-$P\left(A_1^{(1)*}\right)$ equation we take the equation that first was appeared in [**GRO03**] as an asymmetric q-$P_{\rm IV}$ equation, and we use the variables as in Sakai's paper [**Sak07**]. We again consider d-$P\left(A_1^{(1)*}\right)$ to be a birational map $\varphi : \mathbb{P}^1 \times \mathbb{P}^1 \dashrightarrow \mathbb{P}^1 \times \mathbb{P}^1$ with parameters $b, b_1, \ldots, b_8$,

$$\begin{pmatrix} b & b_1 & b_2 & b_3 & b_4 \\ & b_5 & b_6 & b_7 & b_8 \end{pmatrix}; f, g \mapsto \begin{pmatrix} \bar{b} & \bar{b}_1 & \bar{b}_2 & \bar{b}_3 & \bar{b}_4 \\ & \bar{b}_5 & \bar{b}_6 & \bar{b}_7 & \bar{b}_8 \end{pmatrix}; \bar{f}, \bar{g}, \qquad \begin{array}{l} \bar{b} = b - \delta, \\ \bar{b}_i = b_i, \quad i = 1, \ldots, 8, \end{array}$$

$\delta = b_1 + \cdots + b_8$, and $\bar{f}$ and $\bar{g}$ are given by the equations

(3.11)
$$\begin{cases} \dfrac{(g + f - 2b)(g + \bar{f} - b - \bar{b})}{(g + f)(g + \bar{f})} = \dfrac{\prod_{i=1}^4 (g - b + b_i)}{\prod_{i=5}^8 (g - b_i)} \\[4mm] \dfrac{(g + \bar{f} - b - \bar{b})(\bar{g} + \bar{f} - 2\bar{b})}{(g + \bar{f})(\bar{g} + \bar{f})} = \dfrac{\prod_{i=1}^4 (\bar{f} - \bar{b} - b_i)}{\prod_{i=5}^8 (\bar{f} + b_i)} \end{cases}.$$

It is convenient to introduce the notation

$$G_{14} = G_{14}(g) = \prod_{i=1}^4 (g - b + b_i), \quad G_{14}^i = G_{14}^i(g) = \prod_{j=1, j \neq i}^4 (g - b + b_j),$$

and similarly for $G_{58}$, $F_{14}$, and $F_{58}$. The maps $\varphi$ is then given by the sequence $(f, g) \to (\bar{f}, g) \to (\bar{f}, \bar{g})$ described by the equations

(3.12)
$$\bar{f} = -\frac{(g - b - \bar{b})(f + g - 2b)G_{58} - g(f + g)G_{14}}{(f + g - 2b)G_{58} - (f + g)G_{14}},$$

(3.13)
$$\bar{g} = -\frac{(\bar{f} - 2\bar{b})(\bar{f} + g - b - \bar{b})F_{58} - \bar{f}(\bar{f} + g)F_{14}}{(\bar{f} + g - b - \bar{b})F_{58} - (\bar{f} + g)F_{14}},$$

and $\varphi^{-1}$ is given by $(\bar{f}, \bar{g}) \to (\bar{f}, g) \to (f, g)$ given by

$$g = -\frac{(\bar{f} - b - \bar{b})(\bar{f} + \bar{g} - 2\bar{b})F_{58} - \bar{f}(\bar{f} + \bar{g})F_{14}}{(\bar{f} + \bar{g} - 2\bar{b})F_{58} - (\bar{f} + \bar{g})F_{14}},$$

$$f = -\frac{(g - 2b)(\bar{f} + g - b - \bar{b})G_{58} - g(\bar{f} + g)G_{14}}{(\bar{f} + g - b - \bar{b})G_{58} - (\bar{f} + g)G_{14}}.$$

It is easy to see that the indeterminate points of the first map $\bar{f} = \bar{f}(f, g)$ are either given by the conditions $f + g = 2b$ and $G_{14} = 0$, or by the conditions $f + g = 0$ and $G_{58} = 0$. Thus, we get 8 indeterminate points lying on two curves of bi-degree $(1, 1)$, $C_b : f + g = 2b$ and $C_0 : f + g = 0$:

$$p_i(b + b_i, b - b_i), \quad i = 1, \ldots, 4 \qquad \text{and} \qquad p_i(-b_i, b_i), \quad i = 5, \ldots, 8$$

on the $(f, g)$-plane. It is also easy to see that these are also the indeterminate points of all of the other maps (with $b$ changed to $\bar{b}$ for $(\bar{f}, \bar{g})$-coordinates). We then get the following blowup diagram describing the Okamoto space of initial conditions $\mathcal{X}_{\mathbf{b}}$ on Figure 7.

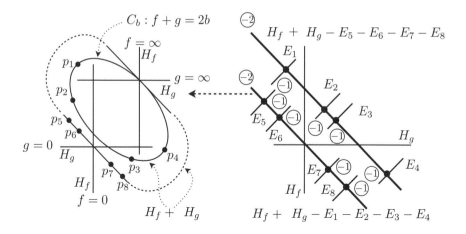

FIGURE 7. Okamoto surface $\mathcal{X}_{\mathbf{b}}$ for the model form of d-$P(A_1^{(1)*})$.

In $\text{Pic}(\mathcal{X}_{\mathbf{b}}) = \mathbb{Z}H_f \oplus \mathbb{Z}H_g \oplus \bigoplus_{i=1}^{8} \mathbb{Z}E_i$, the anti-canonical divisor again decomposes uniquely as the sum of two connected components,

$$-K_{\mathcal{X}} = 2H_f + 2H_g - \sum_{i=1}^{8} E_i = D_0 + D_1, \qquad \text{where}$$
$$D_0 = H_f + H_g - E_1 - E_2 - E_3 - E_4,$$
$$D_1 = H_f + H_g - E_5 - E_6 - E_7 - E_8,$$

and $D_1^2 = D_2^2 = -D_1 \bullet D_2 = -2$. Thus, the configuration of components $D_i$ is described by the Dynkin diagram of type $A_1^{(1)}$. To this diagram again correspond two different types of surfaces, the generic one corresponding to divisors $D_1$ and $D_2$ intersecting at two points gives a multiplicative system of type $A_1^{(1)}$, and the degenerate configuration corresponding to two components touch at one point gives an additive system denoted by $A_1^{(1)*}$, which is our case, see Figure 8.

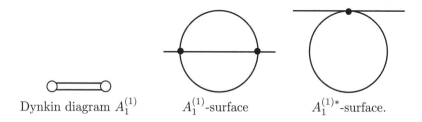

Dynkin diagram $A_1^{(1)}$     $A_1^{(1)}$-surface     $A_1^{(1)*}$-surface.

FIGURE 8. Configurations of type $A_1^{(1)}$

The symmetry sub-lattice $R^{\perp} = \text{Span}_{\mathbb{Z}}\{\alpha_0, \dots, \alpha_7\}$ is of type $E_7^{(1)}$, where the basis of $\alpha_i$ is given on Figure 9.

To compute the action of $\varphi_*$ on $\text{Pic}(\mathcal{X}_{\mathbf{b}})$, we decompose $\varphi = \varphi_2 \circ \varphi_1$, where $\varphi_1 : (f, g) \to (\bar{f}, g)$ is given by (3.12), and $\varphi_2 : (\bar{f}, g) \to (\bar{f}, \bar{g})$ is given by (3.13). Then $(\varphi_1)_*(H_g) = H_g$, it is straightforward to see that $(\varphi_1)_*(E_i) = H_g - E_i$ and

$$\alpha_1 = E_3 - E_4, \qquad \alpha_5 = E_5 - E_6,$$
$$\alpha_2 = E_2 - E_3, \qquad \alpha_6 = E_6 - E_7,$$
$$\alpha_3 = E_1 - E_2, \qquad \alpha_0 = E_7 - E_8,$$
$$\alpha_4 = H_f - E_1 - E_5, \qquad \alpha_7 = H_g - H_f,$$

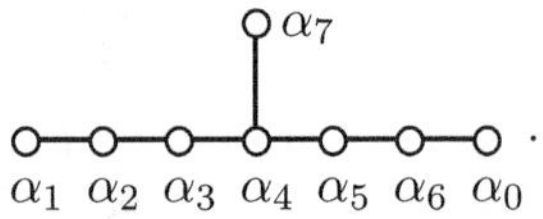

FIGURE 9. Symmetry sub-lattice for d-$P\left(A_1^{(1)*}\right)$

that $(\varphi_1)_*(D_0) = D_1$, $(\varphi_1)_*(D_1) = D_0$, which gives $(\varphi_1)_*(H_f) = H_f + 4H_g - E$, where $E = \sum_{i=1}^{8} E_i$. The situation with $\varphi_2$ is completely symmetric, $(\varphi_2)_*(H_f) = H_f$, $(\varphi_2)_*(E_i) = H_f - E_i$, and $(\varphi_2)_*(H_g) = H_g + 4H_f - E$. Composing these two linear maps, we get the action of $\varphi_*$:

$$H_f \mapsto 9H_f + 4H_g - 3E$$
$$H_g \mapsto 4H_f + H_g - E$$
$$E_i \mapsto 3H_f + H_g - E + E_i, \qquad i = 1, \ldots, 8,$$

and so the induced action $\varphi_*$ on the sub-lattice $R^\perp$ is given by the following translation:

$$(3.14) \qquad (\alpha_0, \alpha_1, \alpha_2, \alpha_3, \alpha_4, \alpha_5, \alpha_6, \alpha_7) \mapsto (\alpha_0, \alpha_1, \alpha_2, \alpha_3, \alpha_4, \alpha_5, \alpha_6, \alpha_7) +$$
$$(0, 0, 0, 0, 1, 0, 0, -2)(-K_{\mathcal{X}}).$$

3.2.2. *Schlesinger Transformations.* We start with a $4 \times 4$ Fuchsian system of the spectral type $(1111, 1111, 22)$. In this case it is convenient to have all singular points to be finite, since we need to consider two different kinds of elementary Schlesinger transformations — one between two points with non-repeating eigenvalues, and the other when at one point we have an eigenvalue of multiplicity two. As before, we can use scalar gauge transformations to make some of the eigenvalues to vanish, and so, putting $\theta_3 = \theta_3^1 = \theta_3^2$ we take our Riemann scheme to be

$$\left\{ \begin{array}{ccc} z_1 & z_2 & z_3 \\ \theta_1^1 & \theta_2^1 & \theta_3 \\ \theta_1^2 & \theta_2^2 & \theta_3 \\ \theta_1^3 & \theta_2^3 & 0 \\ 0 & \theta_2^4 & 0 \end{array} \right\}, \qquad \theta_1^1 + \theta_1^2 + \theta_1^3 + \theta_2^1 + \theta_2^2 + \theta_2^3 + \theta_2^4 + 2\theta_3 = 0.$$

We first consider an elementary Schlesinger transformation $\left\{ {}^1_1 {}^2_1 \right\}$ for which $\bar{\theta}_1^1 = \theta_1^1 - 1$ and $\bar{\theta}_2^1 = \theta_2^1 + 1$. The multiplier matrix for this transformation is

$$\mathbf{R}(z) = \mathbf{I} + \frac{z_1 - z_2}{z - z_1}\mathbf{P}, \qquad \text{where } \mathbf{P} = \frac{\mathbf{b}_{2,1}\mathbf{c}_1^{1\dagger}}{\mathbf{c}_1^{1\dagger}\mathbf{b}_{2,1}} \text{ and we put } \mathbf{Q} = \mathbf{I} - \mathbf{P}.$$

Since this transformation does not involve the point $z_3$ with multiple eigenvalues, the dynamic is again given by equations (2.13–2.19) that now take the form

$$(3.15)$$

$$\bar{\mathbf{b}}_{1,1} = \frac{1}{c_1^1}\mathbf{b}_{2,1}, \qquad \bar{\mathbf{b}}_{1,j} = \frac{1}{c_1^j}\left(\mathbf{I} - \frac{\mathbf{P}}{\theta_1^1 - \theta_1^j - 1}\left(\mathbf{A}_2 + \frac{z_2 - z_1}{z_3 - z_1}\mathbf{A}_1\right)\right)\mathbf{b}_{1,j} \qquad (j = 2, 3);$$

$$\bar{\mathbf{b}}_{2,1} = \frac{1}{c_2^1}\left((\theta_2^1+1)\mathbf{I} + \mathbf{Q}\left(\mathbf{I} + \frac{\mathbf{b}_{2,2}\mathbf{c}_2^{2\dagger}}{\theta_2^1-\theta_2^2+1} + \frac{\mathbf{b}_{2,3}\mathbf{c}_2^{3\dagger}}{\theta_2^1-\theta_2^3+1} + \frac{\mathbf{b}_{2,4}\mathbf{c}_2^{4\dagger}}{\theta_2^1-\theta_2^4+1}\right)\right) \times$$

$$\left(\mathbf{A}_1 + \frac{z_1-z_2}{z_3-z_2}\mathbf{A}_3\right)\right)\frac{\mathbf{b}_{2,1}}{\mathbf{c}_1^{1\dagger}\mathbf{b}_{2,1}};$$

$$\bar{\mathbf{b}}_{2,j} = \frac{1}{c_2^j}\mathbf{Q}\mathbf{b}_{2,j} \quad (j=2,3,4); \qquad \bar{\mathbf{b}}_{3,j} = \frac{1}{c_3^j}\mathbf{R}(z_3)\mathbf{b}_{3,j} \quad (j=1,2);$$

$$\bar{\mathbf{c}}_1^{1\dagger} = c_1^1\frac{\mathbf{c}_1^{1\dagger}}{\mathbf{c}_1^{1\dagger}\mathbf{b}_{2,1}}\left((\theta_1^1-1)\mathbf{I} + \left(\mathbf{A}_2 + \frac{z_2-z_1}{z_3-z_1}\mathbf{A}_3\right)\times\right.$$

$$\left.\left(\mathbf{I} + \frac{\mathbf{b}_{1,2}\mathbf{c}_1^{2\dagger}}{\theta_1^1-\theta_1^2-1} + \frac{\mathbf{b}_{1,3}\mathbf{c}_1^{3\dagger}}{\theta_1^1-\theta_1^3-1}\right)\mathbf{Q}\right), \quad \bar{\mathbf{c}}_1^{j\dagger} = c_1^j\mathbf{c}_1^{j\dagger}\mathbf{Q} \quad (j=2,3);$$

$$\bar{\mathbf{c}}_2^{1\dagger} = c_2^1\mathbf{c}_1^{1\dagger}, \quad \bar{\mathbf{c}}_2^{j\dagger} = c_2^j\mathbf{c}_2^{j\dagger}\left(\mathbf{I} - \left(\mathbf{A}_1 + \frac{z_1-z_2}{z_3-z_2}\mathbf{A}_3\right)\frac{\mathbf{P}}{\theta_2^1-\theta_2^j+1}\right) \quad (j=2,3,4);$$

$$\bar{\mathbf{c}}_3^{j\dagger} = c_3^j\mathbf{c}_3^{j\dagger}\mathbf{R}^{-1}(z_3),$$

where $c_i^j$ are again arbitrary non-zero constants.

Similarly to the previous example, we parameterize the matrices as

$$\mathbf{B}_1 = \begin{bmatrix} 1 & 0 & 0 \\ 0 & 1 & 0 \\ 0 & 0 & 1 \\ 0 & 0 & 0 \end{bmatrix}, \qquad \mathbf{C}_1^\dagger = \begin{bmatrix} \theta_1^1 & 0 & 0 & \alpha \\ 0 & \theta_1^2 & 0 & \beta \\ 0 & 0 & \theta_1^3 & \gamma \end{bmatrix},$$

$$\mathbf{B}_3 = \begin{bmatrix} 0 & 1 \\ 0 & 1 \\ 0 & 1 \\ 1 & 1 \end{bmatrix}, \qquad \mathbf{C}_3^\dagger = \begin{bmatrix} -(x+\theta_3) & 0 & x & \theta_3 \\ 0 & \theta_3-y & y & 0 \end{bmatrix},$$

$$\mathbf{A}_2 = -(\mathbf{A}_1 + \mathbf{A}_3).$$

Using the condition that the eigenvalues of $\mathbf{A}_2$ are $\theta_2^1, \ldots, \theta_2^4$ we get a system of three linear equations on $\alpha$, $\beta$, and $\gamma$ with coefficients depending on $x$ and $y$, which gives us rational functions $\alpha(x,y)$, $\beta(x,y)$, and $\gamma(x,y)$ (again, the resulting expressions are quite large and we omit them). Thus, the space of accessory parameters for Fuchsian systems of this type is two-dimensional and $x$ and $y$ are some coordinates on this space.

The resulting mapping $\psi : (x,y) \to (\bar{x},\bar{y})$ becomes very complicated (and computing it requires a Computer Algebra System; in our work we have used **Mathematica**) and so we omit equations describing the map. Nevertheless, it is possible to do a complete geometric analysis of the mapping.

We find that the indeterminate points of $\psi$ are

$$p_1\left(\frac{(\theta_1^1+\theta_2^1+\theta_3)(\theta_1^3+\theta_2^1)}{\theta_1^1-\theta_1^3}, -\frac{(\theta_1^2+\theta_2^1+\theta_3)(\theta_1^3+\theta_2^1)}{\theta_1^2-\theta_1^3}\right), \quad p_5(0,0),$$

$$p_2\left(\frac{(\theta_1^1+\theta_2^2+\theta_3)(\theta_1^3+\theta_2^2)}{\theta_1^1-\theta_1^3}, -\frac{(\theta_1^2+\theta_2^2+\theta_3)(\theta_1^3+\theta_2^2)}{\theta_1^2-\theta_1^3}\right), \quad p_6(-\theta_3,\theta_3),$$

$$p_3 \left( \frac{(\theta_1^1 + \theta_2^3 + \theta_3)(\theta_1^3 + \theta_2^3)}{\theta_1^1 - \theta_1^3}, \; -\frac{(\theta_1^2 + \theta_2^3 + \theta_3)(\theta_1^3 + \theta_2^3)}{\theta_1^2 - \theta_1^3} \right),$$

$$p_4 \left( \frac{(\theta_1^1 + \theta_2^4 + \theta_3)(\theta_1^3 + \theta_2^4)}{\theta_1^1 - \theta_1^3}, \; -\frac{(\theta_1^2 + \theta_2^4 + \theta_3)(\theta_1^3 + \theta_2^4)}{\theta_1^2 - \theta_1^3} \right),$$

as well as the sequence of infinitely close points

$$p_7 \left( \frac{1}{x} = 0, \frac{1}{y} = 0 \right) \longleftarrow p_8 \left( \frac{1}{x} = 0, \frac{x}{y} = -\frac{(\theta_1^1 + 1)(\theta_1^2 - \theta_1^3)}{(\theta_1^2 + 1)(\theta_1^1 - \theta_1^3)} \right).$$

Note also that the points $p_1, \ldots, p_7$ all lie on a $(2,2)$-curve $Q$ given by the equation

$$(3.16) \quad \left( (\theta_1^3 - \theta_1^1)x + (\theta_1^3 - \theta_1^2)y \right)^2$$
$$+ (\theta_1^1 - \theta_1^2)\left( (\theta_1^3 - \theta_1^2 - \theta_3)(\theta_1^3 - \theta_1^1)x + (\theta_1^3 - \theta_1^1 - \theta_3)(\theta_1^3 - \theta_1^2)y \right) = 0.$$

Resolving indeterminate points of this map using blow-ups gives us the Okamoto surface $\mathcal{X}_\theta$ pictured on Figure 10. Note that the $-2$ curves $D_0 = 2H_x + 2H_y - F_1 - F_2 - F_3 - F_4 - F_5 - F_6 - 2F_7$ and $D_1 = F_7 - F_8$ touch at the point with coordinates $\left( \frac{1}{x} = 0, \frac{x}{y} = -\frac{\theta_1^2 - \theta_1^3}{\theta_1^1 - \theta_1^3} \right)$. Thus, we immediately see that this is indeed a surface of type $A_1^{(1)*}$.

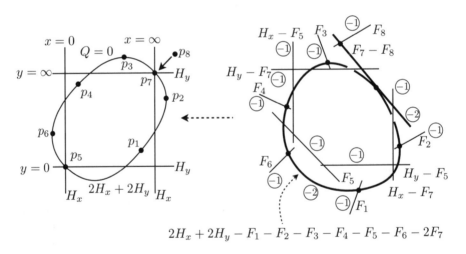

$$2H_x + 2H_y - F_1 - F_2 - F_3 - F_4 - F_5 - F_6 - 2F_7$$

FIGURE 10. Okamoto surface $\mathcal{X}_\theta$ for the Schlesinger transformations reduction to d-$P\left( A_1^{(1)*} \right)$.

3.2.3. *Reduction to the standard form.* We now proceed to match the surface $\mathcal{X}_\theta$ described by the blow-up diagram on Figure 10 with the surface $\mathcal{X}_\mathbf{b}$ described by diagram on Figure 7. As in the previous example, we look for rational classes $\mathcal{H}_f, \mathcal{H}_g, \mathcal{E}_1, \ldots, \mathcal{E}_8$ in $\mathrm{Pic}(\mathcal{X}_\theta)$ such that

$$\mathcal{H}_f \bullet \mathcal{H}_g = 1, \; \mathcal{E}_i^2 = -1, \; \mathcal{H}_f^2 = \mathcal{H}_g^2 = \mathcal{H}_f \bullet \mathcal{E}_i = \mathcal{H}_g \bullet \mathcal{E}_i = \mathcal{E}_i \bullet \mathcal{E}_j = 0, \; 1 \le i \ne j \le 8,$$

and the resulting configuration matches diagram on Figure 7. This time the (virtual) genus formula $g(C) = (C^2 + K_\mathcal{X} \bullet C)/2 + 1$ suggests we see that we should look for classes of rational curves of self-intersection zero among $H_x$, $H_y$, or

$H_x + H_y - F_i - F_j$ and for classes of rational curves of self-intersection $-1$ among $F_i$, $H_x - F_i$, $H_y - F_i$, or $H_x + H_y - F_i - F_j - F_k$.

It is again convenient to start by comparing the $-2$-curves on both diagrams,

$$D_0 = 2H_x + 2H_y - F_1 - F_2 - F_3 - F_4 - F_5 - F_6 - 2F_7$$
$$= \mathcal{H}_f + \mathcal{H}_g - \mathcal{E}_1 - \mathcal{E}_2 - \mathcal{E}_3 - \mathcal{E}_4,$$
$$D_1 = F_7 - F_8 = \mathcal{H}_f + \mathcal{H}_g - \mathcal{E}_5 - \mathcal{E}_6 - \mathcal{E}_7 - \mathcal{E}_8.$$

Given the uniformity of the coordinates of $p_i$, $i = 1, \ldots, 4$, we see that it makes sense to choose $\mathcal{E}_i = F_i$ for $i = 1, \ldots, 4$, and also we can put $\mathcal{E}_8 = F_8$. This results in $\mathcal{H}_f + \mathcal{H}_g = 2H_x + 2H_y - F_5 - F_6 - 2F_7$, which suggests taking $\mathcal{H}_f = H_x + H_y - F_5 - F_7$ and $\mathcal{H}_y = H_x + H_y - F_6 - F_7$ (and so $\mathcal{H}_f \bullet \mathcal{H}_g = 1$ and $\mathcal{H}_f^2 = \mathcal{H}_g^2 = 0$). Then $\mathcal{E}_5 + \mathcal{E}_6 + \mathcal{E}_7 = 2H_x + 2H_y - F_5 - F_6 - 3E_7$, so we take $\mathcal{E}_5 = H_y - F_7$, $\mathcal{E}_6 = H_x - F_7$, and $\mathcal{E}_7 = H_x + H_y - F_5 - F_6 - F_7$. It is not very hard to show that such choice satisfies all of our requirements and moreover, it is essentially unique (up to a permutation of the indices of exceptional divisors). To summarize, we get the following identification:

$$\mathcal{H}_f = H_x + H_y - F_5 - F_7, \qquad \mathcal{E}_1 = F_1, \quad \mathcal{E}_3 = F_3, \quad \mathcal{E}_5 = H_y - F_7,$$
$$\mathcal{H}_g = H_x + H_y - F_6 - F_7, \qquad \mathcal{E}_2 = F_2, \quad \mathcal{E}_4 = F_4, \quad \mathcal{E}_6 = H_x - F_7,$$
$$\mathcal{E}_7 = H_x + H_y - F_5 - F_6 - F_7, \quad \mathcal{E}_8 = F_8.$$

Let us now define the base coordinates $f$ and $g$ of the linear systems $|\mathcal{H}_f|$ and $|\mathcal{H}_g|$ that will map the exceptional fibers of the divisors $\mathcal{E}_i$ to the points $\pi_i$ such that $\pi_1, \ldots, \pi_4$ are on a line $f + g = $ const, and $\pi_5, \ldots, \pi_8$ are on the line $f + g = 0$. Since the pencil $|\mathcal{H}_f|$ consists of $(1,1)$ curves on $\mathbb{P}^1 \times \mathbb{P}^1$ passing through $p_5(0,0)$ and $p_7(\infty, \infty)$,

$$|\mathcal{H}_f| = |H_x + H_y - F_5 - F_7| = \{axy + bx + cy + d = 0 \mid a = d = 0\}$$
$$= \{bx + cy = 0\},$$

and so we can initially define the base coordinate as $f_1 = y/x$. Similarly, the pencil $|\mathcal{H}_g|$ consists of $(1,1)$ curves on $\mathbb{P}^1 \times \mathbb{P}^1$ passing through $p_6(-\theta_3, \theta_3)$ and $p_7(\infty, \infty)$, so

$$|\mathcal{H}_g| = |H_x + H_y - F_6 - F_7| = \{axy + bx + cy + d = 0 \mid a = 0, (c-b)\theta_3 + d = 0\}$$
$$= \{b(x + \theta_3) + c(y - \theta_3) = 0\},$$

and $g_1 = (y - \theta_3)/(x + \theta_3)$. Next we will do a series of affine change of variables to arrange that the points $p_i$, $i = 1, \ldots, 4$, are on a line $f + g = $ const. We have (below $i = 1, \ldots, 4$)

$$f_1(\pi_i) = -\frac{(\theta_1^1 - \theta_1^3)(\theta_1^2 + \theta_2^2 + \theta_3)}{(\theta_1^2 - \theta_1^3)(\theta_1^1 + \theta_2^i + \theta_3)}, \qquad g_1(\pi_i) = -\frac{(\theta_1^1 - \theta_1^3)(\theta_1^2 + \theta_2^i)}{(\theta_1^2 - \theta_1^3)(\theta_1^1 + \theta_2^i)},$$

and therefore, it makes sense to put

$$f_2 = \frac{(\theta_1^2 - \theta_1^3)}{(\theta_1^1 - \theta_1^3)} f_1 + 1, \qquad\qquad g_2 = \frac{(\theta_1^2 - \theta_1^3)}{(\theta_1^1 - \theta_1^3)} g_1 + 1.$$

We then get

$$f_2(\pi_i) = -\frac{(\theta_1^1 - \theta_1^2)}{(\theta_1^1 + \theta_2^i + \theta_3)}, \qquad\qquad g_2(\pi_i) = -\frac{(\theta_1^1 - \theta_1^2)}{(\theta_1^1 + \theta_2^i)},$$

and so we put

$$f_3 = \frac{(\theta_1^1 - \theta_1^2)}{f_2}, \qquad\qquad g_3 = \frac{(\theta_1^1 - \theta_1^2)}{g_2}$$

to get

$$f_3(\pi_i) = \theta_1^1 + \theta_2^i + \theta_3, \qquad\qquad g_3(\pi_i) = \theta_1^1 + \theta_2^i.$$

Thus, our final change of coordinates is

$$(3.17) \qquad f = f_3 - \theta_1^1 = -\frac{\theta_1^2(\theta_1^3 - \theta_1^1)x - \theta_1^1(\theta_1^2 - \theta_1^3)y}{(\theta_1^3 - \theta_1^1)x - (\theta_1^2 - \theta_1^3)y},$$

$$(3.18) \qquad g = \theta_1^1 - g_3 = \frac{\theta_1^2(\theta_1^3 - \theta_1^1)(x + \theta_3) - \theta_1^1(\theta_1^2 - \theta_1^3)(y - \theta_3)}{(\theta_1^3 - \theta_1^1)(x + \theta_3) - (\theta_1^2 - \theta_1^3)(y - \theta_3)},$$

and for $i = 1, \ldots, 4$, $f(\pi_i) = f(p_i) = \theta_2^i + \theta_3$, $g(\pi_i) = g(p_i) = -\theta_2^i$ and so these points lie on the line $f + g = \theta_3$. We also get the identification of some of the parameters, $\theta_2^i = b_i - b$ for $i = 1, \ldots, 4$, and $\theta_3 = 2b$. It remains to verify that this change of variables puts points $\pi_5, \ldots, \pi_8$ on the line $f + g = 0$ and identify the remaining parameters. The exceptional divisor $\mathcal{E}_5$ in the $(x, y)$-coordinates corresponds to the line $y = \infty$, and so $(f, g)(\pi_5) = (-\theta_1^1, \theta_1^1)$. Similarly, $(f, g)(\pi_6) = (-\theta_1^2, \theta_1^2)$. The exceptional divisor $\mathcal{E}_7$ corresponds to the line $x + y = 0$, and so $(f, g)(\pi_7) = (-\theta_1^3, \theta_1^3)$. Finally, $\mathcal{E}_8$ is given by $x = y = \infty$, $\frac{x}{y} = -\frac{(\theta_1^1 + 1)(\theta_1^2 - \theta_1^3)}{(\theta_1^2 + 1)(\theta_1^1 - \theta_1^3)}$, and so $(f, g)(\pi_8) = (1, -1)$. Thus, we see that indeed $\pi_5, \ldots, \pi_8$ lie on the line $f + g = 0$ and the remaining identification between the parameters is $\theta_1^1 = b_5$, $\theta_1^2 = b_6$, $\theta_1^3 = b_7$, and $b_8 = -1$. To summarize, we get the following matching of parameters

$$(3.19) \qquad \begin{array}{llll} b_1 = \theta_2^1 + \theta_3/2, & b_3 = \theta_2^3 + \theta_3/2, & b_5 = \theta_1^1 & b_7 = \theta_1^3 \\ b_2 = \theta_2^2 + \theta_3/2 & b_4 = \theta_2^4 + \theta_3/2, & b_6 = \theta_1^2 & b_8 = -1 \end{array}$$

We are now in the position to compare the dynamic given by an elementary Schlesinger transformation with the dynamic of our model example of d-$P\left(A_1^{(1)*}\right)$. As in the previous example, there are two different ways to do so. First, we can compute the corresponding translation vector. It is not very difficult to show that the action of $\psi_*$ of an elementary Schlesinger transformation $\{\begin{smallmatrix} 1 & 2 \\ 1 & 1 \end{smallmatrix}\}$ on the classes $\mathcal{H}_f$, $\mathcal{H}_g$, and $\mathcal{E}_i$ is

$$\mathcal{H}_f \mapsto 4\mathcal{H}_f + 3\mathcal{H}_g - 3\mathcal{E}_1 - \mathcal{E}_2 - \mathcal{E}_3 - \mathcal{E}_4 - 2\mathcal{E}_6 - 2\mathcal{E}_7 - 2\mathcal{E}_8,$$
$$\mathcal{H}_g \mapsto 3\mathcal{H}_f + 4\mathcal{H}_g - 3\mathcal{E}_1 - \mathcal{E}_2 - \mathcal{E}_3 - \mathcal{E}_4 - 2\mathcal{E}_6 - 2\mathcal{E}_7 - 2\mathcal{E}_8,$$
$$\mathcal{E}_1 \mapsto \mathcal{E}_5,$$

$$\mathcal{E}_2 \mapsto 2\mathcal{H}_f + 2\mathcal{H}_g - 2\mathcal{E}_1 - \mathcal{E}_3 - \mathcal{E}_4 - \mathcal{E}_6 - \mathcal{E}_7 - \mathcal{E}_8,$$
$$\mathcal{E}_3 \mapsto 2\mathcal{H}_f + 2\mathcal{H}_g - 2\mathcal{E}_1 - \mathcal{E}_2 - \mathcal{E}_4 - \mathcal{E}_6 - \mathcal{E}_7 - \mathcal{E}_8,$$
$$\mathcal{E}_4 \mapsto 2\mathcal{H}_f + 2\mathcal{H}_g - 2\mathcal{E}_1 - \mathcal{E}_2 - \mathcal{E}_3 - \mathcal{E}_6 - \mathcal{E}_7 - \mathcal{E}_8,$$
$$\mathcal{E}_5 \mapsto 3\mathcal{H}_f + 3\mathcal{H}_g - 2\mathcal{E}_1 - \mathcal{E}_2 - \mathcal{E}_3 - \mathcal{E}_4 - 2\mathcal{E}_6 - 2\mathcal{E}_7 - 2\mathcal{E}_8,$$
$$\mathcal{E}_6 \mapsto \mathcal{H}_f + \mathcal{H}_g - \mathcal{E}_1 - \mathcal{E}_7 - \mathcal{E}_8,$$
$$\mathcal{E}_7 \mapsto \mathcal{H}_f + \mathcal{H}_g - \mathcal{E}_1 - \mathcal{E}_6 - \mathcal{E}_8,$$
$$\mathcal{E}_8 \mapsto \mathcal{H}_f + \mathcal{H}_g - \mathcal{E}_1 - \mathcal{E}_6 - \mathcal{E}_7.$$

Comparing the action of $\psi_*$ with the standard action $\varphi_*$ given by (3.14) on the symmetry sub-lattice, we see that the translation vectors are different:

$$\psi_* : (\alpha_0, \alpha_1, \alpha_2, \alpha_3, \alpha_4, \alpha_5, \alpha_6, \alpha_7) \mapsto (\alpha_0, \alpha_1, \alpha_2, \alpha_3, \alpha_4, \alpha_5, \alpha_6, \alpha_7)+$$
$$(0, 0, 0, -1, 0, 1, 0, 0)(-K_\mathcal{X}),$$
$$\varphi_* : (\alpha_0, \alpha_1, \alpha_2, \alpha_3, \alpha_4, \alpha_5, \alpha_6, \alpha_7) \mapsto (\alpha_0, \alpha_1, \alpha_2, \alpha_3, \alpha_4, \alpha_5, \alpha_6, \alpha_7)+$$
$$(0, 0, 0, 0, 1, 0, 0, -2)(-K_\mathcal{X}).$$

To get more insight into the relationship between Schlesinger transformations and the standard d-$P(A_1^{(1)*})$ dynamic, it is better to compute the action of d-$P\left(A_1^{(1)*}\right)$ on the Riemann scheme of our Fuchsian system using the above identification of parameters:

$$\left\{ \begin{matrix} z_1 & z_2 & z_3 \\ \theta_1^1 & \theta_2^1 & \theta_3 \\ \theta_1^2 & \theta_2^2 & \theta_3 \\ \theta_1^3 & \theta_2^3 & 0 \\ 0 & \theta_2^4 & 0 \end{matrix} \right\} \overset{\left\{ \begin{smallmatrix} 1 & 2 \\ 1 & 1 \end{smallmatrix} \right\}}{\longmapsto} \left\{ \begin{matrix} z_1 & z_2 & z_3 \\ \theta_1^1 - 1 & \theta_2^1 + 1 & \theta_3 \\ \theta_1^2 & \theta_2^2 & \theta_3 \\ \theta_1^3 & \theta_2^3 & 0 \\ 0 & \theta_2^4 & 0 \end{matrix} \right\},$$

$$\left\{ \begin{matrix} z_1 & z_2 & z_3 \\ \theta_1^1 & \theta_2^1 & \theta_3 \\ \theta_1^2 & \theta_2^2 & \theta_3 \\ \theta_1^3 & \theta_2^3 & 0 \\ 0 & \theta_2^4 & 0 \end{matrix} \right\} \overset{\text{d-}P(A_1^{(1)*})}{\longmapsto} \left\{ \begin{matrix} z_1 & z_2 & z_3 \\ \theta_1^1 & \theta_2^1 - 1 & \theta_3 + 2 \\ \theta_1^2 & \theta_2^2 - 1 & \theta_3 + 2 \\ \theta_1^3 & \theta_2^3 - 1 & 0 \\ 0 & \theta_2^4 - 1 & 0 \end{matrix} \right\}.$$

Thus, we see that the standard d-$P\left(A_1^{(1)*}\right)$ dynamic changes the multiple eigenvalue and so it requires the use of rank-two elementary Schlesinger transformations,

$$\text{d-}P\left(A_1^{(1)*}\right) = \left\{ \begin{smallmatrix} 2 & 3 \\ 3 & 1 \\ 4 & 2 \end{smallmatrix} \right\} \circ \left\{ \begin{smallmatrix} 2 & 3 \\ 1 & 1 \\ 2 & 2 \end{smallmatrix} \right\}.$$

Thus, consider the transformation $\left\{ \begin{smallmatrix} 2 & 3 \\ 1 & 1 \\ 2 & 2 \end{smallmatrix} \right\}$ changing the characteristic indices by $\bar{\theta}_2^1 = \theta_2^1 - 1$, $\bar{\theta}_2^1 = \theta_2^1 - 1$, and $\bar{\theta}_3 = \theta_3 + 1$. This transformation is given by equations (2.24–2.30) that now take the form

(3.20)    $\bar{\mathbf{b}}_{1,j} = \dfrac{1}{c_1^j}\mathbf{R}(z_1)\mathbf{b}_{1,j}$   $(j = 1, 2, 3)$;    $\bar{\mathbf{c}}_1^{j\dagger} = c_1^j \mathbf{c}_1^{j\dagger}\mathbf{R}^{-1}(z_i)$   $(j = 1, 2, 3)$;

$\bar{\mathbf{b}}_{2,1} = \dfrac{1}{c_2^1}\mathbf{Q}_2\mathbf{b}_{3,1}$,    $\bar{\mathbf{b}}_{2,2} = \dfrac{1}{c_2^2}\mathbf{Q}_1\mathbf{b}_{3,2},$

$$\bar{\mathbf{b}}_{2,j} = \frac{1}{c_2^j}\left(\mathbf{I} - \left(\frac{\mathcal{P}_1}{\theta_2^1 - \theta_2^j - 1} + \frac{\mathcal{P}_2}{\theta_2^2 - \theta_2^j - 1}\right)\left(\frac{z_3 - z_2}{z_1 - z_2}\mathbf{A}_1 + \mathbf{A}_3\right)\right)\mathbf{b}_{2,j} \quad (j = 3, 4);$$

$$\bar{\mathbf{c}}_2^{1\dagger} = c_2^1\frac{\mathbf{c}_2^{1\dagger}}{\mathbf{c}_2^{1\dagger}\mathbf{Q}_2\mathbf{b}_{3,1}}\left((\theta_2^1 - 1)\mathbf{I} + \left(\frac{z_3 - z_2}{z_1 - z_2}\mathbf{A}_1 + \mathbf{A}_3\right)\times\right.$$
$$\left.\left(\mathbf{I} + \frac{\mathbf{b}_{2,3}\mathbf{c}_2^{3\dagger}}{\theta_2^1 - \theta_2^3 - 1} + \frac{\mathbf{b}_{2,4}\mathbf{c}_2^{4\dagger}}{\theta_2^1 - \theta_2^4 - 1}\right)\mathcal{Q}\right),$$

$$\bar{\mathbf{c}}_2^{2\dagger} = c_2^2\frac{\mathbf{c}_2^{2\dagger}}{\mathbf{c}_2^{2\dagger}\mathbf{Q}_1\mathbf{b}_{3,2}}\left((\theta_2^2 - 1)\mathbf{I} + \left(\frac{z_3 - z_2}{z_1 - z_2}\mathbf{A}_1 + \mathbf{A}_3\right)\times\right.$$
$$\left.\left(\mathbf{I} + \frac{\mathbf{b}_{2,3}\mathbf{c}_2^{3\dagger}}{\theta_2^2 - \theta_2^3 - 1} + \frac{\mathbf{b}_{2,4}\mathbf{c}_2^{4\dagger}}{\theta_2^2 - \theta_2^4 - 1}\right)\mathcal{Q}\right),$$

$$\bar{\mathbf{c}}_2^{j\dagger} = c_2^j\mathbf{c}_2^{j\dagger}\mathcal{Q} \quad (j = 2, 3);$$

$$\bar{\mathbf{b}}_{3,1} = \frac{1}{c_3^1}\left((\theta_3 + 1)\mathbf{I} + \mathcal{Q}\left(\frac{z_2 - z_3}{z_1 - z_3}\mathbf{A}_1 + \mathbf{A}_2\right)\right)\frac{\mathbf{b}_{3,1}}{\mathbf{c}_2^{1\dagger}\mathbf{Q}_2\mathbf{b}_{3,1}},$$

$$\bar{\mathbf{b}}_{3,2} = \frac{1}{c_3^2}\left((\theta_3 + 1)\mathbf{I} + \mathcal{Q}\left(\frac{z_2 - z_3}{z_1 - z_3}\mathbf{A}_1 + \mathbf{A}_2\right)\right)\frac{\mathbf{b}_{3,2}}{\mathbf{c}_2^{2\dagger}\mathbf{Q}_1\mathbf{b}_{3,2}};$$

$$\bar{\mathbf{c}}_3^{1\dagger} = c_3^1\mathbf{c}_2^{1\dagger}\mathbf{Q}_2, \quad \bar{\mathbf{c}}_3^{2\dagger} = c_3^2\mathbf{c}_2^{2\dagger}\mathbf{Q}_1,$$

where

$$\mathbf{P}_1 = \frac{\mathbf{b}_{3,1}\mathbf{c}_2^{1\dagger}}{\mathbf{c}_2^{1\dagger}\mathbf{b}_{3,1}}, \quad \mathbf{Q}_1 = \mathbf{I} - \mathbf{P}_1, \quad \mathbf{P}_2 = \frac{\mathbf{b}_{3,2}\mathbf{c}_2^{2\dagger}}{\mathbf{c}_2^{2\dagger}\mathbf{b}_{3,2}}, \quad \mathbf{Q}_2 = \mathbf{I} - \mathbf{P}_2;$$

$$\mathcal{P}_1 = \frac{\mathbf{Q}_2\mathbf{P}_1}{\text{Tr}(\mathbf{Q}_2\mathbf{P}_1)}, \quad \mathcal{P}_2 = \frac{\mathbf{Q}_1\mathbf{P}_2}{\text{Tr}(\mathbf{Q}_1\mathbf{P}_2)}, \quad \mathcal{P} = \mathcal{P}_1 + \mathcal{P}_2, \quad \mathcal{Q} = \mathbf{I} - \mathcal{P};$$

$$\mathbf{R}(z) = \mathbf{I} + \frac{z_2 - z_3}{z - z_2}\mathcal{P}.$$

The Okamoto surface for this dynamic is the same as before (it depends only on the Fuchsian system rather than a particular transformation), and its action $\psi_*^{12}$ on $\text{Pic}(\mathcal{X}_\theta)$ is given by

$$\mathcal{H}_f \mapsto 6\mathcal{H}_f + 3\mathcal{H}_g - \mathcal{E}_1 - \mathcal{E}_2 - 3\mathcal{E}_3 - 3\mathcal{E}_4 - 2\mathcal{E}_5 - 2\mathcal{E}_6 - 2\mathcal{E}_7 - 2\mathcal{E}_8,$$
$$\mathcal{H}_g \mapsto 3\mathcal{H}_f + 2\mathcal{H}_g - 2\mathcal{E}_3 - 2\mathcal{E}_4 - \mathcal{E}_5 - \mathcal{E}_6 - \mathcal{E}_7 - \mathcal{E}_8,$$
$$\mathcal{E}_1 \mapsto 3\mathcal{H}_f + 2\mathcal{H}_g - \mathcal{E}_2 - 2\mathcal{E}_3 - 2\mathcal{E}_4 - \mathcal{E}_5 - \mathcal{E}_6 - \mathcal{E}_7 - \mathcal{E}_8,$$
$$\mathcal{E}_2 \mapsto 3\mathcal{H}_f + 2\mathcal{H}_g - \mathcal{E}_1 - 2\mathcal{E}_3 - 2\mathcal{E}_4 - \mathcal{E}_5 - \mathcal{E}_6 - \mathcal{E}_7 - \mathcal{E}_8,$$
$$\mathcal{E}_3 \mapsto \mathcal{H}_f - \mathcal{E}_4,$$
$$\mathcal{E}_4 \mapsto \mathcal{H}_f - \mathcal{E}_3,$$
$$\mathcal{E}_5 \mapsto 2\mathcal{H}_f + \mathcal{H}_g - \mathcal{E}_3 - \mathcal{E}_4 - \mathcal{E}_6 - \mathcal{E}_7 - \mathcal{E}_8,$$
$$\mathcal{E}_6 \mapsto 2\mathcal{H}_f + \mathcal{H}_g - \mathcal{E}_3 - \mathcal{E}_4 - \mathcal{E}_5 - \mathcal{E}_7 - \mathcal{E}_8,$$
$$\mathcal{E}_7 \mapsto 2\mathcal{H}_f + \mathcal{H}_g - \mathcal{E}_3 - \mathcal{E}_4 - \mathcal{E}_5 - \mathcal{E}_6 - \mathcal{E}_8,$$
$$\mathcal{E}_8 \mapsto 2\mathcal{H}_f + \mathcal{H}_g - \mathcal{E}_3 - \mathcal{E}_4 - \mathcal{E}_5 - \mathcal{E}_6 - \mathcal{E}_7,$$

and the action on the symmetry sub-lattice is

$$\psi_*^{12} : (\alpha_0, \alpha_1, \alpha_2, \alpha_3, \alpha_4, \alpha_5, \alpha_6, \alpha_7) \mapsto (\alpha_0, \alpha_1, \alpha_2, \alpha_3, \alpha_4, \alpha_5, \alpha_6, \alpha_7) +$$
$$(0, 0, 1, 0, 0, 0, 0, -1)(-K_\mathcal{X}).$$

Similarly, the action $\psi_*^{34}$ of the rank-two transformation $\left\{ \begin{smallmatrix} 2 & 3 \\ 3 & 1 \\ 4 & 2 \end{smallmatrix} \right\}$ on $\mathrm{Pic}(\mathcal{X}_\theta)$ is given by

$$\mathcal{H}_f \mapsto 6\mathcal{H}_f + 3\mathcal{H}_g - 3\mathcal{E}_1 - 3\mathcal{E}_2 - \mathcal{E}_3 - \mathcal{E}_4 - 2\mathcal{E}_5 - 2\mathcal{E}_6 - 2\mathcal{E}_7 - 2\mathcal{E}_8,$$
$$\mathcal{H}_g \mapsto 3\mathcal{H}_f + 2\mathcal{H}_g - 2\mathcal{E}_1 - 2\mathcal{E}_2 - \mathcal{E}_5 - \mathcal{E}_6 - \mathcal{E}_7 - \mathcal{E}_8,$$
$$\mathcal{E}_1 \mapsto \mathcal{H}_f - \mathcal{E}_2,$$
$$\mathcal{E}_2 \mapsto \mathcal{H}_f - \mathcal{E}_1,$$
$$\mathcal{E}_3 \mapsto 3\mathcal{H}_f + 2\mathcal{H}_g - 2\mathcal{E}_1 - 2\mathcal{E}_2 - \mathcal{E}_4 - \mathcal{E}_5 - \mathcal{E}_6 - \mathcal{E}_7 - \mathcal{E}_8,$$
$$\mathcal{E}_4 \mapsto 3\mathcal{H}_f + 2\mathcal{H}_g - 2\mathcal{E}_1 - 2\mathcal{E}_2 - \mathcal{E}_3 - \mathcal{E}_5 - \mathcal{E}_6 - \mathcal{E}_7 - \mathcal{E}_8,$$
$$\mathcal{E}_5 \mapsto 2\mathcal{H}_f + \mathcal{H}_g - \mathcal{E}_1 - \mathcal{E}_2 - \mathcal{E}_6 - \mathcal{E}_7 - \mathcal{E}_8,$$
$$\mathcal{E}_6 \mapsto 2\mathcal{H}_f + \mathcal{H}_g - \mathcal{E}_1 - \mathcal{E}_2 - \mathcal{E}_5 - \mathcal{E}_7 - \mathcal{E}_8,$$
$$\mathcal{E}_7 \mapsto 2\mathcal{H}_f + \mathcal{H}_g - \mathcal{E}_1 - \mathcal{E}_2 - \mathcal{E}_5 - \mathcal{E}_6 - \mathcal{E}_8,$$
$$\mathcal{E}_8 \mapsto 2\mathcal{H}_f + \mathcal{H}_g - \mathcal{E}_1 - \mathcal{E}_2 - \mathcal{E}_5 - \mathcal{E}_6 - \mathcal{E}_7,$$

and its action on the symmetry sub-lattice is

$$\psi_*^{34} : (\alpha_0, \alpha_1, \alpha_2, \alpha_3, \alpha_4, \alpha_5, \alpha_6, \alpha_7) \mapsto (\alpha_0, \alpha_1, \alpha_2, \alpha_3, \alpha_4, \alpha_5, \alpha_6, \alpha_7) +$$
$$(0, 0, -1, 0, 1, 0, 0, -1)(-K_\mathcal{X}).$$

Thus,

$$\psi_*^{34} \circ \psi_*^{12} : (\alpha_0, \alpha_1, \alpha_2, \alpha_3, \alpha_4, \alpha_5, \alpha_6, \alpha_7) \mapsto (\alpha_0, \alpha_1, \alpha_2, \alpha_3, \alpha_4, \alpha_5, \alpha_6, \alpha_7) +$$
$$(0, 0, 1, 0, 0, 0, 0, -1)(-K_\mathcal{X}) +$$
$$(0, 0, -1, 0, 1, 0, 0, -1)(-K_\mathcal{X})$$
$$= (\alpha_0, \alpha_1, \alpha_2, \alpha_3, \alpha_4, \alpha_5, \alpha_6, \alpha_7) +$$
$$(0, 0, 0, 0, 1, 0, 0, -2)(-K_\mathcal{X}) = \varphi_*.$$

Finally, using a Computer Algebra System we can verify by a direct calculation that

$$\text{d-}P(A_1^{(1)*}) = \left\{ \begin{smallmatrix} 2 & 3 \\ 3 & 1 \\ 4 & 2 \end{smallmatrix} \right\} \circ \left\{ \begin{smallmatrix} 2 & 3 \\ 1 & 1 \\ 2 & 2 \end{smallmatrix} \right\}$$

holds on the level of equations as well. We again summarize our results in the form of a Theorem.

THEOREM 3.2. *Consider a Fuchsian system of the spectral type* $(1111, 1111, 22)$ *of the form*

$$\frac{d\mathbf{Y}}{dz} = \left( \frac{\mathbf{A}_1}{z - z_1} + \frac{\mathbf{A}_2}{z - z_2} + \frac{\mathbf{A}_3}{z - z_3} \right) \mathbf{Y},$$

*where the matrices* $\mathbf{A}_i$ *are normalized by gauge and similarity transformations to be of the form*

$$\mathbf{A}_1 = \begin{bmatrix} 1 & 0 & 0 \\ 0 & 1 & 0 \\ 0 & 0 & 1 \end{bmatrix} \begin{bmatrix} \theta_1^1 & 0 & 0 & \alpha \\ 0 & \theta_1^2 & 0 & \beta \\ 0 & 0 & \theta_1^3 & \gamma \end{bmatrix}, \quad \mathbf{A}_3 = \begin{bmatrix} 0 & 1 \\ 0 & 1 \\ 0 & 1 \\ 1 & 1 \end{bmatrix} \begin{bmatrix} -x - \theta_3 & 0 & x & \theta_3 \\ 0 & \theta_3 - y & y & 0 \end{bmatrix},$$

*where $\alpha = \alpha(x, y)$, $\beta = \beta(x, y)$, $\gamma = \gamma(x, y)$ and $\mathbf{A}_2 = -\mathbf{A}_1 - \mathbf{A}_3$. Do the change of variables inverse to (3.17–3.18):*

$$(3.21) \qquad x = -\frac{\theta_3(f + \theta_1^1)(g - \theta_1^3)}{(\theta_1^1 - \theta_1^3)(f + g)}, \qquad y = \frac{\theta_3(f + \theta_1^2)(g - \theta_1^3)}{(\theta_1^2 - \theta_1^3)(f + g)}.$$

*Then in the coordinates $(f, g)$ the Okamoto surface for the Schlesinger transformation dynamic $\mathcal{X}_\theta$ coincides with the Okamoto surface for the model example dynamic $\mathcal{X}_{\mathbf{b}}$ under the parameter identification (3.19), and the model d-P $\left(A_1^{(1)*}\right)$ dynamic can be represented as a composition of elementary Schlesinger transformations of rank two:*

$$d\text{-}P(A_1^{(1)*}) = \left\{\begin{smallmatrix} 2 & 3 \\ 3 & 1 \\ 4 & 2 \end{smallmatrix}\right\} \circ \left\{\begin{smallmatrix} 2 & 3 \\ 1 & 1 \\ 2 & 2 \end{smallmatrix}\right\}.$$

## 4. Conclusion

In this work we further develop a theory of discrete Schlesinger evolution equations that correspond to elementary Schlesinger transformations of ranks one and two of Fuchsian systems. We showed how to obtain difference Painlevé equations of types d-P $\left(A_2^{(1)*}\right)$ and d-P $\left(A_1^{(1)*}\right)$ as reductions of elementary Schlesinger transformations. We also tried to make our computations very detailed in order to illustrate general techniques on how to study discrete Painlevé equations geometrically.

One interesting observation is that standard examples of difference Painlevé equations of these types in both cases can be represented as compositions of elementary Schlesinger transformations. Thus, Schlesinger dynamic should in principle be simpler and it would be interesting to find a nice and simple form of equations giving this dynamic.

## References

[AB06]    D. Arinkin and A. Borodin, *Moduli spaces of d-connections and difference Painlevé equations*, Duke Math. J. **134** (2006), no. 3, 515–556, arXiv:math/0411584 [math.AG]. MR2254625 (2007h:14047)

[BK90]    É. Brézin and V. A. Kazakov, *Exactly solvable field theories of closed strings*, Phys. Lett. B **236** (1990), no. 2, 144–150, DOI 10.1016/0370-2693(90)90818-Q. MR1040213 (91k:81116)

[Boa09]   Philip Boalch, *Quivers and difference Painlevé equations*, Groups and symmetries, CRM Proc. Lecture Notes, vol. 47, Amer. Math. Soc., Providence, RI, 2009, pp. 25–51. MR2500553 (2011g:39028)

[Bor04]   Alexei Borodin, *Isomonodromy transformations of linear systems of difference equations*, Ann. of Math. (2) **160** (2004), no. 3, 1141–1182, DOI 10.4007/annals.2004.160.1141. MR2144976 (2007b:37149)

[Con99a]  Robert Conte, *The Painlevé approach to nonlinear ordinary differential equations*, The Painlevé property, CRM Ser. Math. Phys., Springer, New York, 1999, pp. 77–180. MR1713577 (2000h:34014)

[Con99b]  *The Painlevé property*, CRM Series in Mathematical Physics, Springer-Verlag, New York, 1999. One century later; Edited by Robert Conte. MR1713574 (2000e:34001)

[DST13]   Anton Dzhamay, Hidetaka Sakai, and Tomoyuki Takenawa, *Discrete Schlesinger transformations, their Hamiltonian formulation, and Difference Painlevé equations*, arXiv:1302.2972v2 [math-ph], 2013, pp. 1–29.

[FIKN06]  Athanassios S. Fokas, Alexander R. Its, Andrei A. Kapaev, and Victor Yu. Novokshenov, *Painlevé transcendents: The Riemann-Hilbert approach*, Mathematical Surveys and Monographs, vol. 128, American Mathematical Society, Providence, RI, 2006. MR2264522 (2010e:33030)

[FN80]    Hermann Flaschka and Alan C. Newell, *Monodromy- and spectrum-preserving deformations. I*, Comm. Math. Phys. **76** (1980), no. 1, 65–116. MR588248 (82g:35103)

[Fuc05]   R. Fuchs, *Sur quelques equations differentielles lineaires du second odre*, Comptes Rendus de l'Acade mie des Sciences Paris **141** (1905), 555–558.

[Fuc07]   Richard Fuchs, *Über lineare homogene Differentialgleichungen zweiter Ordnung mit drei im Endlichen gelegenen wesentlich singulären Stellen* (German), Math. Ann. **63** (1907), no. 3, 301–321, DOI 10.1007/BF01449199. MR1511408

[Gam10]   B. Gambier, *Sur les équations différentielles du second ordre et du premier degré dont l'intégrale générale est a points critiques fixes* (French), Acta Math. **33** (1910), no. 1, 1–55, DOI 10.1007/BF02393211. MR1555055

[Gar26]   R. Garnier, *Solution du problème de riemann pour les systemès différentiels linéaires du second odre*, Annales scientifiques de l'É.N.S. 3e série **43** (1926), 177–307.

[GM90]    David J. Gross and Alexander A. Migdal, *A nonperturbative treatment of two-dimensional quantum gravity*, Nuclear Phys. B **340** (1990), no. 2-3, 333–365, DOI 10.1016/0550-3213(90)90450-R. MR1068087 (91j:81078)

[GR04]    B. Grammaticos and A. Ramani, *Discrete Painlevé equations: a review*, Discrete integrable systems, Lecture Notes in Phys., vol. 644, Springer, Berlin, 2004, pp. 245–321. MR2087743 (2005g:39032)

[GR14]    ———, *Discrete Painlevé equations: an integrability paradigm*, Phys. Scr. **89** (2014), no. 038002, 13.

[GRO03]   B. Grammaticos, A. Ramani, and Y. Ohta, *A unified description of the asymmetric $q$-$P_V$ and d-$P_{IV}$ equations and their Schlesinger transformations*, J. Nonlinear Math. Phys. **10** (2003), no. 2, 215–228, DOI 10.2991/jnmp.2003.10.2.5. MR1976382 (2004c:34273)

[GRP91]   B. Grammaticos, A. Ramani, and V. Papageorgiou, *Do integrable mappings have the Painlevé property?*, Phys. Rev. Lett. **67** (1991), no. 14, 1825–1828, DOI 10.1103/PhysRevLett.67.1825. MR1125950 (92f:58081)

[HF07]    Yoshishige Haraoka and Galina Filipuk, *Middle convolution and deformation for Fuchsian systems*, J. Lond. Math. Soc. (2) **76** (2007), no. 2, 438–450, DOI 10.1112/jlms/jdm064. MR2363425 (2009e:34262)

[IKSY91]  Katsunori Iwasaki, Hironobu Kimura, Shun Shimomura, and Masaaki Yoshida, *From Gauss to Painlevé: A modern theory of special functions*, Aspects of Mathematics, E16, Friedr. Vieweg & Sohn, Braunschweig, 1991. MR1118604 (92j:33001)

[IN86]    Alexander R. Its and Victor Yu. Novokshenov, *The isomonodromic deformation method in the theory of Painlevé equations*, Lecture Notes in Mathematics, vol. 1191, Springer-Verlag, Berlin, 1986. MR851569 (89b:34012)

[JM81]    Michio Jimbo and Tetsuji Miwa, *Monodromy preserving deformation of linear ordinary differential equations with rational coefficients. II*, Phys. D **2** (1981), no. 3, 407–448, DOI 10.1016/0167-2789(81)90021-X. MR625446 (83k:34010b)

[JM82]    Michio Jimbo and Tetsuji Miwa, *Monodromy preserving deformation of linear ordinary differential equations with rational coefficients. III*, Phys. D **4** (1981/82), no. 1, 26–46, DOI 10.1016/0167-2789(81)90003-8. MR636469 (83k:34010c)

[JMMS80]  Michio Jimbo, Tetsuji Miwa, Yasuko Môri, and Mikio Sato, *Density matrix of an impenetrable Bose gas and the fifth Painlevé transcendent*, Phys. D **1** (1980), no. 1, 80–158, DOI 10.1016/0167-2789(80)90006-8. MR573370 (84k:82037)

[JMU81]   Michio Jimbo, Tetsuji Miwa, and Kimio Ueno, *Monodromy preserving deformation of linear ordinary differential equations with rational coefficients. I. General theory and $\tau$-function*, Phys. D **2** (1981), no. 2, 306–352, DOI 10.1016/0167-2789(81)90013-0. MR630674 (83k:34010a)

[JS96]    Michio Jimbo and Hidetaka Sakai, *A $q$-analog of the sixth Painlevé equation*, Lett. Math. Phys. **38** (1996), no. 2, 145–154, DOI 10.1007/BF00398316. MR1403067 (97m:39002)

[Kat96]   Nicholas M. Katz, *Rigid local systems*, Annals of Mathematics Studies, vol. 139, Princeton University Press, Princeton, NJ, 1996. MR1366651 (97e:14027)

[Kos01]   Vladimir Petrov Kostov, *The Deligne-Simpson problem for zero index of rigidity*, Perspectives of complex analysis, differential geometry and mathematical physics (St. Konstantin, 2000), World Sci. Publ., River Edge, NJ, 2001, pp. 1–35. MR1877365 (2002j:14013)

[Mur04]    Mikio Murata, *New expressions for discrete Painlevé equations*, Funkcial. Ekvac. **47** (2004), no. 2, 291–305, DOI 10.1619/fesi.47.291. MR2108677 (2005j:39014)

[NP91]     F. W. Nijhoff and V. G. Papageorgiou, *Similarity reductions of integrable lattices and discrete analogues of the Painlevé* II *equation*, Phys. Lett. A **153** (1991), no. 6-7, 337–344, DOI 10.1016/0375-9601(91)90955-8. MR1098879 (92b:58103)

[Pai02]    P. Painlevé, *Sur les équations différentielles du second ordre et d'ordre supérieur dont l'intégrale générale est uniforme* (French), Acta Math. **25** (1902), no. 1, 1–85, DOI 10.1007/BF02419020. MR1554937

[Pai73]    Paul Painlevé, *Oeuvres de Paul Painlevé. Tome I, II, III.*, Éditions du Centre National de la Recherche Scientifique,Paris, 1973, Preface by René Garnier, Compiled by Raymond Gérard, Georges Reeb and Antoinette Sec. MR0532682 (58 #27154a)

[RGH91]    A. Ramani, B. Grammaticos, and J. Hietarinta, *Discrete versions of the Painlevé equations*, Phys. Rev. Lett. **67** (1991), no. 14, 1829–1832, DOI 10.1103/PhysRevLett.67.1829. MR1125951 (92j:39011)

[Sak01]    Hidetaka Sakai, *Rational surfaces associated with affine root systems and geometry of the Painlevé equations*, Comm. Math. Phys. **220** (2001), no. 1, 165–229, DOI 10.1007/s002200100446. MR1882403 (2003c:14030)

[Sak07]    Hidetaka Sakai, *Problem: discrete Painlevé equations and their Lax forms*, Algebraic, analytic and geometric aspects of complex differential equations and their deformations. Painlevé hierarchies, RIMS Kôkyûroku Bessatsu, B2, Res. Inst. Math. Sci. (RIMS), Kyoto, 2007, pp. 195–208. MR2310030 (2008c:33020)

[Sch12]    L. Schlesinger, *Über eine Klasse von Differentialsystemen beliebiger Ordnung mit festen kritischen Punkten*, J. Reine Angew. Math. **141** (1912), 96–145.

SCHOOL OF MATHEMATICAL SCIENCES, THE UNIVERSITY OF NORTHERN COLORADO, CAMPUS BOX 122, 501 20TH STREET, GREELEY, COLORDAO 80639

*E-mail address*: adzham@unco.edu

FACULTY OF MARINE TECHNOLOGY, TOKYO UNIVERSITY OF MARINE SCIENCE AND TECHNOLOGY, 2-1-6 ETCHU-JIMA, KOTO-KU, TOKYO, 135-8533, JAPAN

*E-mail address*: takenawa@kaiyodai.ac.jp

Contemporary Mathematics
Volume **651**, 2015
http://dx.doi.org/10.1090/conm/651/13036

# Beta Ensembles, Quantum Painlevé Equations and Isomonodromy Systems

## Igor Rumanov

ABSTRACT. This is a review of recent developments in the theory of beta ensembles of random matrices and their relations with conformal filed theory (CFT). There are (almost) no new results here. equations of Belavin-Polyakov-Zamolodchikov (BPZ) type occupy the main stage. This article can serve as a guide on appearances and studies of quantum Painlevé and more general multidimensional linear equations of Belavin-Polyakov-Zamolodchikov (BPZ) type in literature. We demonstrate how BPZ equations of CFT arise from $\beta$-ensemble eigenvalue integrals. Quantum Painlevé equations are relatively simple instances of BPZ or confluent BPZ equations, they are PDEs in two independent variables ("time" and "space"). While CFT is known as quantum integrable theory, here we focus on the appearing links of $\beta$-ensembles and CFT with *classical* integrable structure and isomonodromy systems. The central point is to show on the example of quantum Painlevé II (QPII) [**94**] how classical integrable structure can be extended to general values of $\beta$ (or CFT central charge $c$), beyond the special cases $\beta = 2$ ($c = 1$) and $c \to \infty$ where its appearance is well-established. We also discuss an á priori very different important approach, the ODE/IM correspondence giving information about complex quantum integrable models, e.g. CFT, from some stationary Schrödinger ODEs. Solution of the ODEs depends on (discrete) symmetries leading to functional equations for Stokes multipliers equivalent to discrete integrable Hirota-type equations. The separation of "time" and "space" variables, a consequence of our integrable structure, also leads to Schrödinger ODEs and thus may have a connection with ODE/IM methods.

## 1. Introduction

Beta ensembles of random matrices (RM), introduced by Dyson [**40**] in 1962, find more and more applications in physics and mathematics, e.g. in conformal and integrable quantum field theories (QFT), theory of second order phase transitions, condensed matter theory in connection with conduction in disordered wires, Quantum Hall effect, anyons and fractional (exclusion) statistics. A rather comprehensive treatment can be found in [**48**]. In fact, their wide applicability originates in their very definition [**40**] as Coulomb gas (fluid) of particles-eigenvalues on the real line or in complex plane. Here we would like to concentrate on their properties related to integrability and integrable systems – both classical and quantum, with the aim toward the exact solvability of the related problems, which partly explains

2010 *Mathematics Subject Classification.* Primary 60B20, 37K10, 81T40.
The author was supported in part by NSF Grant #DMS-0905779.

the term integrability here – it is first and foremost exact solvability for us. What we mean by exact solvability should become clearer below in section 4.

There are two kinds of integrable systems (IS), often considered separately and studied by quite different methods – classical and quantum IS. The former usually involve functions and differential equations they satisfy while the latter usually deal with non-commutative algebras of operators, their eigenvalues and eigenfunctions, where the main tools are often representation-theoretic. The common integrable structure, however, is clearly seen in the existence of Lax matrices and compatibility conditions. Albeit, while classical IS usually deal with number-valued Lax matrices, for quantum IS the Lax matrices are operator-valued. A general paradigm coming from physics is to consider a quatum IS as quantization of a classical IS. While arisen historically to become a common lore, quantization procedure is often not well defined mathematically and ambiguous due to operator ordering issues. Integrability cures many of these issues, and quantum integrable theories are often better defined than other quantum theories (this is especially true of QFT).

The relationship between quantum and classical IS, however, goes deeper than just the latter being a classical limit of the former. An exact, without any limit, correspondence between quantum transfer matrices and classical discrete Hirota equations was first found in [69], and later in many other cases, see e.g. [71] and references therein. In another related venue, the ODE/IM (ordinary differential equations/integrable (quantum) models) correspondence was revealed in [15, 16, 34], see [33] for a review. There the solution of Baxter equations for quantum transfer matrices of quantum spin chains, or transfer operators in CFT [14], and their eigenvalues hinged upon finding energy spectra of certain stationary Schrödinger equations (SE). This led to the algebraic Bethe equations for the energy eigenvalues and again to discrete Hirota equations for the Stokes multipliers identified with the eigenvalues of quantum transfer matrices. We are aiming to demonstrate that Dyson beta ensembles and related canonically quantized Painlevé or Garnier equations are excellent natural models for further insight into these deep connections. In this review, we describe important related results focusing on the relatively simple examples.

Plan of the paper is as follows. Section 2 introduces the main objects of study. Simple derivation of BPZ-type equations for $\beta$-ensembles is given in section 3. In the central section 4, several representative examples of exact integrability in conformal field theory (CFT) and $\beta$-ensemble problems for general Dyson index $\beta$ are outlined. The main conjecture based on the results of [94] is that this integrability can be generalized to achieve everything mentioned in section 5 considering special cases $\beta = 2$, $\beta = 0$ and $\beta \to \infty$ where classical integrability is well-known. Section 6 briefly touches on the appearance of more general PDEs involving multidimensional diffusion-drift operators, e.g. of quantized Garnier type. Section 7 contains some concluding remarks.

**1.1. Work of Dyson and further developments.** Dyson considered [40] the Brownian Motion (BM) of eigenvalues of Gaussian $\beta$-ensemble defined by eigenvalue distribution

$$\mathbb{P}(x_1, \ldots, x_n) \sim |\Delta|^\beta \exp\left(-\frac{\beta \sum_i x_i^2}{2a^2}\right) = e^{-\beta W}, \qquad (1.1)$$

where the potential $W$ is

$$W(x_1, \ldots, x_n) = -\sum_{i<j} \ln |x_i - x_j| + \sum_i \frac{x_i^2}{2a^2}.$$

This is equivalent to the dynamics of particles in Brownian motion with positions $x_i$, subjected to an electric force $E(x_i) = -\partial_{x_i} W$ and friction with strength $f$, such that at temperature $T$ during a small time interval $\delta t$ changes in particle positions $\delta x_i$ are given by

$$f\langle \delta x_i \rangle = E(x_i)\delta t, \quad f\langle \delta x_i^2 \rangle = 2T\delta t,$$

and all higher moments are zero. The joint probability density (p.d.f.) $\mathbb{P}(x_1, \ldots, x_n; t)$ then satisfies the Fokker-Planck (FP) equation

$$f\frac{\partial \mathbb{P}}{\partial t} = \sum_i \left( T\frac{\partial^2 \mathbb{P}}{\partial x_i^2} - \frac{\partial}{\partial x_i}(E(x_i)\mathbb{P}) \right).$$

Its unique stationary solution is (1.1), i.e. the original random matrix (RM) eigenvalue distribution. The respective Brownian motinon of matrix elements is defined for $\beta = 1, 2$ or $4$ only[1] and satisfies another FP equation,

$$f\frac{\partial \mathbb{P}}{\partial t} = \sum_i \left( \frac{T}{2}g_i \frac{\partial^2 \mathbb{P}}{\partial M_i^2} + \frac{1}{a^2}\frac{\partial}{\partial M_i}(M_i\mathbb{P}) \right).$$

For an initial condition as $M = M'$ at $t = 0$, the unique solution of the last equation is

$$\mathbb{P}(M; t) = \frac{c_0}{(1-c^2)^{N/2}} \exp\left( -\frac{\text{Tr}(M - cM')^2}{2a^2 k_B T(1-c^2)} \right),$$

where $c_0$ is a numerical constant and $c = e^{-\frac{t}{a^2 f}}$. I.e. here $\beta = 1/T$. The Brownian Motion (BM) is invariant under symmetry-preserving unitary transformations of $M$. Dyson showed that the two BMs correspond to each other in the sense that the first is the BM of eigenvalues for the matrix whose entries are involved in the second one. In applications, e.g. in studies of conductivity of wires with impurities creating disorder, a similar model often arises, that of a random matrix $H$ consisting of random and non-random components,

$$H = |1 - e^{-2\tau}|^{1/2} H_{rand.} + e^{-\tau} H_0,$$

where $H_0$ is non-random, $\tau$ is the strength of disorder playing the role of time for Dyson BM [48, 97].

These FP equations are autonomous and therefore admit simple solutions. It turns out that the whole related nonlinear integrable structure can be derived from *nonautonomous* FP equations [94, 107], which is one of the main subjects of the present paper. Besides, one is more often interested in the integrals of the p.d.f. rather than the p.d.f. itself. The integrals sometimes also satisfy certain FP equations which we consider in section 3.

---

[1] Nowadays, however, starting with the seminal work [39], many matrix models leading to general $\beta$ eigenvalue distributions are known [8, 37, 70].

## 2. Beta ensembles, Virasoro constraints and quantum integrable models

Consider the integrals defining $\beta$-ensembles,

$$I_\beta([t]) = \int \cdots \int \prod_{i=1}^{M} dx_i |\Delta(x)|^\beta e^{\sum_{k=0}^{\infty} t_k \sum_{i=1}^{M} x_i^k}, \qquad \Delta(x) = \prod_{i<j}(x_i - x_j). \quad (2.1)$$

The integration here can be considered over (subsets of) $\mathbb{R}^M$. The integral satisfies the so-called Virasoro constraints derived for general $\beta$ in [9], exposing the connection of $\beta$-ensembles and conformal filed theory (CFT) with central charge

$$c = 1 - 6\frac{(1-\kappa)^2}{\kappa}. \quad (2.2)$$

The following identity expresses the Virasoro constarints

$$0 = \int \cdots \int \prod_{i=1}^{M} dx_i \sum_{i=1}^{M} \frac{\partial}{\partial x_i} \left( x_i^{n+1} |\Delta(x)|^\beta e^{\sum_{k=0}^{\infty} t_k \sum_{i=1}^{M} x_i^k} \right) = L_n I_\beta([t]), \quad (n = -1, 0, 1, \dots),$$

$$(2.3)$$

where

$$L_n = \kappa \sum_{m=0}^{n} \frac{\partial^2}{\partial t_m \partial t_{n-m}} + \sum_{m=1}^{\infty} m t_m \frac{\partial}{\partial t_{n+m}} + (1-\kappa)(n+1)\frac{\partial}{\partial t_n} \quad (2.4)$$

are the Virasoro generators making an infinite subalgebra ($m, n \geq -1$) of the Virasoro algebra with commutation relations

$$[L_n, L_m] = (n-m)L_{n+m} + \frac{c}{12}n(n^2 - 1)\delta_{n,-m}. \quad (2.5)$$

The connection of integral (2.1) with CFT is best demonstrated by introducing so called free collective field operators, or holomorphic (*chiral*) bosons [9, 68]:

$$\partial\phi(z) = \sqrt{\beta} \sum_{n=0}^{\infty} z^{-n-1} \frac{\partial}{\partial t_n} + \frac{1}{\sqrt{\beta}} \sum_{n=1}^{\infty} n t_n z^{n-1}. \quad (2.6)$$

Then the generators $L_n$ in (2.4) are the Laurent (or Fourier) modes of the holomorphic component of the CFT energy-momentum tensor [19, 36, 68]

$$T(z) = \sum_{-\infty}^{\infty} \frac{L_n}{z^{n+2}} = \frac{1}{2} : (\partial\phi(z))^2 : -\frac{1-\kappa}{\sqrt{2\kappa}}\partial^2\phi(z), \quad (2.7)$$

where here and further on $\kappa = \beta/2$, colons denote the necessary normal ordering (i.e. putting all the $t_n$ derivatives to the right) to make square of the operator generating series (2.6) well-defined.

There is also a direct relation of certain integrals of the form (2.1) with quantum Calogero-Sutherland model (CSM) [9, 48] defined by Hamiltonian for $N$ particles,

$$\mathcal{H} = -\frac{1}{2}\sum_{k=1}^{N} \frac{\partial^2}{\partial q_k^2} + \frac{\pi^2}{2L^2}\sum_{j\neq k}^{N} \frac{\kappa(\kappa-1)}{\sin^2 \pi(q_j - q_k)/L}, \quad (2.8)$$

Following [**9**, **48**], denote $y_j = e^{2\pi i q_j/L}$ and consider another multidimensional integral, the Selberg-Aomoto integral,

$$S_{M,N}(a,b,\gamma,\mu;[y]) = \int_{[0,1]^M} \prod_{i=1}^{M} dx_i \cdot \prod_{i=1}^{M}\prod_{k=1}^{N}(1-x_i y_k)^\alpha \prod_{i=1}^{M} x_i^a(1-x_i)^b \prod_{i<j}^{M} |x_i - x_j|^\gamma,$$

$$(2.9)$$

which is essentially an averaged power $\alpha$ of the product of $N$ characteristic polynomials of size $M$ Jacobi $\beta$-ensemble. The correspondence between (2.9) and (2.1) is established via the transformation of symmetric variables $y_j$ into the power sums $t_k = \sum_{j=1}^{N} y_j^k / k$,

$$\prod_{i=1}^{M}\prod_{n=1}^{N}(1-x_i y_n)^\alpha = \prod_{i=1}^{M} e^{-\alpha \sum_{k=1}^{\infty} t_k x_i^k}. \qquad (2.10)$$

Thus, for special points in the infinite space of couplings $\{t_k\}$, such that only the first $N$ of them are independent, the integral (2.1) reduces to the one of the form (2.9) or (3.1) below.

On the other hand, the CSM Hamiltonian (2.8) can be transformed as

$$\tilde{\Delta}(y)^{-\kappa}\mathcal{H}\tilde{\Delta}(y)^\kappa = 2\left(\frac{\pi}{L}\right)^2 \tilde{H} + E_0, \quad \tilde{\Delta}(y) = \prod_{i<j}^{N} \sin\pi(q_i - q_j)/L \sim \prod_i y_i^{-(N-1)/2} \prod_{i<j}^{N}(y_i - y_j),$$

$$(2.11)$$

where

$$\tilde{H} = \sum_{i=1}^{N}(y_i \partial_{y_i})^2 + \kappa \sum_{i<j}^{N} \frac{y_i + y_j}{y_i - y_j}(y_i \partial_{y_i} - y_j \partial_{y_j}), \qquad (2.12)$$

and $E_0 = (\pi/L)^2 \kappa^2 (N^3 - N)/6$ is the eigenvalue of the ground state $\tilde{\Delta}(y)^\kappa$. The eigenfunctions of $\tilde{H}$ are Jack polynomials, see e.g. [**48**, **76**]. The integral (2.9) can be expanded in a (convergent) infinite linear combination of Jack polynomials in $y$-variables with $\kappa = 2\alpha^2/\gamma$, and so is an eigenfunction of the operator $\tilde{H}$. It satisfies a multivariate generalization of hypergeometric differential equation when $\alpha = 1$ or $\alpha = -\gamma/2$, see e.g. [**48**], a special case of PDEs considered in the next section.

## 3. BPZ differential equations for $\beta$-ensembles and CFT [**4**, **10**, **78**–**80**]

As an example for PDE derivation, consider an integral

$$Z = \int \cdots \int \prod_{i=1}^{N} dx_i (z - x_i)^\alpha \Delta^\beta(x) e^{-\sum_{k=1}^{N} V(x_k)}. \qquad (3.1)$$

Denoting by $\langle\ \rangle$ the integration with the measure defined by (3.1) and differentiationg $Z$ (so that e.g. $Z = \langle\rangle$), one has

$$\partial_z Z = \left\langle \sum_{k=1}^{N} \frac{\alpha}{z - x_k} \right\rangle, \qquad (3.2)$$

$$\partial_{zz} Z = \left\langle \sum_{k=1}^{N}\left(\frac{\alpha^2 - \alpha}{(z - x_k)^2} + \sum_{j\neq k}^{N} \frac{2\alpha^2}{(z - x_k)(x_k - x_j)}\right)\right\rangle. \qquad (3.3)$$

On the other hand, consider the identity (it can be considered as a generating function of the Virasoro constraints (2.3), (2.4))

$$0 = \sum_{k=1}^{N} \int \cdots \int \prod_{i=1}^{N} dx_i \frac{\partial}{\partial x_k} \left( \frac{1}{z - x_k} \prod_{i=1}^{N} (z - x_i)^\alpha \Delta^\beta(x) e^{-\sum_{k=1}^{N} V(x_k)} \right) =$$

$$= \left\langle \sum_{k=1}^{N} \left( \frac{1 - \alpha}{(z - x_k)^2} + \sum_{j \neq k}^{N} \frac{\beta}{(z - x_k)(x_k - x_j)} - \frac{V'(x_k)}{z - x_k} \right) \right\rangle. \qquad (3.4)$$

Comparing (3.4) with (3.3), one finds another identity (by subtracting $\alpha \cdot$(3.4) from (3.3)):

$$\partial_{zz}\langle \ \rangle - \left\langle \sum_{k=1}^{N} \sum_{j \neq k}^{N} \frac{\alpha(\beta + 2\alpha)}{(z - x_k)(x_k - x_j)} \right\rangle + \left\langle \sum_{k=1}^{N} \frac{\alpha V'(x_k)}{z - x_k} \right\rangle = 0. \qquad (3.5)$$

Besides, taking another linear combination, $\beta \cdot$(3.2)$+2\alpha^2 \cdot$(3.4) leads to one more identity,

$$\beta \partial_{zz}\langle \ \rangle + \left\langle \sum_{k=1}^{N} \frac{\alpha(1 - \alpha)(\beta + 2\alpha)}{(z - x_k)^2} \right\rangle - \left\langle \sum_{k=1}^{N} \frac{2\alpha^2 V'(x_k)}{z - x_k} \right\rangle = 0. \qquad (3.6)$$

The identities (3.5) and (3.6) can be further simplified by taking the appropriate special values of $\alpha$ and killing the complicated averaged sums. E.g. taking $\alpha = 1$ in (3.6) and using (3.2) yields

$$\frac{\beta}{2} \partial_{zz}\langle \ \rangle - V'(z)\partial_z\langle \ \rangle + \left\langle \sum_{k=1}^{N} \frac{V'(z) - V'(x_k)}{z - x_k} \right\rangle = 0. \qquad (3.7)$$

Similarly, taking $\alpha = -\beta/2$ in (3.5) leads to

$$\partial_{zz}\langle \ \rangle + V'(z)\partial_z\langle \ \rangle + \frac{\beta}{2} \left\langle \sum_{k=1}^{N} \frac{V'(z) - V'(x_k)}{z - x_k} \right\rangle = 0. \qquad (3.8)$$

The last term in (3.7) or (3.8) for many important potentials can be written explicitly in terms of derivatives of $Z$ w.r.t. the coupling parameters. For instance, the so-called multi-Penner potential,

$$V(x) = C(N, \beta) \cdot \sum_{l=1}^{n} m_l \ln(x - w_l). \qquad (3.9)$$

gives

$$\left\langle \sum_{k=1}^{N} \frac{V'(z) - V'(x_k)}{z - x_k} \right\rangle = C(N, \beta) \sum_{l=1}^{n} \frac{m_l}{z - w_l} \left\langle \sum_{k=1}^{N} \frac{1}{x_k - w_l} \right\rangle = \sum_{l=1}^{n} \frac{1}{z - w_l} \frac{\partial}{\partial w_l} \langle \ \rangle, \qquad (3.10)$$

where we used that

$$\partial_{w_l} e^{-C \sum_{k=1}^{N} m_l \ln(x_k - w_l)} = C m_l \sum_{k=1}^{N} 1/(x_k - w_l) e^{-C \sum_{k=1}^{N} m_l \ln(x_k - w_l)}.$$

Thus, we obtain the nonlinear PDEs satisfied by $Z$ for the two special values of $\alpha$, $\alpha = 1$ and $\alpha = -\beta/2$, respectively:

$$\left( \frac{\beta}{2} \partial_{zz} - C(N,\beta) \cdot \sum_{l=1}^{n} \frac{m_l}{z - w_l} \partial_z + \sum_{l=1}^{n} \frac{1}{z - w_l} \frac{\partial}{\partial w_l} \right) \langle \, \rangle_1 = 0, \qquad (3.11)$$

$$\left( \frac{2}{\beta} \partial_{zz} + \frac{2}{\beta} C(N,\beta) \cdot \sum_{l=1}^{n} \frac{m_l}{z - w_l} \partial_z + \sum_{l=1}^{n} \frac{1}{z - w_l} \frac{\partial}{\partial w_l} \right) \langle \, \rangle_{-\beta/2} = 0. \qquad (3.12)$$

These Fuchsian PDEs are essentially (up to a change due to multiplication of $Z$ (3.1) by a simple factor, see (6.1)) Belavin-Polyakov-Zamolodchikov (BPZ) equations of CFT [19] corresponding to the insertion of two types of *degenerate primary fields* (degenerate primary vertex operators) into a CFT correlation function of a product of $n + 1$ non-degenerate primaries (the $\beta$-ensemble integral with $n$ general $w_l$ corresponds to $n$ CFT primaries at their positions and another one implied to be at $\infty$). Our derivation followed [4, 78], and a similar one was used in [80] to obtain the connection between all quantum Painlevé equations and $\beta$-ensembles with certain potentials. Another important case is that of a polynomial potential $V$, $V(x) = \sum_{l=1}^{n} t_l x^l$. Then

$$\left\langle \sum_{k=1}^{N} \frac{V'(z) - V'(x_k)}{z - x_k} \right\rangle = \sum_{l=1}^{n} l t_l \left\langle \sum_{k=1}^{N} x_k^{l-1} \right\rangle = \sum_{l=1}^{n} l t_l \frac{\partial}{\partial t_{l-1}} \langle \, \rangle, \qquad (3.13)$$

which leads to the *confluent* BPZ equations related to *irregular* conformal blocks [61] now realized to be important in asymptotically free gauge theories [50].

Taking $n = 3$ and considering two of the $w_l$ as fixed (usually at 0 and 1) in (3.11) and (3.12) leads to two quantum Painlevé VI equations with two different dual values for the Planck constant, $\hbar = \beta/2$ and $\hbar = 2/\beta$, respectively. The connection of the first of them with $\beta$-ensembles was demonstrated e.g. in [80]. This explains why the different dual values appeared in different contexts: the $\beta$-ensembles of Nagoya [80] contain the averaged characteristic polynomials, i.e. there $\alpha = 1$, which leads to PDEs of type (3.11) with $\hbar = \beta/2$. On the contrary, the FP equations derived in the various large $N$ limits of $\beta$-ensembles by probabilists [22, 23, 41, 89–91] were related to the ensembles with external matrix source (or with "spikes" in statistics terminology). Here the formula obtained by Forrester [49] becomes very illuminating. He found the formula for the probability density of finite $N$ Laguerre (Wishart) $\beta$-ensemble with one spike, i.e. only one eigenvalue of the external source matrix different from 1,

$$P_{\beta,a,\delta}(l_1,\ldots,l_n) \sim \prod_{i<j} |l_i - l_j|^\beta \cdot \prod_{k=1}^{n} l_k^{(a+1)\beta/2-1} e^{-\beta l_k/2} \int_{-\infty}^{+\infty} e^{it} \prod_{k=1}^{n} \left( it - \frac{\delta - 1}{2\delta} l_k \right)^{-\beta/2} dt,$$

$$(3.14)$$

where $\delta$ is the spike parameter. Comparing (3.14) with the expression under the integral of (3.1) one immediately sees that the special power $-\beta/2$ corresponding to the second choice of degenerate Virasoro primary like in (3.12) is present here. Therefore this integral and all its large $N$ limits satisfy PDEs of type (3.12), i.e. with $\hbar = 2/\beta$ as was found in the just cited references by probabilistic methods. Moreover, we see now that before taking any limit, the finite Laguerre (or Gaussian) $\beta$-ensembles in external field should satisfy a PDE of this type if we manage to express the corresponding left-hand side of (3.10) in terms of derivatives w.r.t. the coupling parameters.

To summarize, introducing into a $\beta$-ensemble integral the factors corresponding to insertion of degenerate primary fields in CFT amounts to choosing a convenient generating function to probe the dependence of the integral on coupling parameters like $w_k$ in (3.1) or $t_l$ in (3.13), so that the modified integral (generating function) satisfies a relatively simple linear PDE of BPZ/FP/non-stationary Schrödinger type.

A key to revealing various symmetries and relations among different $\beta$-ensembles could become the remarkable duality formulas of Desrosiers [31],

$$
e^{-p_2(f)} \left\langle \prod_{j=1}^{n} \prod_{k=1}^{N} \left( s_j \pm i\sqrt{\frac{2}{\beta}} x_k \right) {}_0\mathcal{F}_0^{(2/\beta)}(x, 2f) \right\rangle \bigg|_{x \in G\beta E_N} =
$$

$$
= e^{-p_2(s)} \left\langle \prod_{j=1}^{n} \prod_{k=1}^{N} \left( y_j \pm i\sqrt{\frac{2}{\beta}} f_k \right) {}_0\mathcal{F}_0^{(\beta/2)}(y, 2s) \right\rangle \bigg|_{y \in G4/\beta E_n} , \qquad (3.15)
$$

$$
e^{-p_2(f)} \left\langle \prod_{j=1}^{n} \prod_{k=1}^{N} (s_j \pm x_k)^{-\beta/2} {}_0\mathcal{F}_0^{(2/\beta)}(x, 2f) \right\rangle \bigg|_{x \in G\beta E_N} =
$$

$$
= e^{-p_2(s)} \left\langle \prod_{j=1}^{n} \prod_{k=1}^{N} (y_j \pm f_k)^{-\beta/2} {}_0\mathcal{F}_0^{(2/\beta)}(y, 2s) \right\rangle \bigg|_{y \in G\beta E_n} , \qquad (3.16)
$$

where ${}_0\mathcal{F}_0^{(\gamma)}(X, Y)$ is the hypergeometric function of two matrix arguments $X$ and $Y$ (see e.g. [76] or [48]), $p_2(s) = \sum_j s_j^2$ etc., relating averaged products of $n$ characteristic polynomials in $s_j$-variables over Gaussian $\beta$-ensemble of size $N$ with external matrix source (external field) $f$ with eigenvalues $f_k$ to the ones of degree $N$ over dual $4/\beta$-ensemble of size $n$ with vectors $s$ and $f$ interchanged. There are similar formulas for Laguerre $\beta$-ensembles [31] involving function ${}_0\mathcal{F}_1$. Applying the transformation (2.10) one sees that in fact these formulas connect also various $\beta$-ensembles with non-Gaussian polynomial potentials in external fields. It would be interesting to extend these formulas e.g. to the largest eigenvalue distributions of the corresponding RME, e.g.

$$
e^{-p_2(f)} \int_{-\infty}^{t} \cdots \int_{-\infty}^{t} \prod_{j=1}^{n} \prod_{k=1}^{N} \left( s_j \pm i\sqrt{\frac{2}{\beta}} x_k \right) {}_0\mathcal{F}_0^{(2/\beta)}(x, 2f) \Delta^\beta(x) \prod_{i=1}^{N} e^{-x_i^2} dx_i .
$$
$$(3.17)$$

Shifting the integration variables $x_i \to x_i + t$ one obtains an integral of the form

$$
e^{-p_2(f+t)} \int_{-\infty}^{0} \cdots \int_{-\infty}^{0} \prod_{j=1}^{n} \prod_{k=1}^{N} \left( \left( s_j \mp it\sqrt{\frac{2}{\beta}} \right) \pm i\sqrt{\frac{2}{\beta}} x_k \right) {}_0\mathcal{F}_0^{(2/\beta)}(x, 2(f+t)) \Delta^\beta(x) \prod_{i=1}^{N} e^{-x_i^2} dx_i .
$$
$$(3.18)$$

Then, redefining the characteristic polynomial and external source variables by the opposite shift, $s_j \to s_j - t$, $f_k \to f_k - t$, one returns to the integral of the original form but over the intervals $(-\infty, 0]$ instead of $(-\infty, t]$. Thus, the problem reduces to the possibility of extension of the dualities between the integrals over the whole $\mathbb{R}^N$ to the ones over the single $N$-dimensional orthant. The idea of such shift of integration variables (in the opposite direction) was recently used in [86] to obtain information on gauge theory instanton partition function and a phase transition for it from the Gaussian unitary ensemble (GUE) largest eigenvalue probability and the Tracy-Widom distribution [102] in its large matrix size limit.

Integrals of the form (2.1) also recently appeared [4, 50, 51] in the string theory and AGT correspondence [6] between 2-dimensional Liouville CFT and 4-dimensional supersymmetric gauge theories, can be obtained from those with potential (3.9) by the limiting confluence procedure, when the Penner singularities $w_l$ merge. In CFT this corresponds to the introduction of degenerate or confluent primary fields, see e.g. [61]. This is quite similar to the confluence process of Garnier systems [58], where the regular singular points merge leading from the Garnier systems to the confluent Garnier systems with irregular singularities [64, 66, 96].

## 4. Results and conjectures for general $\beta$

### 4.1. Stationary Schrödinger equations and ODE/IM correspondence.
As a representative example, let us consider the system of [15, 16]. In [15], the authors proved the extended version of the earlier conjecture [34] about the relation between vacuum eigenvalues of CFT $Q$-operators (analogs of Baxter $Q$-operators for integrable lattice models and quantum spin chains, see e.g. [33] and references therein) introduced in [14], part II, and spectral determinants of one-dimensional Schrödinger equations (SE) on the half-line $x > 0$ with homogeneous potential,

$$\left(-\partial_{xx} + x^{2\alpha} + \frac{l(l+1)}{x^2}\right)\Psi(x) = E\Psi(x). \tag{4.1}$$

The vacuum eigenvalues of $Q$-operators $Q_\pm(\lambda)$ are their eigenvalues at the highest weight state $|p\rangle$ of a representation of Virasoro algebra parameterized by a momentum-like parameter $p$, the highest weight $\Delta$ is given by $\Delta = p^2/\kappa + (c-1)/24$, and $c = 1 - 6(1-\kappa)^2/\kappa$ is the central charge of the CFT. I.e.

$$\langle p|Q_\pm(\lambda)|p\rangle = \lambda^{\pm 2\pi i p/\kappa} A_\pm(\lambda, p), \tag{4.2}$$

where $\lambda$ is a spectral parameter. The analytical properties of $A_\pm(\lambda, p)$ as functions of $\lambda$ and $p$ were studied in [14]. They turned out [15, 34, 35] to correspond to the properties of the spectral determinants

$$D_\pm(E, l) = \prod_{n=1}^{\infty}\left(1 - \frac{E}{E_n^\pm}\right) \tag{4.3}$$

of (4.1), where the "plus" and "minus" signs correspond to the even and odd eigenfunctions with the ordered eigenvalues $E_n^\pm$, respectively, with the degree and the "angular momentum" parameter equal to

$$\alpha = \frac{1}{\kappa} - 1, \qquad l = \frac{2p}{\kappa} - \frac{1}{2}. \tag{4.4}$$

The relation shown to hold in [15] reads:

$$A_\pm(\lambda, p) = D_\pm(\rho\lambda^2, 2p/\kappa - 1/2), \qquad \rho = \left(\frac{2}{\kappa}\right)^{2-2\kappa}\Gamma^2(1-\kappa). \tag{4.5}$$

It holds for $\alpha > 1$, i.e. for $\kappa < 1/2$ (the range $1 \leq \beta \leq 2$, i.e. $\beta = 1, 2$, should be reached by analytic continuation from the domain $\beta < 1$, and the spectral determinants $D_\pm$ should then be defined differently from just (4.3), using the additional Weierstrass exponential factors in the infinite products, see e.g. [3] or [104]). The proof of [15] proceeds as follows.

Assuming $\Re l > -3/2$, a solution $\psi^+(x, E) = \psi(x, E, l)$ of (4.1) is uniquely specified by the asymptotic at $x \to 0$,

$$\psi(x, E, l) \to \sqrt{\frac{2\pi}{1+\alpha}} \frac{x^{l+1}}{(2+2\alpha)^{(2l+1)/(2+2\alpha)}} \Gamma(1 + (2l+1)/(2+2\alpha)) + O(x^{l+3}). \quad (4.6)$$

It can be analytically continued outside $\Re l > -3/2$, and the function $\psi^-(x, E) = \psi(x, E, -l-1)$ is a linearly independent solution of the same equation for generic $l$ since their Wronskian $W[\psi^+, \psi^-]$ is equal to

$$W[\psi^+, \psi^-] = \psi^+ \partial_x \psi^- - \psi^- \partial_x \psi^+ = 2i(q^{l+1/2} - q^{-l-1/2}), \qquad q = e^{i\pi\kappa} = e^{\frac{i\pi}{1+\alpha}}. \quad (4.7)$$

Then the asymptotics of (4.1) as $x \to +\infty$ are considered. There is a unique solution $\chi(x, E, l)$ which decays at large $x$, e.g. normalized so that

$$\chi(x, E, l) \to \frac{1}{x^{\alpha/2}} \exp\left(-\frac{x^{1+\alpha}}{1+\alpha} + O(x^{1-\alpha})\right). \quad (4.8)$$

Now the crucial role is played by the discrete symmetries of (4.1) given by the two transformations:

$$\hat{\Lambda}: \quad x \to x, \ E \to E, \ l \to -l-1; \qquad \hat{\Omega}: x \to qx, \ E \to q^{-2}E, \ l \to l. \quad (4.9)$$

The transformation $\hat{\Omega}$ applied to $\chi(x, E, l)$ yields another linearly independent solution. With the choice of [15], $\chi^-(x, E, l) = iq^{-1/2}\chi(qx, q^{-2}E, l)$, their Wronskian

$$W[\chi, \chi^-] = 2. \quad (4.10)$$

Then one matches the two asymptotics expanding the solutions $\psi^+$ and $\psi^-$ into the basis $\chi, \chi^-$, e.g.

$$\psi^+ = C(E, l)\chi + D(E, l)\chi^-, \quad (4.11)$$

where the notation $D(E, l)$ anticipates the result to be seen shortly. It is easy to see, that the action of the transformations (4.9) on the four solutions introduced is the following:

$$\hat{\Lambda}\psi^\pm = \psi^\mp; \qquad \hat{\Lambda}\chi^\pm = \chi^\pm, \quad (4.12)$$

$$\hat{\Omega}\psi^\pm = q^{1/2\pm(l+1/2)}\psi^\pm; \qquad \hat{\Omega}\chi^+ = -iq^{1/2}\chi^-, \qquad \hat{\Omega}\chi^- = -iq^{1/2}\chi^+ + u\chi^-, \quad (4.13)$$

with some coefficient $u = u(E, l)$. From (4.13) and (4.11) it follows that

$$C(E, l) = -iq^{-l-1/2}D(q^{-2}E, l), \quad (4.14)$$

and applying (4.12) to (4.11) gives

$$\psi^- = D(E, -l-1)\chi^- - iq^{l+1/2}D(q^{-2}E, -l-1)\chi^+. \quad (4.15)$$

At last, taking the Wronskian $W[\chi^+, \psi^+]$ and using (4.11) and (4.10) one finds

$$D(E, l) = \frac{1}{2}W[\chi^+, \psi^+]. \quad (4.16)$$

By (4.16), if $D(E, l) = 0$, it means that $\psi^+ \sim \chi^+$, which is an eigenfunction of discrete spectrum of (4.1), i.e. $E$ must belong to the zeros of $D^+(E, l)$ (4.3), and vice versa. Both are entire functions of $E$. Therefore also $\log(D^+(E, l)/D(E, l))$ is entire. By semiclassical asymptotics as $E \to \infty$ one finds that $\log(D^+(E, l)/D(E, l)) \to$

0 in this limit and hence $D^+(E, l) = D(E, l)$. From all the above facts one can establish (4.5) by showing that $D(E, l)$ satisfy the properties uniquely defining functions $A_+$ [14]. The bilinear relation

$$q^{l+1/2}D(q^2 E, l)D(E, -l-1) - q^{-l-1/2}D(E, l)D(q^2 E, -l-1) = q^{l+1/2} - q^{-l-1/2} \tag{4.17}$$

follows from combining (4.11), (4.14), (4.15), (4.7) and (4.10). It exactly corresponds the so-called quantum Wronskian relation which $A_\pm(\lambda, p)$ satisfy [14]. Their matching analiticity conditions can be derived from (4.16) and WKB analysis of (4.1) [15, 35], we skip the details.

The correspondence between one-dimensional SE and quantum integrable models was extended to the excited states of the Baxter $Q$-operators in [16]. Then the SE $(-\partial_{xx} + V(x))\Psi(x) = E\Psi(x)$ arises with the following modified potential:

$$V(x) = x^{2\alpha} + \frac{l(l+1)}{x^2} - 2\frac{d^2}{dx^2} \sum_{k=1}^{L} \ln(x^{2\alpha+2} - z_k), \tag{4.18}$$

where $z_k$ are $L$ pairwise different complex numbers satisfying $L$ algebraic Bethe ansatz type equations

$$\sum_{j \neq k}^{L} \frac{z_k(z_k^2 + (3+\alpha)(1+2\alpha)z_k z_j + \alpha(1+2\alpha)z_j^2)}{(z_k - z_j)^3} - \frac{\alpha z_k}{4(1+\alpha)} + \frac{(2l+1)^2 - 4\alpha^2}{16(\alpha+1)} = 0. \tag{4.19}$$

The SE with potential (4.18) is in fact equivalent to another one involving the "circular polygon potentials" (see e.g. [3] about these potentials) under the natural change of variable already considered in [74],

$$z = x^{2\alpha+2} = x^{2/\kappa} \implies x = z^{\kappa/2}.$$

In terms of variable $z$ the SE of [16] becomes

$$\left(-\partial_{zz} + \frac{\kappa - 2}{2z}\partial_z + \frac{\kappa^2 l(l+1)}{4z^2} + \frac{\kappa^2}{4z} + \sum_{k=1}^{L}\left(\frac{2}{(z-z_k)^2} + \frac{\kappa-2}{z(z-z_k)}\right) - \frac{\kappa^2 E}{4}z^{\kappa-2}\right)\Psi(x(z)) = 0, \tag{4.20}$$

and changing the wavefunction by the factor

$$\Psi(x(z)) = z^{(\kappa-2)/4}\psi(z)$$

and rearranging terms in (4.20) yields the SE for the new function $\psi(z)$:

$$\left(-\partial_{zz} + \frac{\kappa^2 l(l+1) + (\kappa-2)(6-\kappa)/4}{4z^2} + \left(\frac{\kappa^2}{4} - \sum_{k=1}^{L}\frac{\kappa-2}{z_k}\right)\frac{1}{z} + \right.$$

$$\left. + \sum_{k=1}^{L}\left(\frac{2}{(z-z_k)^2} + \frac{\kappa-2}{z_k(z-z_k)}\right) - \frac{\kappa^2 E}{4}z^{\kappa-2}\right)\psi(z) = 0. \tag{4.21}$$

This equation was obtained in [74] with different notations, e.g. $\kappa = -b^2$ where $b$ is the Liouville theory parameter considered in [74] (the real parameter $b$ corresponds to $\beta < 0$). In (4.21) we see indeed a polygon potential added to the original potential characterizing the ground state. In contrast to the case of nonstationary equation QPII considered below, here the appearance of the polygon potential is unrelated to the value of $\kappa$ which is tied with the degree $2\alpha$ of the ground state potential.

**4.2. Simplest nonstationary cases – quantum Painlevé equations.**
These are Fokker-Planck (FP) or nonstationary Schrödinger equations (NSSE) in
one "time" and one "space" variable, with Hamiltonians being the canonically quantized classical Painlevé Hamiltonians, i.e. $H(x,p) \to H(x, \hbar\partial_x)^2$. Aside from CFT
applications, their connection with $\beta$-ensembles first appeared for general beta from
two very different sources. First, special cases of large matrix size $N$ limits of certain $\beta$-ensembles – so-called soft edge and hard edge limits – were found to satisfy
the quantum PII [**22, 41, 90**] and quantum PIII [**89, 91**] equations, respectively.
In the above papers, however, the linear PDEs obtained were not identified with
quantized Painlevé equations. This was done in the second source – paper [**80**],
which started from the canonical quantization of all Painlevé Hamiltonians and
then showed that eigenvalue integrals of beta ensembles with special potentials
were particular solutions of the quantum Painlevé (QP) equations.

In [**94**], we studied the quantum Painlevé II equation (QPII) from the point
of view of possible exact quantum-classical correspondence – we posed and partly
solved the problem of finding a classical $2 \times 2$ matrix Lax pair such that one of its
eigenvector components satisfies the QPII for general $\beta$ or $\kappa \equiv \beta/2$:

$$\left(\kappa\partial_t + \partial_{xx} + (t - x^2)\partial_x\right)\mathcal{F}(t,x) = 0, \tag{4.22}$$

driven in part by the fact of existence of the pair for two special values $\beta = 2, 4$ (or
$\kappa = 1, 2$) found in [**22**] for the QPII and in [**93**] for hard edge related QPIII,

$$(\kappa t\partial_t + x^2\partial_{xx} + (ax - x^2 - 1/t)\partial_x)\mathcal{F}^H(t,x) = 0.$$

The boundary conditions ensure that the solution $\mathcal{F}(t,x)$ is a probability distribution function. Moreover, its large $x$ asymptotic $F_\beta(t)$,

$$\mathcal{F}(t,x) \to F_\beta(t) \quad \text{as } x \to +\infty, \tag{4.23}$$

is the Tracy-Widom distribution $(TW_\beta)$ for general $\beta$. The celebrated distributions
$TW_2$ [**102**], $TW_1$ and $TW_4$ [**103**] of RMT with unitary, orthogonal and symplectic
symmetry, respectively, giving it the name universally appear in certain asymptotics
of random particle system processes, see e.g. [**45**] and references therein.

We set up to find a Lax pair of the form

$$\partial_x \begin{pmatrix} \mathcal{F} \\ G \end{pmatrix} = L \begin{pmatrix} \mathcal{F} \\ G \end{pmatrix}, \qquad \partial_t \begin{pmatrix} \mathcal{F} \\ G \end{pmatrix} = B \begin{pmatrix} \mathcal{F} \\ G \end{pmatrix}, \tag{4.24}$$

where we denoted

$$L = \begin{pmatrix} L_1 & L_+ \\ L_- & L_2 \end{pmatrix}, \qquad B = \begin{pmatrix} B_1 & B_+ \\ B_- & B_2 \end{pmatrix}.$$

Then, eliminating $G$ from the first components of these equations, one obtains
another, first-order, PDE for $\mathcal{F}$:

$$\partial_t\mathcal{F} - b_+\partial_x\mathcal{F} + b_1\mathcal{F} = 0, \tag{4.25}$$

where we denoted

$$b_+ = \frac{B_+}{L_+}, \qquad b_1 = b_+L_1 - B_1. \tag{4.26}$$

---

[2]the usual ordering ambiguity of quantization is cured here by shifts of the free parameters
of Painlevé equations

Eliminating $\partial_t \mathcal{F}$ from (4.22) and (4.25), one sees that $\mathcal{F}$ satisfies also an ODE in $x$:

$$(\partial_{xx} + (t - x^2 + \kappa b_+)\partial_x - \kappa b_1))\mathcal{F} = 0, \qquad (4.27)$$

which amounts to the effective separation of variables in QPII. All this is in fact consistent in a greater generality [94], i.e. for an FP equation with general differentiable coefficients in place of QPII. It depends only upon the solvability of a closed governing system of two nonlinear PDEs for the functions $P = -\kappa b_+$ and $b = \kappa b_1$, which in QPII case reads[3] ($v(t, x) = t - x^2$):

$$\kappa \partial_t (P - v) + \partial_x (\partial_x P + P(P - v) + 2b) = 0, \qquad (4.28)$$

$$\kappa \partial_t b + \partial_{xx} b + v \partial_x b = -2b \partial_x P. \qquad (4.29)$$

We found [94] an explicit solution of this system for all integer $\kappa$ (i.e. even $\beta$), using an intuition from the known simplest cases $\kappa = 1, 2$ [22,93] and from a consideration of quite similar problem for all Painlevé equations, but not considering general $\beta$, in [107]. In [107], the importance of poles in zero-curvature equations was stressed and solutions with one and two poles, corresponding to Painlevé equations were found, starting from an FP equation like (4.22). Their solutions in fact correspond to $\kappa = 1, 2$, respectively. Our solution for all $\kappa \in \mathbb{N}$ has $\kappa$ poles:

$$b_+(t, x) = -\frac{1}{\kappa} \sum_{k=1}^{\kappa} \frac{1}{x - Q_k(t)}, \qquad (4.30)$$

$$2b_1(t, x) = \frac{1}{\kappa} \sum_{k=1}^{\kappa} \frac{\kappa Q_k' + t - Q_k^2 - 2R_k}{x - Q_k} - \frac{1}{\kappa} \sum_{k=1}^{\kappa} Q_k - \frac{1}{\kappa}\left(\frac{t^2}{2} + U(t)\right), \qquad (4.31)$$

where we denoted

$$R_k = \sum_{j \neq k}^{N} \frac{1}{Q_k - Q_j}, \qquad (4.32)$$

function $U(t)$ is defined by

$$\kappa U'(t) = -\sum_{k=1}^{\kappa} Q_k^2, \qquad (4.33)$$

and the poles $Q_k(t)$ satisfy equations of motion with Calogero interaction in external time-dependent cubic ("Painlevé II implying") potential,

$$\kappa^2 Q_k'' = -2Q_k(t - Q_k^2) + \kappa - 2 - \sum_{j \neq k} \frac{8}{(Q_k - Q_j)^3}, \qquad (4.34)$$

which, considered together with (4.33), have $\kappa$ first integrals

$$\frac{(\kappa Q_k')^2}{2} + t Q_k^2 - \frac{Q_k^4}{2} - (\kappa - 2)Q_k - \sum_{j \neq k}^{\kappa} \frac{2}{(Q_k - Q_j)^2} + U(t) -$$

$$- \sum_{j \neq k}^{\kappa} \frac{\kappa Q_k' + \kappa Q_j'}{Q_k - Q_j} + \sum_{j \neq k}^{\kappa} \sum_{l \neq k,j}^{\kappa} \frac{2}{(Q_k - Q_j)(Q_j - Q_l)} = 0, \qquad (4.35)$$

(which can be written more concisely if one uses the exchange operators between $Q_k$ particles [94]). The functions $Q_k(t)$ are in many respects similar to the co-ordinates of Garnier system [58], e.g. in their origin as apparent singularities of

---

[3]This system with $v = 0$ to the best of our knowledge first appeared in [24] where it was used to find similarity solutions to the heat equation.

equation (4.27). By comparing with the considerations of the previous subsection, one observes that the equations (4.35) play here the role of *non-stationary Bethe ansatz equations*.

Thus, the exact quantum-classical correspondence is established for quantum Painlevé II – explicitly for $\kappa \in \mathbb{N}$ and conjecturally, but plausibly, since the system (4.28), (4.29) has a Laurent series solution for all $\kappa$. The found integrable structure has already allowed us to obtain a Painlevé II representation for $TW_6$ ($\kappa = 3$) [95] which was beyond the classical integrable theory before. **In fact, see sections 3, 4.3, 6, in [95] we implicitly used the irregular "5-point conformal block with one degenerate primary"** $\sim \mathcal{F}(t, x)$ **to gain information on the irregular "4-point conformal block"** $\sim \mathcal{F}_0(t)$. Here we want to present some further facts and conjectures about the system for various $\kappa$. Recall the ODE in $x$ (4.27) that a solution of QPII must satisfy if the governing system has a solution. For $\kappa \in \mathbb{N}$ this is certainly the case. The ODE can be brought to a form of stationary Schrödinger equation by making the first derivative drop out:

$$\mathcal{F} = \Psi Y^{1/2} e^{-1/2(tx - x^3/3)} = \Psi e^{1/2 \int^x P(t,x)dx} e^{-1/2(tx - x^3/3)},$$

so that function $\Psi(x; t)$ satisfies

$$\partial_{xx}\Psi - V\Psi = 0, \qquad V = b + \frac{(P - t + x^2)^2}{4} - \frac{\partial_x(P - t + x^2)}{2}, \qquad (4.36)$$

which is explicit for $\kappa \in \mathbb{N}$:

$$V(x) = \frac{3}{4} \sum_{k=1}^{\kappa} \frac{1}{(x - Q_k)^2} + \frac{1}{2} \sum_{k=1}^{\kappa} \frac{\kappa Q_k' - R_k}{x - Q_k} - \frac{U(t)}{2} + \frac{(\kappa - 2)x}{2} - \frac{tx^2}{2} + \frac{x^4}{4}. \quad (4.37)$$

This potential is a sum of quartic potential for non-symmetric anharmonic oscillator (so has an irregular singularity at $x = \infty$) and the potential arising in studies of conformal mappings of circular polygons [3]. The appearance of the circular polygon potentials seems to be generic for conformal theory related problems, recall e.g. the higher order eigenvalues of quantum transfer operators [16] from the previous subsection. This genericity finds a natural explanation in the theory of isomonodromic deformations where the vertices of the polygons appear as *apparent singularities* i.e. the singular points of the coefficients of an ODE in the $x$-complex plane such that its solutions are meromorphic at these points [38, 47, 58]. Indeed, all the $Q_k$ in (4.37) are apparent singularities. Moreover, for general $\kappa$, all the possible singularities of the function $P(t, x)$ are simple poles and they lead to such apparent singularities for the equation (4.36).

For general $\kappa$, the governing system (4.28), (4.29) can be rewritten in terms of $P$ and the Schrödinger potential $V$ (4.36) as follows:

$$\kappa \partial_t V + P \partial_x V + 2V \partial_x P - \frac{\partial_{xxx}P}{2} = 0, \qquad (4.38)$$

$$\kappa \partial_t P + 2\partial_{xx}P + P \partial_x P + 2x(t - x^2) - (\kappa - 2) + 2\partial_x V = 0. \qquad (4.39)$$

Comparing with CSM-related governing systems arising in CFT, see section 5.2, one can see that the main difference is in the additional time-dependent free term $2x(t - x^2) - (\kappa - 2)$ here reflecting the fact that we deal with Calogero-like system in time-dependent external field.

Also, for general $\kappa$, the logarithmic $x$-anti-derivative of $P$, let us denote it $Y$, i.e.

$$P = \frac{\partial_x Y}{Y},$$

satisfies a bilinear PDE of Hirota-like form,

$$Y(\kappa \partial_t + \partial_{xx})^2 Y - ((\kappa \partial_t + \partial_{xx})Y)^2 - 2\partial_x Y \partial_x (\kappa \partial_t + \partial_{xx})Y + 2\partial_{xx} Y(\kappa \partial_t + \partial_{xx})Y +$$
$$+ Y(\kappa \partial_t f(v) \cdot Y + \partial_x f(v) \cdot \partial_x Y) + 2f(v)(Y \partial_{xx} Y - (\partial_x Y)^2) = 0, \qquad (4.40)$$

where

$$f(v) = -\int_0^x \partial_t v(t, z) dz - \partial_x v - \frac{v^2}{2} = -(\kappa - 2)x - \frac{(t - x^2)^2}{2}.$$

For $\kappa \in \mathbb{N}$, it has polynomial solutions, $Y = \prod_{k=1}^{\kappa}(x - Q_k(t))$, in terms of the above functions $Q_k(t)$. The explicit Lax pair found for $\kappa \in \mathbb{N}$, reads[4] [**94**]

$$L = \begin{pmatrix} \frac{1}{2}(-v + L_d) & Y \\ -\frac{1}{2Y}(\kappa B_d + \partial_x L_d + L_d^2/2 + f_v) & \frac{1}{2}(-v - L_d) \end{pmatrix},$$

$$B = \begin{pmatrix} \frac{1}{2}\left(-x + \frac{U(t) + t^2/2}{\kappa} + B_d\right) & -\frac{\partial_x Y}{\kappa} \\ -\frac{2L - \partial_x Y + \kappa \partial_t L_d - \kappa \partial_x B_d}{2\kappa Y} & \frac{1}{2}\left(-x + \frac{U(t) + t^2/2}{\kappa} - B_d\right) \end{pmatrix},$$

where $v = t - x^2$,

$$L_d = -Y \cdot \sum_{k=1}^{\kappa} \frac{\kappa Q'_k - 2R_k}{(x - Q_k)\prod_{j \neq k}(Q_k - Q_j)} = -\sum_{k=1}^{\kappa}(\kappa Q'_k - 2R_k)\prod_{j \neq k}\frac{x - Q_j}{Q_k - Q_j},$$

$$\kappa B_d = \sum_{k=1}^{\kappa}\frac{\kappa Q'_k - 2R_k}{x - Q_k}\left(\sum_{l=1}^{\kappa}\frac{\prod_{j \neq l}^{\kappa}(x - Q_j)}{\prod_{j \neq k}^{\kappa}(Q_k - Q_j)} - 1\right),$$

$$f_v = -\frac{x^4}{2} - tx^2 + (\kappa - 2)x - U(t).$$

Another example with similarities and important differences from the foregoing, is the system [**54**] originating from NSSE

$$i\partial_t \psi = \frac{1}{2}\left(-\partial_{xx} + x^6 - \nu x^2 + \frac{l(l+1)}{x^2}\right)\psi = 0. \qquad (4.41)$$

Here, unlike above, the potential of the NSSE does not depend on time $t$, an important difference. Starting with this equation and using Darboux transformations by its quasirational solutions of the form $\psi^{(0)} = \sum_j^M c_j \psi_j(x) \exp(-i\lambda_j t/2)$, where $\psi_j$ is quasirational solution of the corresponding SE with eigenvalue $\lambda_j$, for special values of $\nu$ and $l$, the authors derived a new NSSE with an additional potential having $m$ double poles, so that the total (time-dependent) potential is

$$V(t, x) = \frac{1}{2}\left(x^6 - \nu x^2 + \frac{l(l+1)}{x^2}\right) - 2\partial_{xx} \ln W_m(\psi_n^{(0)})$$

in terms of the Wronskian $W_m$ of $m$ independent quasirational solutions $\psi_n^{(0)}$, $n = 1, \ldots, m$, of (4.41) found in [**54**]. The function

$$\psi^{(m)}(t, x) = x^\mu \frac{\prod_{j=1}^{n_z}(x - x_j(t))^{\alpha_j}}{\prod_{k=1}^{n_p}(x - y_k(t))^{\gamma_k}}e^{-x^4/4}\exp(if(t)), \qquad \mu = -l \text{ or } l + 1,$$

---

[4]The Lax pair is not unique, in general it has two arbitrary functions of $x$ and $t$, but the presented form seems convenient for general $\kappa \in \mathbb{N}$, being polynomial in $x$.

where all $\alpha_j \in \mathbb{N}$, $\gamma_k \in \mathbb{N}$, similar in form to the solutions of the SE, solves the obtained NSSE for $\mu \in \mathbb{Z}$, $\nu = 3 + 2\mu + 2N$, $N \in \mathbb{Z}$, if $\sum_j \alpha_j - \sum_k \gamma_k = N - 3m - \mu$, the constraints $\sum_j \alpha_j x_j = \sum_k \gamma_k y_k$ and $\sum_j \alpha_j / x_j = \sum_k \gamma_k / y_k$ are satisfied, and, finally, the functions $x_j(t)$ and $y_k(t)$ satisfy the time-dependent Bethe ansatz type equations ensuring also the absence of monodromy for the solutions $\psi^{(m)}$,

$$-i\dot{x}_j + \sum_{k \neq j}^{n_z} \frac{\alpha_k}{x_j - x_k} - \sum_{k=1}^{n_p} \frac{\gamma_k}{x_j - y_k} - x_j^3 + \frac{\mu}{x_j} = 0,$$

$$i\dot{y}_k - \sum_{j=1}^{n_z} \frac{\alpha_j}{y_k - x_j} - \sum_{j \neq k}^{n_p} \frac{\gamma_j}{y_k - y_j} + \gamma_k y_k^3 - \frac{\mu}{y_j} = 0,$$

$$\sum_{j=1}^{n_z} \alpha_j x_j^2 - \sum_{k=1}^{n_p} \gamma_k y_k^2 + \dot{f} = 0$$

(compare with (4.35) and (4.33) !). In the case $\mu = 0$, $\alpha_j = \gamma_k = 1$ for all $j$ and $k$, the $x_j$ and $y_k$ decouple into similar independent subsystems of equations of motion,

$$\ddot{x}_j(t) = -3x_j^5 - (2(n_z - n_p) - 3)x_j + \sum_{l \neq j}^{n_z} \frac{2}{(x_j - x_l)^3},$$

$$\ddot{y}_k(t) = -3y_k^5 - (2(n_z - n_p) + 3)y_k + \sum_{l \neq k}^{n_p} \frac{2}{(y_k - y_l)^3}.$$

A similarity with our equations (4.34) here is especially transparent as is the important difference due to the absence of terms explicitly depending on $t$. This system in fact can be considered as a peculiar "remnant" of quantum Painlevé IV (QPIV). Indeed, starting with the form of QPIV written down e.g. in [80] and making simple change of $x$-variable $x \to x^2$ there and then changing the dependent function $\Psi$ by $\Psi = x^{1/2}\psi$ to remove the first $x$-derivative, one arrives at the equation of the form

$$\left( \kappa\partial_t + \frac{1}{4}\partial_{xx} - \frac{x^6}{4} - t\frac{x^4}{2} + (n - \kappa - 1/2 - t^2/4)x^2 + \frac{\kappa(2 - \kappa) - 3/4}{4x^2} - (\alpha + \kappa/2)t \right) \psi = 0,$$

$$(4.42)$$

with some parameters $n$ and $\alpha$ ($n$ is the integer number of a $\beta$-ensemble eigenvalues in [80]), which contains the substantial additional term $\sim tx^4$ as compared to (4.41). This implies that the system of [54] might be considered from the more general QPIV point of view. The exact relation between them, however, needs a further investigation.

**4.3. Quantum Painlevé equations in CFT.** The quantum Painlevé (QP) equations appeared naturally in the studies of quantum integrable systems and CFT. The first detailed consideration of QPVI (quantum Painlevé VI) in this context seems to date back to 1993, where it appeared in [42] in representation-theoretic studies of quantum affine algebras and Wess-Zumino-Novikov-Witten (WZNW) quantum field theory. More recently QPVI surfaced in the studies of Liouville CFT correlators in [44] where both rational and elliptic parameterizations were considered and explicit transformation between them written down. Moreover, exact solutions of QPVI for special values of the parameters were found, related with the elliptic or algebraic solutions of Painlevé VI. Liouville CFT considered by [44] has central charge $c > 25$ (i.e. $\beta < 0$), but results for CFT with $c < 1$ related to the $\beta$-ensembles can be obtained from it by careful analytic continuation.

By the principles of CFT based on representation theory of Virasoro algebra (2.5), the general 4-point correlation function of primary operators has an expression

$$\langle V_{\alpha_1}(z_1, \bar{z}_1)V_{\alpha_2}(z_2, \bar{z}_2)V_{\alpha_3}(z_3, \bar{z}_3)V_{\alpha_4}(z_4, \bar{z}_4)\rangle =$$

$$= \prod_{i<j}|z_i - z_j|^{\gamma_{ij}} \int C(\alpha_1, \alpha_2, iP + Q/2)C(Q/2 - iP, \alpha_3, \alpha_4) |\mathcal{F}_P(\{\alpha_i\}; t)|^2 \frac{dP}{2},$$

$$(4.43)$$

where $Q = b + 1/b$ with $b$ being the coupling parameter of the Liouville CFT ($b^2 = -\beta$), $\gamma_{ij}$ are certain combinations of conformal dimensions (weights), $C(\alpha, \gamma, \delta)$ are the structure constants of conformal algebra (which determine the CFT 3-point correlation functions), $\mathcal{F}_P(\{\alpha_i\}; t)$ called conformal blocks are in general known only as infinite series in the cross-ratio

$$t = \frac{(z_1 - z_2)(z_3 - z_4)}{(z_1 - z_3)(z_2 - z_4)}$$

invariant w.r.t. the global conformal i.e. projective transformations, and integration is w.r.t. the momentum parameter $P$ which determines the intermediate scaling dimensions $\Delta = P^2 + Q^2/4$ in the given scattering cross-channel. If one of the four primaries is the so-called degenerate field, i.e. $V_\alpha$ with a scaling dimension of the form

$$\alpha = \alpha_{mn} = -\frac{mb}{2} - \frac{n}{2b} \qquad (4.44)$$

with integer $m$ and $n$, then the correlator (4.43) satisfies a PDE which is of second order e.g. for $m = 1, n = 0$ (how it arises we saw with the $\beta$-ensemble model in section 3) and in the considered case is equivalent to a Gauss hypergeometric equation. The authors of [44] considered the case $n = 0$ and general $m$, which they studied with the help of a 5-point correlator with one degenerate field $V_{1/2b}$,

$$\langle V_{1/2b}(z)V_{\alpha_1}(0)V_{\alpha_2}(1)V_{\alpha_3}(\infty)V_{\alpha_4}(t)\rangle,$$

where the antiholomorphic dependence identical to the holomorphic one is suppressed in the notation. The last function satisfies a QPVI equation, which in the elliptic parameterization takes Schrödinger form

$$\left(\frac{4i}{\pi b^2}\partial_\tau + \partial_{uu} - \sum_{k=1}^{4} s_k(s_k + 1)\mathcal{P}(u - \omega_k) + \sum_{k=1}^{3} s_k(s_k + 1)\mathcal{P}(\omega_k)\right)\Psi(u, \tau), \quad (4.45)$$

where $\mathcal{P}(y)$ is the Weierstrass elliptic function,

$$\tau = i\frac{K(1-t)}{K(t)}, \qquad u = \frac{\pi}{4K(t)}\int_0^{\frac{z-t}{t(z-1)}} \frac{dr}{\sqrt{r(1-r)(1-tr)}}, \qquad (4.46)$$

$K(t)$ is the elliptic integral of the first kind,

$$K(t) = \frac{1}{2}\int_0^1 \frac{dr}{\sqrt{r(1-r)(1-tr)}},$$

$\omega_k$ are half-periods

$$\omega_1 = \frac{\pi}{2}, \qquad \omega_2 = \frac{\pi\tau}{2}, \qquad \omega_3 = \omega_1 + \omega_2, \qquad \omega_4 = 0,$$

$s_k$ are related to parameters $\alpha_k$ as

$$\alpha_k = \frac{Q}{2} - \frac{b}{2}\left(s_k + \frac{1}{2}\right), \qquad (4.47)$$

and the 5-point correlator is given in terms of $\Psi$ and several factors containing elliptic $\Theta_1$ function,

$$\langle V_{1/2b}(z)V_{\alpha_1}(0)V_{\alpha_2}(1)V_{\alpha_3}(\infty)V_{\alpha_4}(t)\rangle$$

$$= (z(z-1))^{1/b^2}\frac{(z(z-1)(z-t))^{1/4}}{(t(t-1))^{\frac{8\Delta(\alpha_4)+1}{12}}}\frac{(\Theta_1(u))^{1/b^2}}{(\Theta'(0))^{(1+1/b^2)/3}}\Psi(u,\tau).$$

The authors noted that for special values of parameters $s_k$,

$$s_k = m_k + \frac{2n_k}{b^2},\quad m_k, n_k \in \mathbb{Z},$$

the general solution of the QPVI (4.45) can be obtained from the general solution of the corresponding heat equation, i.e. (4.45) without the elliptic potential. Using this, they found explicitly exact solutions of QPVI for $s_1 = s_2 = s_3 = 0, s_4 = m + 2n/b^2$ in terms of $m+n$-dimensional integrals of elliptic functions and obtained some 4-point conformal blocks as limits from them. This QPVI for integer values of all $s_k$ also appeared in the studies of special symmetry point of eight-vertex model [17,18] where also some special solutions have been obtained in terms of elliptic functions and their relation with elliptic solutions of Painlevé VI was revealed. For integer $s_k$ the potential is a finite gap potential. More general family of special solutions for the whole 4-dimensional lattice of integer $s_k$ and the corresponding Painlevé VI $\tau$-functions have later been found in [92]. Thus, for QPVI with any value of $\kappa$ (or $b$), the special values of Painlevé VI related four parameters admit relatively simple exact solutions, just like for classical Painlevé VI itself.

Exactly as with classical Painlevé equations, it is possible to obtain all the other QP equations from QPVI by confluence of singularities procedure. This way e.g. the exponential form of QPIII was obtained from (4.45) in [78].

## 5. Special cases $\beta = 2(c = 1)$ and $c \to \infty$ ($\beta \to 0$ or $\beta \to \infty$)

In these two cases many stronger results have been obtained than for general $\beta$ (or central charge $c$). The main point we wish to stress is that most of the results described below must be possible to extend to the general $\beta$ and related theories by means of the exact quantum-classical correspondence like the one we revealed in [94] and discuss in section 4.2. Indeed, Garnier systems (both regular and confluent) can be naturally considered as compatibility conditions of linear systems of PDEs. The linear PDEs involved are equivalent to the BPZ (or Knizhnik-Zamolodchikov (KZ) [67]) equations of a (confluent) CFT for special values of central charge $c$, i.e. effectively $\beta$, usually for $\beta = 2$, but also in the limits as $\beta \to \infty$ or $\beta \to 0$. We propose that a similar equivalence may hold for general values of $\beta$, with some type of (modified) Garnier or more general isomonodromy systems. A concrete example of this we encountered studying quantum Painlevé II equation (QPII) related to the soft edge large size limit of $\beta$-ensembles, for all even integer $\beta$ [94] and conjecturally (not completely explicitly yet) for the other $\beta$, see section 4.2. What leads us to this conjecture is the existence of classical integrable structure for general $\beta$ implied by the results of [94] as well as the appearance and crucial importance of a number of *apparent singular points* of differential equations, both in BPZ equations and in isomonodromy systems [38,47,58]. The addition of a sufficient number of apparent singularities ensures the solvability of a general Riemann-Hilbert problem of finding a Fuchsian ODE with given singularities and their monodromies [47,58].

**5.1. Results for $\beta = 2$ ($c = 1$) – traditional exact classical integrability.**
The isomonodromic deformations of Fuchsian ODEs leading to the Garnier systems
are excellently described in the book [**58**]. E.g. Heun equation with one additional
apparent singularity at the value of the *accessory parameter q*,

$$\partial_{xx}y + \left(\frac{c}{x} + \frac{d}{x-1} + \frac{e}{x-t} - \frac{1}{x-q}\right)\partial_x y + \left(\frac{ab}{x(x-1)} + \frac{ht(t-1)}{x(x-1)(x-t)} + \frac{pq(q-1)}{x-q}\right),$$
$$(5.1)$$

where $p(t)$ is the classical momentum for the coordinate $q(t)$ and $h(q,p,t)$ is the
associated Painlevé VI Hamiltonian,

$$h = \frac{q(q-1)(q-t)p^2 + ((c-1)(q-1)(q-t) + (d-1)q(q-t) + eq(q-1))p + ab(q-t)}{t(t-1)},$$

describes the isomonodromic deformations of Heun equation itself leading to Painlevé
VI satisfied by function $q(t)$. Equation (5.1) is the equation of the type (4.27) for
QPVI with $\hbar = \kappa = 1$ which itself is the nonstationary FP equation

$$\hbar\partial_t y = h(x, \hbar\partial_x, t)y$$

with $h(q,p,t)$ above. All types of the confluent Heun equations with one apparent
pole added similarly lead to the other Painlevé equations as written out e.g. in [**98**].
As is clear now from the results of [**94**], this is, for QPVI in place of QPII, exactly
the simplest case $\kappa = \beta/2 = 1$ (i.e. $\hbar = 1$) of the series of cases with $\kappa \in \mathbb{N}$ where the
classical Lax pairs related to ODEs with $\kappa$ such apparent singularities are available.
Some confluent Garnier systems, with two or more "time" variables like $t$ above,
were studied e.g. in [**64, 66, 96**]. Generic determinantal solutions of Garnier system
have been recently found in [**77**].

The idea of quantized Painlevé equations in the sense considered here, i.e. as
FP or NSSE equations in one space and one time dimensions, appeared in the
papers of Suleimanov (e.g. [**99**]), long before the recent prominent applications.
He used the scalar Garnier Lax pair for Painlevé II (PII) to work out the connection of QPII with classical integrability and PII itself. A restricted version of
governing system [**94**] also appeared there. The connection of QPII with PII works
this way only for $\hbar = 1$ ($\beta = 2$). Later the relations between Painlevé equations
and NSSE satisfied by solutions of the associated linear problem were developed in
much detail in the works of Zabrodin and Zotov [**106, 107**], using various parameterizations (trigonometric/exponential and elliptic besides usual rational) and ideas
of Painlevé-Calogero correspondence [**73**] bringing the classical Painlevé equations
into the form of standard Newton equations of motion in special Calogero-like but
time-dependent potentials. There again effectively $\hbar = \kappa = 1$, except for the case
of two poles in [**107**], which may correspond to $\kappa = 2$, see section 4.2, or $\kappa = 1/2$
depending on the type of PDE like in section 3.

Other type of relevant results for $\beta = 2$ came from random matrix theory
and its applications in statistics studying large random samples of large number of
(correlated) variables. E.g. sample covariance matrix eigenvalue p.d.f. leads to the
integrals similar in form to the left-hand side of (3.17) with $n = 0$. For $\beta = 2$, phase
transition from distribution $TW_2$ through critical 2-variable distribution $\mathcal{F}(t,x)_{\beta=2}$
of section 4.2 to Gaussian distribution for several largest eigenvalues was rigorously
shown and quantified in [**11, 12**]. Using integrable structure, e.g. Darboux transformations, [**20**] extended these considerations to express multiparameter critical
distributions, related to certain determinantal random point processes and multi-
matrix Dyson BMs, in terms of simpler distributions like $\mathcal{F}(t,x)_{\beta=2}$. Some results,

e.g. the BBP[12] phase transition and description of critical distributions for all $\beta$, have already been achieved for the integrals of the type (3.15) [32], which is another indication that further generalizations to all $\beta$ are possible.

Generating functions of entries of generic Hankel determinant solutions to Painlevé II and IV were considered in [62, 63]. They were shown to be simply related to an eigenvector component of the corresponding Lax pairs. E.g. function $\mathcal{F}(t,x)$ of eq. (4.22) with $\kappa = 1$ appeared in [62] where log-derivative of the first eigenvector component $Y_1$ of traceless Lax pair for Painlevé II [46, 59] was identified as such a generating function. In our normalization $\mathcal{F}(t,x) = F_2(t)\exp(-1/2(tx - x^3/3))Y_1$. It is an interesting open question if the Hankel determinant structure coming from Toda equations satisfied by Painlevé $\tau$-functions can be generalized to the other values of $\kappa = \beta/2$.

Remarkable expansions of generic $\tau$-functions for Painlevé VI, V and III around their regular (and irregular for Painlevé III) singular points with coefficients given by discrete Fourier transforms of Virasoro conformal blocks were conjectured in [52, 53] using the combinatorial expansion of conformal blocks from the proof of AGT [6] correspondence by [5] via representation theory of tensor product of Virasoro and Heisenberg algebras. The conformal expansions of Painlevé $\tau$-functions have the form

$$\tau(t) = \sum_{n\in\mathbb{Z}} C(\{\theta_i\}; \sigma + n)s^n \mathcal{F}_{c=1}[\{\theta_i\}; \sigma + n](t), \qquad (5.2)$$

where $\theta_i$ are the monodromy exponents at the singular points $i$ (e.g. $\theta_0, \theta_t, \theta_1, \theta_\infty$ for the standard form of Painlevé VI), the parameters $\sigma$ and $s$ correspond to Painlevé integration constants and $\mathcal{F}_{c=1}[\{\theta_i\}; \sigma + n](t)$ are $c = 1$ CFT conformal blocks, e.g. in case of Painlevé VI those related to the holomorphic part of 4-point correlator of the Virasoro primary operators located at $0, t, 1, \infty$. The parameters $\sigma$ and $s$ were studied by Jimbo in [60] where the first terms of expansions (5.2) were found. Vastly extending his results, exact formulas for all the terms were obtained in [52, 53]. This later led to finding exact formulas for the connection coefficients between the expansions of $\tau$-functions at different singular points for Painlevé VI [56] and Painlevé III [57].

Matrix models for $\beta = 2$ were considered from $c = 1$ CFT point of view by [43] where it was shown that classical integrable structure e.g. Lax matrix considerations may give more powerful results than just the application of general but complicated topological recursion of [30]. By showing how local conformal symmetry translates into isomonodromy of linear matrix first order ODE with Lax matrix of coefficients the authors [43] gave a clearer treatment to some of the results of [52]. The classical integrable structure in the form of Lax pairs is exactly what was generalized in [94] to the other $\beta$.

Special solutions in terms of elliptic $\Theta$-functions for multitime $c = 1$ BPZ equations were obtained in [87]. This list of results could be continued – many more exist for $\beta = 2$ and everything is known in principle how to compute here.

**5.2. Results for $c \to \infty$ – the quasiclassical (WKB) limit.** The general ideas and results are well described in a comprehensive review [101]. There are two different types of this limit: $\beta \to 0$ and $\beta \to \infty$. One of them is stationary, corresponding to the Nekrasov-Shatashvili (NS) limit [78, 84] in supersymmetric gauge theories and related to quantum integrable systems like quantum spin chains. Considering dependence of the limiting SE-type equations on their free accessory parameters leads to isomonodromy systems. On the other hand, using the links to quantum (gauge) theories and their NS limits can lead to determination of accessory parameters and solving the related uniformization of Riemann surfaces problem, see e.g. [85, 88] and references therein. A different stationary limit arises when one considers large matrix size $N$ limit together with $\alpha \sim 1/N \to 0$ in the integrals of section 3 [29]. Then one obtains an SE of the form $(\hbar\partial)^2 - V_{eff})\psi = 0$ with $\hbar \sim (\beta/2 - 1)$ [29], i.e. there case $c \to 1$ becomes the further classical limit.

The other $c \to \infty$ limit leads to conventional classical integrable systems directly and is in this respect similar to the case $c = 1(\beta = 2)$, but very different phenomena like Wigner crystallization of eigenvalue positions of $\beta$-ensembles [39, 41] also occur. Which of the limits corresponds to $\beta \to 0$ and which – to $\beta \to \infty$, depends on the type of equation, (3.11) or (3.12). The stationary limit corresponds to strong diffusion, and the nonstationary – to the weak one. In the last limit there are also recent new results on connection problem for Painlevé VI arising as the equation of motion from the classical action, the limit of the logarithm of conformal block [75]. This is one of the simplest cases of the problem of uniformization and finding the accessory parameters, which are on the one hand the derivatives of the classical Liouville theory action [100] and on the other hand the values of the classical Gaudin Hamiltonians translating into the Garnier-type Hamiltonians [101].

An important example of the nonstationary limit comes from the hydrodynamic ("collective field") description of the equations of motion of the classical CSM system (with arbitrary coupling constant $g$) for *any* number of particles including the infinite number of particles limit. It turns out [2] to be equivalent to the complex (or *bidirectional*) Benjamin-Ono (BO) equation, i.e. equation for a complex function $u$,

$$\partial_t u + \partial_x \left( \frac{u^2}{2} + i\frac{g}{2}\partial_x \bar{u} \right) = 0, \tag{5.3}$$

$$u = u_0 + u_1, \qquad \bar{u} = u_0 - u_1,$$

$$u_1 = -i\frac{\pi g}{L} \sum_{j=1}^{N} \cot \frac{\pi}{L}(x - x_j), \qquad u_0 = i\frac{\pi g}{L} \sum_{j=1}^{N} \cot \frac{\pi}{L}(x - y_j),$$

where $x_j$ are the coordinates of the CSM particles, $p_j$ are their momenta,

$$\dot{x}_j(t) = p_j, \qquad \dot{p}_j(t) = -\left(\frac{\pi g}{L}\right)^2 \partial_j \sum_{k \neq j}^{N} \left( \cot \frac{\pi}{L}(x_j - x_k) \right)^2,$$

and the auxiliary complex variables $y_j$ are implicitly defined by the equations

$$-\frac{2\pi}{L}p_j = i\frac{\dot{w}_j}{w_j} = \frac{g}{2}\left(\frac{2\pi}{L}\right)^2 \left( \sum_{k=1}^{N} \frac{w_j + u_k}{w_j - u_k} - \sum_{k \neq j}^{N} \frac{w_j + w_k}{w_j - w_k} \right),$$

where $w_j = e^{2i\pi x_j/L}$ and $u_k = e^{2i\pi y_k/L}$. The coupling constant $g$ of the classical CSM system can be always rescaled to 1 by rescaling $x$, $x_j$ and $t$. This is in contrast to the quantum CSM system considered in the similar way in [1], where the coupling constant is the physically important $\kappa = \beta/2$ parameter on which the properties of the system may crucially depend (and which cannot be removed by rescaling).

If one splits the function $u$ into the real and imaginary parts,

$$u = \mu + ir,$$

and introduces function $V = \partial_x r - r^2$, the complex BO equation can be rewritten as the real system

$$\partial_t V + \mu \partial_x V + 2V \partial_x \mu + \frac{\partial_{xxx}\mu}{2} = 0, \tag{5.4}$$

$$\partial_t \mu + \mathcal{H}\partial_{xx}\mu + \mu\partial_x\mu + \frac{\partial_x V}{2} = 0, \tag{5.5}$$

where the trigonometric Hilbert transform $\mathcal{H}$,

$$(\mathcal{H}\varphi)(z) = v.p. \int_{|w|=1} \frac{dw}{2\pi i w}\varphi(w)\frac{e^{i(z-w)/2} + e^{-i(z-w)/2}}{e^{i(z-w)/2} - e^{-i(z-w)/2}}. \tag{5.6}$$

is used. Comparing now with our governing system for QPII in the form (4.38), (4.39), we see that the first equations are similar while the second ones are different though of the same one-dimensional hydrodynamics with pressure type. The universal equation (5.4) should be the same for all such systems, in geometry it expresses the relation between a Beltrami differential $\mu$ (recall our main governing function, the ratio of upper non-diagonal entries of Lax pair matrices) and a projective connection $V$ (Schrödinger potential, holomorphic part of conformal energy-momentum tensor), see e.g. [21, 85]. The last system of PDEs appeared in the description of quantum integrals of motion whose joint eigenfunctions make the appropriate orthogonal basis for combinatorial expansion of Liouville CFT conformal blocks [5] important for establishing the AGT correspondence [6] with Nekrasov partition functions [83] of supersymmetric gauge theory. Recently these considerations were also generalized [26, 74] to the quantum Intermediate Long Wave (ILW) hierarchy (related to more general versions of AGT correspondence), which continuously interpolates between two important limiting cases of (quantum) KdV and (quantum) BO hierarchies. The first is related to the studies of [14–16], e.g. to the Schrödinger equation (4.1) considered above. The governing system for the semiclassical limit of quantum ILW is similar to (5.4), (5.5) [26, 74], differing only in the generalization of the Hilbert transform from trigonometric to elliptic version using theta-function $\Theta_1$,

$$(\mathcal{H}f)(z) \to (\mathcal{T}f)(z) = v.p. \int_{|w|=1} \frac{dw}{2\pi i w} f(w)\left(\ln\Theta_1((w-z)/2|e^{-\tau})\right)',$$

with $\tau \to 0$ and $\tau \to \infty$ corresponding to the KdV and the BO limits, respectively.

## 6. Multidimensional FP operators

The following integral defines a multi-Penner $\beta$-ensemble and also, up to a simple factor, certain Liouville CFT correlation function [4, 25, 29]:

$$Z_N = \int \cdots \int \prod_{i=1}^{N} dx_i \prod_{i<j} (x_i - x_j)^\beta \prod_{i=1}^{N}\prod_{k=1}^{n}(x_i - w_k)^{-C(N,\beta)m_k} \prod_{j=1}^{l}\prod_{i=1}^{N}(z_j - x_i), \tag{6.1}$$

In Liouville CFT [4, 25, 29] it corresponds to a $\beta$-ensemble with the multi-Penner potential (3.9) with $C = -2b/g_s$, where $g_s$ is the gauge or string coupling constant, but there $\beta = -2b^2 < 0$[5] and central charge $c > 25$, see (2.2). The additional mass

---

[5]so the integrations should be performed over a contour in complex plane where the integrals converge

parameter $m_{n+1}$ of the theory implied to be at $\infty$ in CFT correlators satisfies the constraint

$$\frac{b}{g_s}\sum_{k=1}^{n+1} m_k + \frac{l}{2} = b^2 N - b^2 - 1.$$

Integrals of this type are important in the AGT correspondence [6], see e.g. [25]. Such an integral satisfies the linear partial differential equations (PDEs) derived as in section 3,

$$\left(\frac{\beta}{2}\partial_{z_i z_i} + \sum_{j\neq i}^{l} \frac{1}{z_i - z_j}(\partial_{z_i} - \partial_{z_j}) - C(N,\beta)\sum_{k=1}^{n}\frac{m_k}{z_i - w_k}\partial_{z_i} + \sum_{k=1}^{n}\frac{1}{z_i - w_k}\partial_{w_k}\right) Z_N = 0$$

(6.2)

These are the more general BPZ equations – the equations satisfied by correlation functions of CFT with $l$ "degenerate at level two" primary fields (operator generating functions) at $z_i, i = 1,\ldots,l$ included [19]. This fact was implicitly used immediately after the creation of CFT [19] in the Coulomb gas approach [36], where more general multidimensional integrals over complex eigenvalues appeared, even before the Virasoro constraints were introduced as a main tool to tackle matrix integrals related to gauge QFT and lower-dimensional quantum gravity.

On the one hand, studies of the soft edge limit of $\beta$-ensemble with external matrix source with $r$ non-zero eigenvalues ($r$ *spikes* in statistics community terminology) by probabilistic methods yielded [23] the multidimensional FP equation corresponding to QPII for $r = 1$,

$$\left[r\partial_t + \sum_{k=1}^{r}\left(\frac{2}{\beta}\frac{\partial^2}{\partial x_k^2} + (t - x_k^2)\frac{\partial}{\partial x_k} + \sum_{j\neq k}^{r}\frac{1}{x_k - x_j}\left(\frac{\partial}{\partial x_k} - \frac{\partial}{\partial x_j}\right)\right)\right]\mathcal{F}(t,x) = 0.$$

(6.3)

This implies the presence of $r$ degenerate level two primaries in the corresponding CFT correlation function. We recall that in general an ODE from CFT has first order w.r.t. the locations of non-degenerate primary operators and second order w.r.t. the degenerate level 2 primaries. Up to a change $\beta/2 \to 2/\beta$ in (6.2), the PDE (6.3) can be obtained as the sum from 1 to $r$ of triconfluent limits of equations (6.2) with $n = 3$ and $l = r$ leading to just one "time" variable (the cross-ratio of the $w_k$ including $w_4 = \infty$). Details will appear elsewhere.

A multidimensional PDE of the type (6.3) but nonconfluent (i.e. generalizing QPVI rather than QPII) and in elliptic form like (4.45) was considered by [72] and elliptic solutions for special values of $s_k$-like parameters (see (4.45)) were obtained far extending the result of [44]. It would be interesting to find all possible confluent limits of these solutions.

A linear PDE of different but related type – multidimensional "quantum isomonodromy" equation with one "time" and $N$ Garnier-like coordinates as independent variables –was proposed by Yamada [105] to describe the instanton partition function of superconformal gauge theory with symmetry group $SU(N)$.

Returning to CSM systems, if one introduces in (2.12) the power-sum operators

$$a_n = \frac{\partial}{\partial t_n}, \qquad a_{-n} = a_n^\dagger = \frac{n t_n}{\kappa}, \qquad n \geq 0$$

in terms of $t_n = \sum_j y_j^k/k$, satisfying the Heisenberg algebra $[a_n, a_m^\dagger] = \frac{n}{\kappa}\delta_{n,m}$, $(n, m > 0)$, one can rewrite the modified CSM Hamiltonian (2.12) as

$$\tilde{H}_{CSM} = \kappa \sum_{i,j\geq 1} \left( \kappa i j t_{i+j} \frac{\partial^2}{\partial t_i \partial t_j} + (i+j)t_i t_j \frac{\partial}{\partial t_{i+j}} \right) + \kappa(\kappa - 1) \sum_{i\geq 1} i^2 t_i \frac{\partial}{\partial t_i}. \quad (6.4)$$

Thus, the CSM diffusion-drift operator for infinite number of particles expressed in terms of the power sums of the particle coordinates becomes the $\kappa$-deformed *cut-and-join operator* [55]. The cut-and-join operator corresponding to $\kappa = 1$ is important in enumerative geometry and representation theory of the symmetric group. In this case, certain exponential generating function of Hurwitz numbers was shown [65] to satisfy the infinite-dimensional heat-type equation

$$\frac{\partial e^{H(s,\mathbf{t})}}{\partial s} = \frac{1}{2} \sum_{i,j\geq 1} \left( ijt_{i+j} \frac{\partial^2}{\partial t_i \partial t_j} + (i+j)t_i t_j \frac{\partial}{\partial t_{i+j}} \right) e^{H(s,\mathbf{t})}, \quad (6.5)$$

which makes its solution $e^{H(s,\mathbf{t})}$ a $\tau$-function of the KP hierarchy. Different diffusion operators of this type appear in the theory of so-called z-measures [28] in representation theory of the symmetric group.

Using the operator for general $\kappa$, the quantum CSM Hamiltonian, one can derive the correspondence between CSM with infinite number of particles and quantized Benjamin-Ono (QBO) equation [1, 81],

$$\tilde{H}_{CSM} = \frac{\kappa^3}{3} \sum_{l+m+n=0} : a_l a_m a_n : + \frac{\kappa^2(\kappa-1)}{2} \sum_{m+n=0} |n| : a_m a_n :=$$

$$= \int_{S^1} \frac{dz}{2\pi i z} \left( \frac{1}{3} : \varphi^3(z) : + \frac{\kappa - 1}{2} iz : \varphi'(z)(\mathcal{H}\varphi)(z) : \right) = H_{QBO}, \quad (6.6)$$

where the generating operator function $\varphi(z)$ is the collective field variable denoted $\partial\phi(z)$ in (2.6), and $\mathcal{H}\varphi$ is the Hilbert transform defined by (5.6). In [81, 82] exact generating function of an infinite family of commuting integrals of the quantum BO equation is derived using its quantum Lax matrix in the power-sum representation.

It seems natural to raise the question of generalization of the classical ILW governing system to the quantum case for any $\kappa = \beta/2$ which would give an exact quantum-classical correspondence similar to the one we exposed for the QPII in section 4.2. We conjecture that such a *classical* governing system exists, and its first equation is similar to (5.4), up to the change $\partial_t \to \hbar\partial_t$ as in (4.38) but the second equation – substitute for (5.5) – may change more substantially. Interesting hints about what happens can be found in [1], e.g. here likely $\hbar \sim 2/(\sqrt{\kappa} - 1/\sqrt{\kappa})$, but more study is needed.

## 7. Concluding remarks

Quantum Painlevé equations and their multidimensional generalizations are equivalent to the (confluent) BPZ equations [19] of CFT. They are satisfied by averaged powers of characteristic polynomials of general $\beta$-ensembles of random matrices which are excellent toy models to study various properties of CFT. These linear PDEs themselves carry all information about associated isomonodromic nonlinear integrable systems. Such equations establish an exact quantum-classical correspondence by means of which quantum integrable systems find their equivalent description in terms of classical integrable isomonodromy systems. Thus, these special "quantum conformal" PDEs deserve the most thorough analytic investigation.

We do not consider it but everything here should readily generalize to the case of different symmetry group (here we implicitly considered $GL(2)$) where $W$-algebras substitute the Virasoro algebra and PDEs of higher order arise, and to the $q$-deformed models where Jack functions and CSM operators are replaced by Macdonald functions and difference operators, see e.g. [**27**], and so the $q$-difference equations should naturally arise instead of the considered PDEs.

A number of topics, without which the picture drawn here is very incomplete, remained out of the scope of this paper. They include the relations between BPZ [**19**] and KZ [**67**] equations of CFT, quantum spin chains and their recently discovered connection with classical integrable hierarchies [**7**], the use of Nekrasov functions [**83**] (the other side of AGT correspondence), complex $\beta$-ensembles, Stochastic Löwner Evolutions (SLE) and their integrable description coming from that of CFT and $\beta$-ensembles, the recently emerged field of integrable probability and Macdonald processes [**27**], non-commutative integrable systems, see e.g. [**61**], and their possible description in terms of commutative ones. At last, the unifying link should be provided by the discrete Hirota bilinear equations reviewed in [**71**], hints of which appeared e.g. in [**69**], and in [**33**] and references therein, see section 4.1, and in [**7**]. We plan to address these topics in the second part of this review, in progress.

**Acknowledgments.** The author is grateful to A. Dzhamay, K. Maruno and C. Ormerod for the invitation to write this review for JMM 2014 Proceedings, to M. Ablowitz and R. Maier for useful discussions. Partial support by NSF grant DMS-0905779 is acknowledged.

## References

[1] A. Abanov, P. Wiegmann. Quantum hydrodynamics, quantum Benjamin-Ono equation, and Calogero model. *Phys. Rev. Lett.*, 95:076402, 2005; *arXiv:cond-mat/0504041.*

[2] A. G. Abanov, E. Bettelheim, and P. Wiegmann, *Integrable hydrodynamics of Calogero-Sutherland model: bidirectional Benjamin-Ono equation*, J. Phys. A **42** (2009), no. 13, 135201, 24, DOI 10.1088/1751-8113/42/13/135201. *arXiv:0810.5327.* MR2485800 (2011b:37123)

[3] M. J. Ablowitz and A. S. Fokas, *Complex variables: introduction and applications*, 2nd ed., Cambridge Texts in Applied Mathematics, Cambridge University Press, Cambridge, 2003. MR1989049 (2004f:30001)

[4] M. Aganagic, M. C. N. Cheng, R. Dijkgraaf, D. Krefl, and C. Vafa, *Quantum geometry of refined topological strings*, J. High Energy Phys. **11** (2012), 019, front matter + 52. *arXiv:1105.0630.* MR3036500

[5] V. A. Alba, V. A. Fateev, A. V. Litvinov, and G. M. Tarnopolskiy, *On combinatorial expansion of the conformal blocks arising from AGT conjecture*, Lett. Math. Phys. **98** (2011), no. 1, 33–64, DOI 10.1007/s11005-011-0503-z. *arXiv:1012.1312.* MR2836428 (2012i:81232)

[6] L. F. Alday, D. Gaiotto, and Y. Tachikawa, *Liouville correlation functions from four-dimensional gauge theories*, Lett. Math. Phys. **91** (2010), no. 2, 167–197, DOI 10.1007/s11005-010-0369-5. *arXiv:0906.3219v2.* MR2586871 (2010k:81243)

[7] A. Alexandrov, V. kazakov, S. Leurent, Z. Tsuboi, A. Zabrodin. Classical $\tau$-function for quantum spin chains. *J. High Energy Phys.*, 1309:064, 2013. *arXiv:1112.3310v2.* MR3102207; A. Alexandrov, S. Leurent, Z. Tsuboi, A. Zabrodin. The master $T$-operator for the Gaudin model and the KP hierarchy. *Nucl. Phys. B*, 883:173–223, 2014. *arXiv:1306.1111.* MR3199960.

[8] R. Allez, J.-P. Bouchaud, A. Guionnet. Invariant $\beta$-ensembles and the Gauss-Wigner crossover. *Phys. Rev. Lett.*, 109, id:094102, 2012; *arXiv:1205.3598*

[9] H. Awata, Y. Matsuo, S. Odake, and J. Shiraishi, *Collective field theory, Calogero-Sutherland model and generalized matrix models*, Phys. Lett. B **347** (1995), no. 1-2, 49–55, DOI 10.1016/0370-2693(95)00055-P. *arXiv:hep-th/9411053v3*. MR1322003 (96e:81183)

[10] H. Awata, H. Fuji, H. Kanno, M. Manabe, and Y. Yamada, *Localization with a surface operator, irregular conformal blocks and open topological string*, Adv. Theor. Math. Phys. **16** (2012), no. 3, 725–804. *arXiv:1008.0574v5*. MR3024274

[11] J. Baik, *Painlevé formulas of the limiting distributions for nonnull complex sample covariance matrices*, Duke Math. J. **133** (2006), no. 2, 205–235, DOI 10.1215/S0012-7094-06-13321-5. *arXiv:math/0504606*. MR2225691 (2007a:33019)

[12] J. Baik, G. Ben Arous, and S. Péché, *Phase transition of the largest eigenvalue for non-null complex sample covariance matrices*, Ann. Probab. **33** (2005), no. 5, 1643–1697, DOI 10.1214/009117905000000233. *arXiv:math/0403022*. MR2165575 (2006g:15046)

[13] J. Baik and E. M. Rains, *Limiting distributions for a polynuclear growth model with external sources*, J. Statist. Phys. **100** (2000), no. 3-4, 523–541, DOI 10.1023/A:1018615306992. *arXiv:math/0003130*. MR1788477 (2001h:82067)

[14] V. Bazhanov, S. Lukyanov, A. Zamolodchikov. Integrable Structure of Conformal Field Theory I, II, III. *Comm. Math. Phys.*, 177:381–398, 1996; 190:247–278, 1997; 200:297–324, 1999; *arXiv:hep-th/9412229*; *arXiv:hep-th/9604044*; *arXiv:hep-th/9805008*.

[15] V. V. Bazhanov, S. L. Lukyanov, and A. B. Zamolodchikov, *Spectral determinants for Schrödinger equation and **Q**-operators of conformal field theory*, Proceedings of the Baxter Revolution in Mathematical Physics (Canberra, 2000), J. Statist. Phys. **102** (2001), no. 3-4, 567–576, DOI 10.1023/A:1004838616921. *arXiv: hep-th/9812247v2*. MR1832065 (2002i:81233)

[16] V. V. Bazhanov, S. L. Lukyanov, and A. B. Zamolodchikov, *Higher-level eigenvalues of Q-operators and Schrödinger equation*, Adv. Theor. Math. Phys. **7** (2003), no. 4, 711–725. *arXiv: hep-th/0307108v2*. MR2039035 (2004m:81222)

[17] V. V. Bazhanov and V. V. Mangazeev, *Eight-vertex model and non-stationary Lamé equation*, J. Phys. A **38** (2005), no. 8, L145–L153, DOI 10.1088/0305-4470/38/8/L01. *arXiv: hep-th/0411094*. MR2119173 (2005j:82010)

[18] V. Bazhanov, V. Mangazeev. The eight-vertex model and Painlevé VI. *J. Phys. A*, 39:12235–12243, 2006. *arXiv:hep-th/0602122*, MR2266223 (2008a:82018); The eight-vertex model and Painlevé VI equation II: eigenvector results.*J. Phys. A*, 43:085206, 2010. *arXiv: 0912.2163*. MR2592349 (2012a:82025)

[19] A. A. Belavin, A. M. Polyakov, and A. B. Zamolodchikov, *Infinite conformal symmetry in two-dimensional quantum field theory*, Nuclear Phys. B **241** (1984), no. 2, 333–380, DOI 10.1016/0550-3213(84)90052-X. MR757857 (86m:81097)

[20] M. Bertola, M. Cafasso. Darboux transformations and random point processes. *arXiv:1401.4752*, 2014.

[21] A. Bilal, V. V. Fock, and I. I. Kogan, *On the origin of W-algebras*, Nuclear Phys. B **359** (1991), no. 2-3, 635–672, DOI 10.1016/0550-3213(91)90075-9. MR1116701 (93b:81090)

[22] A. Bloemendal and B. Virág, *Limits of spiked random matrices I*, Probab. Theory Related Fields **156** (2013), no. 3-4, 795–825, DOI 10.1007/s00440-012-0443-2. *arXiv:1011.1877*. MR3078286

[23] A. Bloemendal, B. Virag. Limits of spiked random matrices II. *arXiv:1109.3704*, 2011.

[24] G. W. Bluman and J. D. Cole, *The general similarity solution of the heat equation*, J. Math. Mech. **18** (1968/69), 1025–1042. MR0293257 (45 #2334)

[25] G. Bonelli, K. Maruyoshi, A. Tanzini. Quantum Hitchin systems via $\beta$-deformed matrix models. *arXiv:1104.4016v2*.

[26] G. Bonelli, A. Sciarappa, A. Tanzini, P. Vasko. Six-dimensional supersymmetric gauge theories, quantum cohomology of instanton moduli spaces and $gl(N)$ Quantum Intermediate Long Wave Hydrodynamics. *arXiv:1306.0659*, 2013.

[27] A. Borodin, I. Corwin. Macdonald processes. *Prob. Th. Rel. Fields*, 158:225–400, 2014; *arXiv:1111.4408*, 2011. MR3152785. A. Borodin, I. Corwin, V. Gorin, Sh. Shakirov. Observables of Macdonald processes. *arXiv:1306.0659*, 2013.

[28] A. Borodin, G. Olshanski. Infinite-dimensional diffusions as limits of random walks on partitions. *Prob. Th. Rel. Fields*, 144:281–319, 2009; *arXiv: 0706.1034.* MR2480792 (2010c:60224). G. Olshanski. Anisotropic Young diagrams and infinite-dimensional diffusion processes with Jack parameter. *Intern. Math. Res. Notices*, rnp168:1102–1166, 2010; *arXiv: 0902.3395.*

[29] J.-E. Bourgine, *Large N limit of β-ensembles and deformed Seiberg-Witten relations*, J. High Energy Phys. **8** (2012), 46 pp. *arXiv:1206.1696v2.* MR3006952

[30] B. Eynard, N. Orantin. Invariants of algebraic curves and topological expansion, *Commun. Number Th.*, 1:347–452, 2007; *arXiv:math-ph/0702045v4.* MR2346575 (2008m:14049). L. Chekhov, B. Eynard, O. Marchal. Topological expansion of the Bethe ansatz and quantum algebraic geometry. *arXiv:0911.1664.*

[31] P. Desrosiers, *Duality in random matrix ensembles for all β*, Nuclear Phys. B **817** (2009), no. 3, 224–251, DOI 10.1016/j.nuclphysb.2009.02.019. *arXiv:0801.3438.* MR2522665 (2010i:82078)

[32] P. Desrosiers, D.-Z. Liu. Scaling limits of correlations of characteristic polynomials for the Gaussian β-ensemble with external source. *Int. Math. Res. Not.*, rnu039:1–31, 2014; *arXiv:1306.4058v3.*

[33] P. Dorey, C. Dunning, and R. Tateo, *The ODE/IM correspondence*, J. Phys. A **40** (2007), no. 32, R205–R283, DOI 10.1088/1751-8113/40/32/R01. *arXiv:hep-th/0703066.* MR2370538 (2008m:81074)

[34] P. Dorey and R. Tateo, *Anharmonic oscillators, the thermodynamic Bethe ansatz and non-linear integral equations*, J. Phys. A **32** (1999), no. 38, L419–L425, DOI 10.1088/0305-4470/32/38/102. *arXiv:hep-th/9812211.* MR1733841 (2000h:81061)

[35] P. Dorey and R. Tateo, *On the relation between Stokes multipliers and the T-Q systems of conformal field theory*, Nuclear Phys. B **563** (1999), no. 3, 573–602, DOI 10.1016/S0550-3213(99)00609-4. *arXiv:hep-th/9906219.* MR1731342 (2000m:81183)

[36] Vl. Dotsenko, V. Fateev. Conformal algebra and multipoint correlation functions in 2D statistical models. *Nucl. Phys. B*, 240:312–348, 1984. MR0762194 (85i:82061). Four-point correlation functions and the operator algebra in 2D conformal invariant theories with central charge $c \leq 1$. *Nucl. Phys. B*, 251[FS13]:691–734, 1985. MR789026 (86m:81100a)

[37] A. Dubbs, A. Edelman, P. Koev, and P. Venkataramana, *The beta-Wishart ensemble*, J. Math. Phys. **54** (2013), no. 8, 083507, 20, DOI 10.1063/1.4818304. *arXiv:1305.3561.* MR3135489

[38] B. Dubrovin and M. Mazzocco, *Canonical structure and symmetries of the Schlesinger equations*, Comm. Math. Phys. **271** (2007), no. 2, 289–373, DOI 10.1007/s00220-006-0165-3. *arXiv:math/0311261v4.* MR2287909 (2008a:32015)

[39] I. Dumitriu and A. Edelman, *Matrix models for beta ensembles*, J. Math. Phys. **43** (2002), no. 11, 5830–5847, DOI 10.1063/1.1507823. *arXiv:math-ph/0206043.* MR1936554 (2004g:82044)

[40] F. J. Dyson, *A Brownian-motion model for the eigenvalues of a random matrix*, J. Mathematical Phys. **3** (1962), 1191–1198. MR0148397 (26 #5904)

[41] A. Edelman and B. D. Sutton, *From random matrices to stochastic operators*, J. Stat. Phys. **127** (2007), no. 6, 1121–1165, DOI 10.1007/s10955-006-9226-4. *arXiv:math-ph/0607038.* MR2331033 (2009b:82037)

[42] P. I. Etingof and A. A. Kirillov Jr., *Representations of affine Lie algebras, parabolic differential equations, and Lamé functions*, Duke Math. J. **74** (1994), no. 3, 585–614, DOI 10.1215/S0012-7094-94-07421-8. *arXiv:hep-th/9310083v2.* MR1277946 (96a:81043)

[43] B. Eynard, S. Ribault. Lax matrix solution of $c = 1$ Conformal Field Theory. *arXiv:1307.4865v2.*

[44] V. A. Fateev, A. V. Litvinov, A. Neveu, and E. Onofri, *A differential equation for a four-point correlation function in Liouville field theory and elliptic four-point conformal blocks*, J. Phys. A **42** (2009), no. 30, 304011, 29, DOI 10.1088/1751-8113/42/30/304011. *arXiv:0902.1331.* MR2521330 (2010h:81198)

[45] P. Ferrari. Why random matrices share universal processes with interacting particle systems? *arXiv:1312.1126v2*, 2013.

[46] H. Flaschka and A. C. Newell, *Monodromy- and spectrum-preserving deformations. I*, Comm. Math. Phys. **76** (1980), no. 1, 65–116. MR588248 (82g:35103)

[47] A. S. Fokas, A. R. Its, A. A. Kapaev, and V. Yu. Novokshenov, *Painlevé transcendents: The Riemann-Hilbert approach*, Mathematical Surveys and Monographs, vol. 128, American Mathematical Society, Providence, RI, 2006. MR2264522 (2010e:33030)

[48] P. J. Forrester, *Log-gases and random matrices*, London Mathematical Society Monographs Series, vol. 34, Princeton University Press, Princeton, NJ, 2010. MR2641363 (2011d:82001)

[49] P. Forrester. Probability densities and distributions for spiked Wishart $\beta$-ensembles. *arXiv:1101.2261v2*, 2011.

[50] D. Gaiotto. Asymptotically free $\mathcal{N} = 2$ theories and irregular conformal blocks. *arXiv:0908.0307*.

[51] D. Gaiotto and J. Teschner, *Irregular singularities in Liouville theory and Argyres-Douglas type gauge theories*, J. High Energy Phys. **12** (2012), 50 pp. *arXiv:1203.1052*. MR3045283 *arXiv:1207.0787*, 2012.

[52] O. Gamayun, N. Iorgov, and O. Lisovyy, *Conformal field theory of Painlevé VI*, J. High Energy Phys. **10** (2012), 38 pp. *arXiv:1207.0787*. MR3033880

[53] O. Gamayun, N. Iorgov, and O. Lisovyy, *How instanton combinatorics solves Painlevé VI, V and IIIs*, J. Phys. A **46** (2013), no. 33, 335203, 29, DOI 10.1088/1751-8113/46/33/335203. *arXiv:1302.1832*. MR3093009

[54] J. Gibbons and A. P. Veselov, *On the rational monodromy-free potentials with sextic growth*, J. Math. Phys. **50** (2009), no. 1, 013513, 25, DOI 10.1063/1.3001604. *arXiv:0807.3501*. MR2492623 (2009k:34203)

[55] I. Goulden. A differential operator for symmetric functions and the combinatorics of multiplying transpositions. *Trans. Amer. Math. Soc.*, 344:421–440, 1994. I. Goulden, D. Jackson. Transitive factorizations into transpositions and holomorphic mappings on the sphere. *Proc. Amer. Math. Soc.*, 125:51–60, 1997.

[56] N. Iorgov, O. Lisovyy, and Yu. Tykhyy, *Painlevé VI connection problem and monodromy of c = 1 conformal blocks*, J. High Energy Phys. **12** (2013), 029, front matter+26, DOI 10.1007/JHEP12(2013)029. *arXiv:1308.4092v2*. MR3138677

[57] A. Its, O. Lisovyy, Yu. Tykhyy. Connection problem for the sine-Gordon/Painlevé III tau function and irregular conformal blocks. *arXiv:1403.1235*, 2014.

[58] K. Iwasaki, H. Kimura, S. Shimomura, and M. Yoshida, *From Gauss to Painlevé: A modern theory of special functions*, Aspects of Mathematics, E16, Friedr. Vieweg & Sohn, Braunschweig, 1991. MR1118604 (92j:33001)

[59] M. Jimbo and T. Miwa, *Monodromy preserving deformation of linear ordinary differential equations with rational coefficients. II*, Phys. D **2** (1981), no. 3, 407–448, DOI 10.1016/0167-2789(81)90021-X. MR625446 (83k:34010b)

[60] M. Jimbo, *Monodromy problem and the boundary condition for some Painlevé equations*, Publ. Res. Inst. Math. Sci. **18** (1982), no. 3, 1137–1161, DOI 10.2977/prims/1195183300. MR688949 (85c:58050)

[61] M. Jimbo, H. Nagoya, and J. Sun, *Remarks on the confluent KZ equation for $\mathfrak{sl}_2$ and quantum Painlevé equations*, J. Phys. A **41** (2008), no. 17, 175205, 14, DOI 10.1088/1751-8113/41/17/175205. MR2451670 (2009h:32023)

[62] N. Joshi, K. Kajiwara, and M. Mazzocco, *Generating function associated with the determinant formula for the solutions of the Painlevé II equation* (English, with English and French summaries), Astérisque **297** (2004), 67–78. *arXiv:nlin/0406035*. MR2135675 (2006d:34199)

[63] N. Joshi, K. Kajiwara, and M. Mazzocco, *Generating function associated with the Hankel determinant formula for the solutions of the Painlevé IV equation*, Funkcial. Ekvac. **49** (2006), no. 3, 451–468, DOI 10.1619/fesi.49.451. *arXiv:nlin/0512041*. MR2297948 (2007m:34209)

[64] H. Kawamuko, *On the holonomic deformation of linear differential equations with a regular singular point and an irregular singular point*, Kyushu J. Math. **57** (2003), no. 1, 1–28, DOI 10.2206/kyushujm.57.1. MR2069730 (2005g:34235)

[65] M. Kazarian, *KP hierarchy for Hodge integrals*, Adv. Math. **221** (2009), no. 1, 1–21, DOI 10.1016/j.aim.2008.10.017. *arXiv:math/0809.3263*. MR2509319 (2010b:53158)

[66] H. Kimura, *The degeneration of the two-dimensional Garnier system and the polynomial Hamiltonian structure*, Ann. Mat. Pura Appl. (4) **155** (1989), 25–74, DOI 10.1007/BF01765933. MR1042827 (91b:58096)

[67] V. G. Knizhnik and A. B. Zamolodchikov, *Current algebra and Wess-Zumino model in two dimensions*, Nuclear Phys. B **247** (1984), no. 1, 83–103, DOI 10.1016/0550-3213(84)90374-2. MR853258 (87h:81129)

[68] I. Kostov. Conformal field theory techniques for matrix models. *arXiv:hep-th/9907060*.

[69] I. Krichever, O. Lipan, P. Wiegmann, and A. Zabrodin, *Quantum integrable models and discrete classical Hirota equations*, Comm. Math. Phys. **188** (1997), no. 2, 267–304, DOI 10.1007/s002200050165. *arXiv:hep-th/9604080*. MR1471815 (99c:58076)

[70] M. Krishnapur, B. Rider, B. Virag. Universality of the stochastic Airy operator. *arXiv:1306.4832*.

[71] A. Kuniba, T. Nakanishi, and J. Suzuki, *T-systems and Y-systems in integrable systems*, J. Phys. A **44** (2011), no. 10, 103001, 146, DOI 10.1088/1751-8113/44/10/103001. *arXiv:1010.1344*. MR2773889 (2012c:81102)

[72] E. Langmann and K. Takemura, *Source identity and kernel functions for Inozemtsev-type systems*, J. Math. Phys. **53** (2012), no. 8, 082105, 19, DOI 10.1063/1.4745001. *arXiv:1202.3544*. MR3012626

[73] A. Levin, M. Olshanetsky. Painlevé-Calogero correspondence. *CRM series in Math. Phys.*, pp. 313–332, Springer, 2000; *arXiv:alg-geom/9706010*. MR1843578 (2002k:37106). K. Takasaki. Painlevé-Calogero correspondence revisited. *J. Math. Phys.*, 42:1443–1473, 2001; *arXiv:math/0004118*. MR1814699 (200im:3198)

[74] A. V. Litvinov, *On spectrum of ILW hierarchy in conformal field theory*, J. High Energy Phys. **11** (2013), 155 pp., DOI 10.1007/JHEP11(2013)155. *arXiv:1307.8094*. MR3132164

[75] A. Litvinov, S. Lukyanov, N. Nekrasov, and A. Zamolodchikov, *Classical conformal blocks and Painlevé VI*, J. High Energy Phys. **7** (2014), 144 pp. *arXiv:1309.4700v2*. MR3250114

[76] I. G. Macdonald, *Symmetric functions and Hall polynomials*, 2nd ed., Oxford Mathematical Monographs, The Clarendon Press, Oxford University Press, New York, 1995. With contributions by A. Zelevinsky; Oxford Science Publications. MR1354144 (96h:05207)

[77] T. Mano, *Determinant formula for solutions of the Garnier system and Padé approximation*, J. Phys. A **45** (2012), no. 13, 135206, 14, DOI 10.1088/1751-8113/45/13/135206. MR2904751

[78] A. Marshakov, A. Mironov, and A. Morosov, *On AGT relations with surface operator insertion and a stationary limit of beta-ensembles*, J. Geom. Phys. **61** (2011), no. 7, 1203–1222, DOI 10.1016/j.geomphys.2011.01.012. *arXiv:1011.4491*. MR2788323 (2012d:81294)

[79] K. Maruyoshi and M. Taki, *Deformed prepotential, quantum integrable system and Liouville field theory*, Nuclear Phys. B **841** (2010), no. 3, 388–425, DOI 10.1016/j.nuclphysb.2010.08.008. *arXiv:1006.4505*. MR2720858 (2011k:81139)

[80] H. Nagoya, *Hypergeometric solutions to Schrödinger equations for the quantum Painlevé equations*, J. Math. Phys. **52** (2011), no. 8, 083509, 16, DOI 10.1063/1.3620412. *arXiv:1109.1645*. MR2858065 (2012j:81090)

[81] M. L. Nazarov and E. K. Sklyanin, *Sekiguchi-Debiard operators at infinity*, Comm. Math. Phys. **324** (2013), no. 3, 831–849, DOI 10.1007/s00220-013-1821-z. *arXiv:1212.2781v2*. MR3123538

[82] M. Nazarov and E. Sklyanin, *Integrable hierarchy of the quantum Benjamin-Ono equation*, SIGMA Symmetry Integrability Geom. Methods Appl. **9** (2013), Paper 078, 14. *arXiv:1309.6464*. MR3141546

[83] N. Nekrasov, Seiberg-Witten prepotential from instanton counting. *Adv. Theor. Math. Phys.*, 7:831, 2004; *arXiv:hep-th/0206161*. MR2045303 (2005i:53109). N. Nekrasov, A. Okounkov, Seiberg-Witten theory and random partitions. *Progr. Math.*, 244:525–596, 2006. *arXiv:hep-th/0306238*. MR2181816 (2008a:81227)

[84] N. A. Nekrasov and S. L. Shatashvili, *Quantization of integrable systems and four dimensional gauge theories*, XVIth International Congress on Mathematical Physics, World Sci. Publ., Hackensack, NJ, 2010, pp. 265–289, DOI 10.1142/9789814304634_0015. *arXiv:0908.4052*. MR2730782 (2012b:81109)

[85] N. Nekrasov, A. Rosly, and S. Shatashvili, *Darboux coordinates, Yang-Yang functional, and gauge theory*, Nuclear Phys. B Proc. Suppl. **216** (2011), 69–93, DOI 10.1016/j.nuclphysbps.2011.04.150. *arXiv:1103.3919*. MR2851597

[86] S. Nishigaki, F. Sugino. Tracy-Widom distribution as instanton sum of 2D IIA superstrings. *arXiv:1405.1633*.

[87] D. P. Novikov, *The Schlesinger system with* 2 × 2 *matrices and the Belavin-Polyakov-Zamolodchikov equation* (Russian, with Russian summary), Teoret. Mat. Fiz. **161** (2009), no. 2, 191–203, DOI 10.1007/s11232-009-0135-y; English transl., Theoret. and Math. Phys. **161** (2009), no. 2, 1485–1496. *arXiv:1212.2781v2.* MR2667345 (2011j:34297)

[88] M. Piatek. Classical torus conformal block, $\mathcal{N} = 2^*$ twisted superpotential and the accessory parameter of Lamé equation. *J. High Energy Phys.*, 1403:124, 2014; *arXiv:1309.7672v3.*

[89] J. A. Ramírez and B. Rider, *Diffusion at the random matrix hard edge*, Comm. Math. Phys. **288** (2009), no. 3, 887–906, DOI 10.1007/s00220-008-0712-1. *arXiv:0803.2043v4.* MR2504858 (2010g:47083)

[90] J. A. Ramírez, B. Rider, and B. Virág, *Beta ensembles, stochastic Airy spectrum, and a diffusion*, J. Amer. Math. Soc. **24** (2011), no. 4, 919–944, DOI 10.1090/S0894-0347-2011-00703-0. *arXiv:math/0607331.* MR2813333 (2012c:60022)

[91] J. A. Ramírez, B. Rider, and O. Zeitouni, *Hard edge tail asymptotics*, Electron. Commun. Probab. **16** (2011), 741–752, DOI 10.1214/ECP.v16-1682. *arXiv:1109.4121.* MR2861438 (2012m:60027)

[92] H. Rosengren. Special polynomials related to the supersymmetric eight-vertex model. II. Schrödinger equation. *arXiv:1312.5879v2*, 2013.

[93] I. Rumanov, *Hard edge for β-ensembles and Painlevé III*, Int. Math. Res. Not. IMRN **23** (2014), 6576–6617. *arXiv:1212.5333.* MR3286348

[94] I. Rumanov. Classical integrability for beta-ensembles and general Fokker-Planck equations. *J. Math. Phys.*, 56:013508, 2015; *arXiv:1306.2117.*

[95] I. Rumanov. Painlevé representation of Tracy-Widom$_\beta$ distribution for $\beta = 6$. *arXiv:1408.3779*, 2014.

[96] S. Shimomura, *Pole loci of solutions of a degenerate Garnier system*, Nonlinearity **14** (2001), no. 2, 193–203, DOI 10.1088/0951-7715/14/2/301. MR1819793 (2002a:34132)

[97] B. Simons, P. Lee, B. Altshuler. Matrix models, one-dimensional fermions and quantum chaos. *Phys. Rev. Lett.*, 72:64–67, 1994. Exact results for quantum chaotic systems and one-dimensional fermions from matrix models. *Nucl. Phys. B*, 409[FS]:487–508, 1993. MR1247411 (94k:81071

[98] S. Yu. Slavyanov, *Isomonodromic deformations of Heun equations, and Painlevé equations* (Russian, with Russian summary), Teoret. Mat. Fiz. **123** (2000), no. 3, 395–406, DOI 10.1007/BF02551029; English transl., Theoret. and Math. Phys. **123** (2000), no. 3, 744–753. MR1794008 (2001i:34158)

[99] B. Suleimanov. Hamiltonian structure of Painlevé equations and the method of isomonodromic deformations. *Diff. Eqs.*, 30:726–732, 1994. "Quantizations" of the second Painlevé equation and the problem of the equivalence of its L–A pairs. *Theor. Math. Phys.*, 156:1280–1291, 2008.

[100] L. Takhtajan and P. Zograf, *Hyperbolic 2-spheres with conical singularities, accessory parameters and Kähler metrics on* $\mathcal{M}_{0,n}$, Trans. Amer. Math. Soc. **355** (2003), no. 5, 1857–1867 (electronic), DOI 10.1090/S0002-9947-02-03243-9. *arXiv:math/0112170.* MR1953529 (2003j:32031)

[101] J. Teschner, *Quantization of the Hitchin moduli spaces, Liouville theory and the geometric Langlands correspondence I*, Adv. Theor. Math. Phys. **15** (2011), no. 2, 471–564. *arXiv:1005.2846v5.* MR2924236

[102] C. A. Tracy and H. Widom, *Level-spacing distributions and the Airy kernel*, Comm. Math. Phys. **159** (1994), no. 1, 151–174. *arXiv:hep-th/9211141.* MR1257246 (95e:82003)

[103] C. A. Tracy and H. Widom, *On orthogonal and symplectic matrix ensembles*, Comm. Math. Phys. **177** (1996), no. 3, 727–754. *arXiv:solv-int/9509007.* MR1385083 (97a:82055)

[104] A. Voros. Spectral functions, special functions and the Selberg Zeta function. *Commun. Math. Phys.*, 110:439–465, 1987. MR891947 (89b:58173. Zeta-regularization for exact WKB resolution of a general 1D Schrödinger equation. *arXiv:1202.3200v2*, 2012. MR2970524

[105] Y. Yamada, *A quantum isomonodromy equation and its application to* $\mathcal{N} = 2$ SU($N$) *gauge theories*, J. Phys. A **44** (2011), no. 5, 055403, 9, DOI 10.1088/1751-8113/44/5/055403. *arXiv:1011.0292.* MR2763466 (2012b:81218)

[106] A. Zabrodin, A. Zotov. Quantum Painlevé-Calogero correspondence. *J. Math. Phys.*, 53:073507, 2012; *arXiv:1107.5672.* Quantum Painlevé-Calogero correspondence for Painlevé VI. *J. Math. Phys.*, 53:073508, 2012; *arXiv:1107.5672.*

[107] A. Zabrodin, A. Zotov. Classical-Quantum Correspondence and Functional Relations for Painlevé Equations. Constr. Approx. (2015) 41:385–423 DOI 10.1007/s00365-015-9284-4. *arXiv:1212.5813*, 2012.

DEPARTMENT OF APPLIED MATHEMATICS, CU BOULDER, BOULDER, COLORADO
*E-mail address*: igor.rumanov@colorado.edu

Contemporary Mathematics
Volume **651**, 2015
http://dx.doi.org/10.1090/conm/651/13035

# Inverse Scattering Transform for the Focusing Nonlinear Schrödinger Equation with a One-Sided Non-Zero Boundary Condition

B. Prinari and F. Vitale

ABSTRACT. The inverse scattering transform (IST) as a tool to solve the initial-value problem for the focusing nonlinear Schrödinger (NLS) equation with one-sided non-zero boundary value $q_r(t) \equiv A_r e^{-2iA_r^2 t + i\theta_r}$, $A_r \geq 0$, $0 \leq \theta_r < 2\pi$, as $x \to +\infty$ is presented. The direct problem is shown to be well-defined for NLS solutions $q(x,t)$ such that $[q(x,t) - q_r(t)\vartheta(x)] \in L^{1,1}(\mathbb{R})$ [here and in the following $\vartheta(x)$ denotes the Heaviside function] with respect to $x \in \mathbb{R}$ for all $t \geq 0$, for which analyticity properties of eigenfunctions and scattering data are established. The inverse scattering problem is formulated both via (left and right) Marchenko integral equations and as a Riemann-Hilbert problem on a single sheet of the scattering variable $\lambda_r = \sqrt{k^2 + A_r^2}$, where $k$ is the usual complex scattering parameter in the IST. The direct and inverse problems are also formulated in terms of a suitable uniformization variable that maps the two-sheeted Riemann surface for $k$ into a single copy of the complex plane. The time evolution of the scattering coefficients is then derived, showing that, unlike the case of solutions with the same amplitude as $x \to \pm\infty$, here both reflection and transmission coefficients have a nontrivial (although explicit) time dependence. The results presented in this paper will be instrumental for the investigation of the long-time asymptotic behavior of physically relevant NLS solutions with nontrivial boundary conditions, either via the nonlinear steepest descent method on the Riemann-Hilbert problem, or via matched asymptotic expansions on the Marchenko integral equations.

## 1. Introduction

Nonlinear Schrödinger (NLS) systems have been extensively investigated both mathematically and physically for almost sixty years, and remarkably continue to offer interesting research problems and new venues for applications. Equations of NLS-type have proven over the years to be fundamental for modelling nonlinear wave phenomena in such diverse fields as deep water waves [**4**, **38**], plasma physics [**34**], nonlinear fiber optics [**22**], magnetic spin waves [**16**, **40**], low temperature physics and Bose-Einstein condensates [**29**], just to mention a few. Mathematically, the scalar NLS equation is particularly relevant in view of its universal nature, since most dispersive energy preserving systems reduce to it in appropriate limits. All this clearly explains the keen interest in NLS equations as prototypical

---

2010 *Mathematics Subject Classification.* Primary 35Q55, 37K15, 47J35, 35C08.

integrable systems, and motivates the effort put into advancing our mathematical understanding of this equation.

The inverse scattering transform (IST) as a method to solve the initial-value problem for the scalar NLS equation:

(1.1)                                $iq_t = q_{xx} - 2\sigma |q|^2 q$

(subscripts $x$ and $t$ denote partial differentiation throughout) has been amply studied in the literature, both in the focusing ($\sigma = -1$) and in the defocusing ($\sigma = 1$) dispersion regimes; see, for instance, [2–4, 15, 27, 38] for detailed accounts of the IST in the case of solutions $q(x,t)$ rapidly decaying as $x \to \pm\infty$. The situation is quite different when one is interested in solutions that do not decay at space infinity. As a matter of fact, even though the IST for the focusing NLS equation with rapidly decaying potentials was first proposed more than 40 years ago, and has been subsequently the subject of a vast amount of studies and applications, not nearly as much is available in the literature in the case of nontrivial boundary conditions. The reason for this deficiency is twofold: on one hand, the technical difficulties resulting from the non-zero boundary conditions (NZBCs) significantly complicate the formulation of the IST; on the other hand, the onset of modulational instability, also known as the Benjamin-Feir instability [10, 11] in the context of water waves, was believed to be an obstacle to the development of the IST, or at least to its validity. Nonetheless, direct methods have been extensively used over the years to derive a large number of exact solutions to the focusing NLS equation with NZBCs, known as Peregrine and multi-Peregrine solutions, Akhmediev breathers, and more general solitonic solutions [6–9, 23–26, 30, 33, 35]. Lately these solutions have been the subject of a renewed interest, due to the fact that the development of modulation instability in the governing equation has been recently suggested as a mechanism for the formation of "extreme" (rogue, freak) waves, for which energy density exceeds the mean level by an order of magnitude (see, for instance, [28, 36, 37] in relation to water waves, and [32] regarding the observation of "rogue" waves in optical systems).

In view of these recent developments, it is natural to wonder about the role that soliton solutions play in the nonlinear development of the modulation instability, which makes the study of the long-time asymptotics of NLS solutions of great practical importance, crucial for developing a consistent theory for rogue waves in the ocean, and for extreme events in optical fibers. Likewise, also the investigation of the IST for the focusing case with NZBCs [Eq. (1.1) with $\sigma = -1$], i.e.

(1.2)                                $iq_t = q_{xx} + 2|q|^2 q \,,$

as a means to provide the time evolution of a fairly general initial one-dimensional pulse/wave profile over a nontrivial background, has been receiving a greater deal of attention, since it allows the study of the long-time asymptotic behavior via the nonlinear steepest descent method [14, 17], matched asymptotic expansions [1, 5], or other germane techniques.

The IST for the defocusing NLS equation with NZBCs was first studied in 1973 [39]; the problem was subsequently clarified and generalized in various works (see [18, 20] and references therein). On the other hand, to the best of our knowledge until recently the only results on the IST for the focusing NLS with NZBCs available in the literature could be found in [21, 25], which only partially address the problem since the study was limited to the case of completely symmetric boundary conditions with $\lim_{x\to+\infty} q(x,t) = \lim_{x\to-\infty} q(x,t)$, i.e., only the case in which the potential exhibits no asymptotic phase difference and no amplitude difference is treated. In [12], Biondini & Kovačić have contributed filling this gap by developing the IST for potentials with an arbitrary asymptotic phase difference, although assuming the same amplitude at both space infinities. They also discuss the general behavior of the soliton solutions, as well as the reductions to all special solutions previously known in the literature and mentioned above. The IST for focusing NLS with fully asymmetric NZBCs has been developed in [19], where different asymptotic amplitudes and phases are considered: $q(x,t) \to A_{l/r} e^{-2iA_{l/r}^2 t + i\theta_{l/r}}$ as $x \to \mp\infty$, with $A_r \geq A_l > 0$. This is a nontrivial generalization of [12], and it involves dealing with additional technical difficulties, the most important of which being the fact that when the amplitudes of the NLS solutions as $x \to \pm\infty$ are different, in the spectral domain one cannot introduce a uniformization variable [20] that allows mapping the multiply sheeted Riemann surface for the scattering parameter to a single complex plane, and which provides a remarkable simplification in the study of both direct and inverse problems.

In this work we will develop the IST for the scalar focusing NLS (1.2) with the following one-sided NZBCs:

(1.3a) $\qquad\qquad q(x,t) \to 0 \quad \text{as} \quad x \to -\infty,$

(1.3b) $\qquad\qquad q(x,t) \to q_r(t) = A_r\, e^{-2iA_r^2 t + i\theta_r} \quad \text{as} \quad x \to +\infty,$

where $A_r > 0$ and $0 \leq \theta_r < 2\pi$ are arbitrary constants. Obviously, the case of a nontrivial boundary condition only as $x \to -\infty$ can be treated in a similar fashion.

Such kind of boundary conditions are obviously outside the class considered in [12], where the amplitudes of the background field are taken to be the same at both space infinities. The problem with one-sided non-zero boundary conditions clearly has a physical relevance on its own, and, unlike what happens, for instance, for the Korteweg-de Vries (KdV) equation, for NLS one cannot set without loss of generality one of the boundary conditions to zero by performing suitable rescalings of the field. At the same time, the mathematical motivation for the present work is twofold. On one hand, in [19] we assumed $A_l > 0$, and the limit $A_l \to 0$ is a singular limit, which makes recovering the corresponding results from the fully asymmetric case far from straightforward. On the other hand, the case of one-sided NZBCs presents some specific features that make it deserving a separate investigation. In fact, unlike the case of fully asymmetric boundary conditions and similarly to the same-amplitude case dealt with in [12], with boundary conditions such as the ones in (1.3) it is still possible to introduce a uniformization variable [20] that allows mapping the multiply sheeted Riemann surface for the scattering parameter into a single complex plane. Yet, important differences with respect to the same-amplitude case arise both in the direct and in the inverse problems, and they will be properly highlighted in this work. From the point of view of physical applications, such a work would be particularly significant for the theoretical investigation of rogue waves and perturbed soliton solutions in microstructured fiber optical systems with different background amplitudes enforced at either end of the fiber. This work would also be relevant in clarifying the role that soliton solutions play in the nonlinear development of modulation instability in such systems.

The plan of the paper is outlined below. Sec. 2 is devoted to the study of the direct scattering problem on a single sheet of the scattering variables $k$, $\lambda_r = \sqrt{k^2 + A_r^2}$, where $k$ is the usual complex scattering parameter in the IST. We will show that the direct problem is well defined for potentials $q(x,t)$ such that $[q(x,t) - \vartheta(x)q_r(t)] \in L^{1,1}(\mathbb{R})$ with respect to $x \in \mathbb{R}$ and for all $t \geq 0$, where $L^{1,s}(\mathbb{R})$ is the complex Banach space of all measurable functions $f(x)$ for which $(1 + |x|)^s f(x)$ is integrable and $\vartheta(x)$ is the Heaviside function [i.e., $\vartheta(x) = 1$ for $x \geq 0$, and zero otherwise]. We will then establish analyticity of eigenfunctions and scattering data in $k$, and obtain integral representations for the latter for potentials in this class. In Sec. 3 we will formulate the inverse problem both in terms of Marchenko integral equations, and as a Riemann-Hilbert (RH) problem on a single sheet of the scattering variables $(k, \lambda_r(k))$. Important differences with respect to the symmetric case also arise in the inverse problem, where, in addition to solitons (corresponding to the discrete eigenvalues of the scattering problem), and to radiation (corresponding to the continuous spectrum of the scattering operator, and represented in the inverse problem by the reflection coefficients for $k \in \mathbb{R}$), one also has a nontrivial contribution from additional spectral data for $k \in (-iA_r, iA_r)$, which appears in both formulations of the inverse problem. In particular, this implies that no pure soliton solutions exist, and solitons are always accompanied by a radiative contribution of some sort. As a consequence, unlike the equal-amplitude case dealt with in [12], here no explicit solution can be obtained by simply reducing the inverse problem to a set of algebraic equations. In view of this, the present study provides a very powerful tool for the asymptotic investigation of NLS solutions that cannot be obtained by direct methods. Specifically, the RH formulation of the inverse problem makes it amenable to the study of the long-time asymptotic behavior via the nonlinear steepest descent method, as shown, for instance, in [17] for the modified KdV equation, or in [14] for the focusing NLS with initial condition $q(x,0) = A e^{i\mu|x|}$, $A$ and $\mu$ being positive constants. The Marchenko integral equations provide an alternative setup for the study of the long-time behavior of the solutions by means of matched asymptotics, as was recently done for KdV in [1]. Sec. 4 deals with the time evolution of eigenfunctions and scattering coefficients. In Sec. 5 we develop the direct scattering problem in terms of the uniform variable $z = k + \lambda_r$, and formulate the inverse problem as a Riemann-Hilbert problem in $z \in \mathbb{C}$. Finally, Sec. 6 is devoted to some concluding remarks.

## 2. Direct problem

It is well-known that the focusing NLS equation (1.2) can be associated to the following Lax pair:

(2.1a)
$$\frac{\partial v}{\partial x} = (-ik\sigma_3 + Q)\, v\,,$$

(2.1b)
$$\frac{\partial v}{\partial t} = \left[i(2k^2 - |q|^2 + Q_x)\sigma_3 - 2kQ\right] v\,,$$

where $v(x,k,t)$ is a two component vector, $k \in \mathbb{C}$ the scattering parameter, and

(2.2)  $\sigma_2 = \begin{pmatrix} 0 & -i \\ i & 0 \end{pmatrix}$,     $\sigma_3 = \begin{pmatrix} 1 & 0 \\ 0 & -1 \end{pmatrix}$,     $Q(x,t) = \begin{pmatrix} 0 & q(x,t) \\ -q^*(x,t) & 0 \end{pmatrix}$.

[Here and in the following the asterisk indicates complex conjugates; $\sigma_2$ is given for future reference.] We will consider potentials $q(x,t)$ with one-sided NZBC as in (1.3), where we assume that the limits exist in the standard sense, with asymptotic amplitude $A_r$ positive and time-independent, and asymptotic phase given by $\theta_r(t) = -2A_r^2 t + \theta_r$, to ensure compatibility with the NLS evolution. Furthermore, we assume the following integrability condition

$$(2.3) \qquad (\boldsymbol{H}_s): \qquad \int_{-\infty}^{\infty} dx\, (1 + |x|)^s\, |q(x,t) - q_r(t)\vartheta(x)| < \infty\,,$$

for all $t \geq 0$, where $s = 0, 1$ depending on the situation.

For later convenience, we denote by $Q_r(t)$ the limit of $Q(x,t)$ as $x \to +\infty$ [obviously, $Q(x,t) \to 0_{2\times 2}$ as $x \to -\infty$, according to (1.3)]. We also introduce the "free" potential matrix $Q_f(x,t)$ as follows:

$$(2.4) \qquad Q_f(x,t) = Q_r(t)\,\vartheta(x)\,.$$

In the formulation of the direct problem we will omit to explicitly specify the time-dependence for brevity. It will be clear from the context whether one is considering $t = 0$ or an arbitrary $t > 0$.

It is convenient to introduce the asymptotic scattering operator corresponding to the NZBC as $x \to +\infty$, namely:

$$(2.5a) \qquad \Lambda_r(k) = -ik\sigma_3 + Q_r\,,$$

as well as

$$(2.5b) \qquad \Lambda(x,k) = -ik\sigma_3 + Q_f(x) = -ik\sigma_3 + \vartheta(x)\Lambda_r(k)\,,$$

and define the *fundamental eigensolution* $\tilde{\Psi}(x,k)$ and the eigenfunction $\Phi(x,k)$ as those $2 \times 2$ matrix solutions to (2.1a) which satisfy the asymptotic conditions

$$(2.6a) \qquad \tilde{\Psi}(x,k) = e^{x\,\Lambda_r(k)}[I_2 + o(1)] \qquad x \to +\infty\,,$$

$$(2.6b) \qquad \Phi(x,k) = e^{-ikx\sigma_3}[I_2 + o(1)] \qquad x \to -\infty\,,$$

[here and in the following $I_2$ denotes the $2 \times 2$ identity matrix]. Note that because of the choice of boundary conditions (1.3), $\Phi(x,k)$ coincides with the usual pair of Jost solutions from the left for the scattering problem (cf. Sec. 2.1). On the other hand, as far as the eigensolution from the right is concerned, $e^{x\,\Lambda_r(k)}$ is a bounded group for all $x \in \mathbb{R}$ iff $\Lambda_r(k)$ has only zero or purely imaginary eigenvalues and is diagonalizable, i.e., iff $k \in \mathbb{R} \cup (-iA_r, iA_r)$. For $k = \pm iA_r$ the norm of the group $e^{x\,\Lambda_r(k)}$ grows linearly in $x$ as $x \to +\infty$. Then the following result can be established.

PROPOSITION 2.1. *Let the potential satisfy* $(\boldsymbol{H}_0)$. *Then for* $k \in \mathbb{R}$ *the eigenfunction* $\Phi(x,k)$ *is given by the unique solution to the integral equation*

$$(2.7a) \qquad \Phi(x,k) = e^{-ikx\sigma_3} + \int_{-\infty}^{x} dy\, e^{-ik(x-y)\sigma_3} Q(y)\Phi(y,k)\,,$$

*continuous for* $x \in \mathbb{R}$, *and for all* $k \in \mathbb{R}$. *For* $k \in \mathbb{R} \cup (-iA_r, iA_r)$ *the fundamental eigensolution* $\tilde{\Psi}(x,k)$ *with asymptotic behavior* (2.6a) *can be obtained as the unique solution to the integral equation*

$$(2.7b) \qquad \tilde{\Psi}(x,k) = e^{x\,\Lambda_r(k)} - \int_{x}^{\infty} dy\, e^{(x-y)\Lambda_r(k)}[Q(y) - Q_r]\tilde{\Psi}(y,k)\,.$$

*Moreover, $\tilde{\Psi}(x, k)$ is continuous for $x_0 \leq x$ for any finite $x_0$[1], and, as a function of $k$, for all $k \in \mathbb{R} \cup (-iA_r, iA_r)$. In addition, if the potential satisfies $(\boldsymbol{H}_1)$, then (2.7b) has a unique, continuous solution for $k \in [-iA_r, iA_r]$ (i.e., the continuity result can be extended to include the branch points $k = \pm iA_r$).*

The result for $\Phi(x, k)$ follows from standard iteration arguments for decaying potentials in $L^1$, and for $\tilde{\Psi}(x, k)$ it can be proved as in [**18**] and [**19**].

Assuming $(\boldsymbol{H}_1)$, one can replace the integral equations (2.7) by different ones. To this aim, let us introduce the fundamental matrix $\mathcal{G}(x, y; k)$ as follows:

$$(2.8) \qquad \mathcal{G}(x, y; k) = \vartheta(x)\vartheta(y)\, e^{(x-y)\Lambda_r(k)} + \vartheta(-x)\vartheta(-y)\, e^{-ik\sigma_3(x-y)}$$
$$+ \vartheta(x)\vartheta(-y)\, e^{x\Lambda_r(k)}\, e^{ik\sigma_3 y} + \vartheta(-x)\vartheta(y)\, e^{-ik\sigma_3 x}\, e^{-y\Lambda_r(k)}\,.$$

$\mathcal{G}(x, y; k)$ is a continuous matrix function of $(x, y, k) \in \mathbb{R}^2 \times \mathbb{C}$ which satisfies the initial value problems:

$$(2.9a) \qquad \frac{\partial \mathcal{G}(x, y; k)}{\partial x} = \Lambda(x, k)\mathcal{G}(x, y, k)\,, \quad \mathcal{G}(y, y; k) = I_2\,,$$

$$(2.9b) \qquad \frac{\partial \mathcal{G}(x, y; k)}{\partial y} = -\,\mathcal{G}(x, y, k)\Lambda(y, k)\,, \quad \mathcal{G}(x, x; k) = I_2\,,$$

where $\Lambda(x, k)$ is given by (2.5b). For further details on the fundamental matrix we refer to [**18**, App A], where the analogous problem is considered for the defocusing NLS equation. Then using (2.9) one can easily check that the eigenfunction $\Phi(x, k)$ and the fundamental eigensolution $\tilde{\Psi}(x, k)$ also satisfy the integral equations

$$(2.10a) \qquad \Phi(x, k) = \mathcal{G}(x, 0; k) + \int_{-\infty}^{x} dy\, \mathcal{G}(x, y; k)[Q(y) - Q_f(y)]\Phi(y, k)\,,$$

$$(2.10b) \qquad \tilde{\Psi}(x, k) = \mathcal{G}(x, 0; k) - \int_{x}^{\infty} dy\, \mathcal{G}(x, y; k)[Q(y) - Q_f(y)]\tilde{\Psi}(y, k)\,,$$

where $\mathcal{G}(x, 0; k) = \vartheta(x)e^{x\Lambda_r(k)} + \vartheta(-x)e^{-ik\sigma_3 x}$, according to (2.8), and $Q_f(x)$ is given by (2.4). Note that (2.10a) coincides with (2.7a) for $x \leq 0$, and (2.10b) coincides with (2.7b) for $x \geq 0$. On the other hand, using (2.8) we get

(2.11a)

$$\Phi(x, k) = e^{x\Lambda_r(k)} \left[ I_2 + \int_{-\infty}^{x} dy\, \mathcal{G}(0, y; k)[Q(y) - Q_f(y)]\Phi(y, k) \right]\,, \quad x \geq 0\,,$$

(2.11b)

$$\tilde{\Psi}(x, k) = e^{-ik\sigma_3 x} \left[ I_2 - \int_{x}^{\infty} dy\, \mathcal{G}(0, y; k)[Q(y) - Q_f(y)]\tilde{\Psi}(y, k) \right]\,, \quad x \leq 0\,.$$

Note that in (2.11b) the integral in the right-hand side converges absolutely as $x \to -\infty$, unlike (2.7b).

**2.1. Jost solutions.** Since the asymptotic scattering operator $\Lambda_r(k)$ is traceless, and such that $\Lambda_r^2(k) = -(k^2 + A_r^2)I_2$, we consider the two-sheeted Riemann surface associated with $\lambda_r^2 = k^2 + A_r^2$ by introducing appropriate local polar coordinates, with $r_j \geq 0$ and $-\pi/2 \leq \theta_j < 3\pi/2$ for $j = 1, 2$, and define:
(2.12)
$$\lambda_r = \sqrt{r_1 r_2}\, e^{i(\theta_1 + \theta_2)/2} \quad \text{on Sheet I}\,, \qquad \lambda_r = -\sqrt{r_1 r_2}\, e^{i(\theta_1 + \theta_2)/2} \quad \text{on Sheet II}\,.$$

---

[1]Note that the integral in (2.7b) does not converge absolutely for $x \to -\infty$.

Sheet I

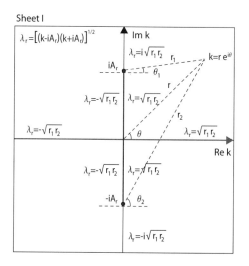

Sheet II

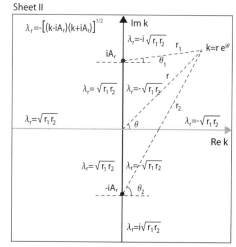

FIGURE 1. The branch cut on the two-sheeted Riemann surface associated with $\lambda_r^2 = k^2 + A_r^2$: we define $\lambda_r = \sqrt{r_1 r_2}\, e^{i(\theta_1 + \theta_2)/2}$ on Sheet I, and $\lambda_r = -\sqrt{r_1 r_2}\, e^{i(\theta_1 + \theta_2)/2}$ on Sheet II, with $r_1 = |k - iA_r|$, $r_2 = |k + iA_r|$ and angles $-\pi/2 \leq \theta_1, \theta_2 < 3\pi/2$ for $j = 1, 2$.

The branch cut is along the imaginary segment $\Sigma_r = [-iA_r, iA_r]$. The Riemann surface is then obtained by gluing together the two copies of the complex plane along the cut $\Sigma_r$ [see Fig. 1].

For the purpose of Secs. 2-4, we will consider a single sheet (Sheet I) of the complex plane for $k$, and denote by $\mathbb{K}_r$ the plane cut along the segment $\Sigma_r$ on the imaginary axis. $\mathbb{C}^\pm$ will denote the open upper/lower complex half planes, and $\mathbb{K}_r^\pm$ the open upper/lower complex half-planes, respectively, cut along $\Sigma_r$.

It is easy to show that $\lambda_r$ provides one-to-one correspondences between the following sets:

- $k \in \mathbb{K}_r^+ \equiv \mathbb{C}^+ \setminus (0, iA_r]$ and $\lambda_r \in \mathbb{C}^+$
- $k \in \partial\mathbb{K}_r^+ \equiv \mathbb{R} \cup \{is - 0^+ : 0 < s < A_r\} \cup \{iA_r\} \cup \{is + 0^+ : 0 < s < A_r\}$ and $\lambda_r \in \mathbb{R}$
- $k \in \mathbb{K}_r^- \equiv \mathbb{C}^- \setminus [-iA_r, 0)$ and $\lambda_r \in \mathbb{C}^-$
- $k \in \partial\mathbb{K}_r^- \equiv \mathbb{R} \cup \{is - 0^+ : -A_r < s < 0\} \cup \{-iA_r\} \cup \{is + 0^+ : -A_r < s < 0\}$ and $\lambda_r \in \mathbb{R}$.

Note that with this choice for the branch cut one has $\lambda_r \sim k$ as $k \to \infty$ in the entire Sheet I, while $\lambda_r \sim -k$ as $k \to \infty$ on Sheet II. In the following, $\lambda_r^\pm(k)$ will denote the boundary values taken by $\lambda_r(k)$ for $k \in \Sigma_r$ from the right/left edge of the cut on Sheet I, with

$$(2.13) \qquad \lambda_r^\pm(k) = \pm\sqrt{A_r^2 - |k|^2}\,, \qquad k = is \pm 0^+, \quad |s| \leq A_r$$

on the right/left edge (cf. Fig. 1).

The eigenvalues of $\Lambda_r(k)$ are $\pm i\lambda_r$, and the eigenvector matrix $W_r(k)$, such that

$$(2.14) \qquad\qquad \Lambda_r(k)W_r(k) = -i\lambda_r W_r(k)\sigma_3\,,$$

can be conveniently chosen as follows:

$$(2.15) \qquad\qquad W_r(k) = I_2 - \frac{i}{\lambda_r + k}\sigma_3 Q_r\,.$$

Note that $\det W_r(k) = \frac{2\lambda_r}{\lambda_r + k}$, and $W_r(k)$ is a nonsingular matrix on either sheet because $A_r > 0$ holds strictly ($\lambda_r + k$ can only vanish on Sheet II, and only in the limit $k \to \infty$).

We can then define the "right" Jost solutions $\Psi(x,k) = \left(\bar{\psi}(x,k)\ \ \psi(x,k)\right)$ in terms of the fundamental eigensolution $\tilde{\Psi}(x,k)$ as:

$$(2.16) \qquad \Psi(x,k) = \left(\bar{\psi}(x,k)\ \ \psi(x,k)\right) := \tilde{\Psi}(x,k)W_r(k)\,,$$

which then satisfies the following boundary condition:

$$(2.17) \qquad\qquad \Psi(x,k) \sim W_r(k)\,e^{-i\lambda_r x\sigma_3}\,, \qquad x \to +\infty\,.$$

The Jost solutions from the right $\bar{\psi}(x,k)$, $\psi(x,k)$ are then defined via a customary asymptotic plane wave behavior (cf. (2.17)) when $\lambda_r \in \mathbb{R}$, i.e., for $k \in \partial\mathbb{K}_r^+ \cup \partial\mathbb{K}_r^-$, and when $k = is \in [-iA_r, iA_r]$ we will denote with a superscript $\pm$ the values on the right/left edge of the cut in both half-planes, i.e.:

$$(2.18) \quad \Psi^\pm(x,is) \equiv \left(\bar{\psi}^\pm(x,is)\ \ \psi^\pm(x,is)\right) := \tilde{\Psi}(x,is)\,W_r(is \pm 0^+) \qquad |s| \le A_r\,,$$

since $\tilde{\Psi}(x,k)$ is single-valued across the cut, and $W_r(k)$ has right/left limits defined by (2.13).

On the other hand, the "left" Jost solutions $\Phi(x,k) = \left(\phi(x,k)\ \ \bar{\phi}(x,k)\right)$ are defined as asymptotic "plane waves" (cf. (2.6b)) for $k \in \mathbb{R}$.

Taking into account (2.7) and (2.16), the Jost solutions can be represented in terms of the following integral equations:

$$(2.19a) \quad e^{i\lambda_r x}\bar{\psi}(x,k) = W_{r,1}(k) - \int_x^\infty dy\,\Xi_r^-(y - x, k)[Q(y) - Q_r]e^{i\lambda_r y}\bar{\psi}(y,k),$$

$$(2.19b) \quad e^{-i\lambda_r x}\psi(x,k) = W_{r,2}(k) - \int_x^\infty dy\,\Xi_r^+(y - x, k)[Q(y) - Q_r]e^{-i\lambda_r y}\psi(y,k),$$

$$(2.19c) \quad e^{ikx}\phi(x,k) = \begin{pmatrix} 1 \\ 0 \end{pmatrix} + \int_{-\infty}^x dy\,\Xi_l^+(x - y, k)\,Q(y)\,e^{iky}\phi(y,k),$$

$$(2.19d) \quad e^{-ikx}\bar{\phi}(x,k) = \begin{pmatrix} 0 \\ 1 \end{pmatrix} + \int_{-\infty}^x dy\,\Xi_l^-(x - y, k)\,Q(y)\,e^{-iky}\bar{\phi}(y,k),$$

where the subscripts $j = 1, 2$ in the matrix $W_r(k)$ denote its $j$-th column, and

(2.20a)    $\Xi_r^-(x, k) = \begin{pmatrix} 1 + \frac{\lambda_r - k}{2\lambda_r}\left[e^{-2i\lambda_r x} - 1\right] & -\frac{iq_r}{2\lambda_r}\left[e^{-2i\lambda_r x} - 1\right] \\ \frac{iq_r^*}{2\lambda_r}\left[e^{-2i\lambda_r x} - 1\right] & e^{-2i\lambda_r x} - \frac{\lambda_r - k}{2\lambda_r}\left[e^{-2i\lambda_r x} - 1\right] \end{pmatrix},$

(2.20b)    $\Xi_r^+(x, k) = \begin{pmatrix} e^{2i\lambda_r x} - \frac{\lambda_r - k}{2\lambda_r}\left[e^{2i\lambda_r x} - 1\right] & \frac{iq_r}{2\lambda_r}\left[e^{2i\lambda_r x} - 1\right] \\ -\frac{iq_r^*}{2\lambda_r}\left[e^{2i\lambda_r x} - 1\right] & 1 + \frac{\lambda_r - k}{2\lambda_r}\left[e^{2i\lambda_r x} - 1\right] \end{pmatrix},$

(2.20c)    $\Xi_l^+(x, k) = \begin{pmatrix} 1 & 0 \\ 0 & e^{2ikx} \end{pmatrix},$

(2.20d)    $\Xi_l^-(x, k) = \begin{pmatrix} e^{-2ikx} & 0 \\ 0 & 1 \end{pmatrix}.$

Note that the integral equations (2.19a) and (2.19b) for the "right" Jost solutions are the same as in the fully asymmetric case [19], while the integral equations (2.19c) and (2.19d) for the "left" Jost solutions are the same as in the vanishing case [3].

The following result then establishes the analyticity properties of the Jost solutions as functions of the scattering parameter $k$.

PROPOSITION 2.2. *Suppose* $(\boldsymbol{H}_1)$ *holds.*[2] *Then, for every* $x \in \mathbb{R}$, *the Jost solution* $\psi(x, k)$ *[resp.* $\bar{\psi}(x, k)$*] extends to a function that is analytic for* $k \in \mathbb{K}_r^+$ *[resp.* $k \in \mathbb{K}_r^-$*], and continuous for* $k \in \mathbb{K}_r^+ \cup \partial\mathbb{K}_r^+ \cup \partial\mathbb{K}_r^-$ *[resp.* $k \in \mathbb{K}_r^- \cup \partial\mathbb{K}_r^- \cup \partial\mathbb{K}_r^+$*]. On the other hand, the Jost solution* $\phi(x, k)$ *[resp.* $\bar{\phi}(x, k)$*] extends to a function that is continuous for* $k \in \mathbb{C}^+ \cup \mathbb{R}$ *[resp.* $k \in \mathbb{C}^- \cup \mathbb{R}$*] and analytic for* $k \in \mathbb{C}^+$ *[resp.* $k \in \mathbb{C}^-$*].*

Note that $\mathbb{K}_r^\pm$ are intended as analytic manifolds, and continuity of the Jost solutions across the cuts is intended as the existence of right/left continuous limits only in the domains that have the branch cut as part of their boundary as an analytic manifold, i.e. $\mathbb{K}_r^+$ for $\psi(x, k)$ and $\mathbb{K}_r^-$ for $\bar{\psi}(x, k)$. In the half-planes where locally there is no analytic continuation off the branch cut, the functions $\psi^\pm(x, k)$, $\bar{\psi}^\pm(x, k)$ are as given in (2.18) with the two choices of $\lambda_r^\pm$, and can be obtained as the unique solutions of the corresponding Volterra integral equations (2.19). The proof of Prop. 2.2 for $\psi(x, k)$ and $\bar{\psi}(x, k)$ can be carried out as in [18, 19], while the result for $\phi(x, k)$ and $\bar{\phi}(x, k)$ can be established following the standard approach for sufficiently rapidly decaying potentials.

In conclusion, for potentials satisfying $(\boldsymbol{H}_1)$ all four Jost solutions are in general simultaneously defined only for $k \in \mathbb{R}$. Note that $\phi^+(x, k) = \phi^-(x, k)$ for $k \in [0, iA_r]$, and $\bar{\phi}^+(x, k) = \bar{\phi}^-(x, k)$ for $k \in [-iA_r, 0]$, where the superscript $\pm$ denotes again the values on the right/left edge of the cut in both half-planes. On the contrary, $\bar{\phi}(x, k)$ is in general not defined for $k \in (0, iA_r]$, and $\phi(x, k)$ is not defined for $k \in [-iA_r, 0)$.

**2.2. Scattering coefficients.** From the integral equations (2.11) for $\Phi(x, k)$ and $\tilde{\Psi}(x, k)$, one can easily find

(2.21a)        $\tilde{\Psi}(x, k) = e^{-ik\sigma_3 x}[B_r(k) + o(1)], \qquad x \to -\infty,$

(2.21b)        $\Phi(x, k) = e^{x\Lambda_r(k)}[B_l(k) + o(1)], \qquad x \to +\infty,$

---

[2]In fact, since the potential decays as $x \to -\infty$, it would be enough to assume $[q(x,t) - \vartheta(x)q_r(t)] \in L^1(\mathbb{R}^-) \cap L^{1,1}(\mathbb{R}^+)$ with respect to $x$ for all $t \geq 0$.

where the coupling matrices

$$(2.22a) \qquad B_r(k) = I_2 - \int_{-\infty}^{\infty} dy\, \mathcal{G}(0, y; k)[Q(y) - Q_f(y)]\tilde{\Psi}(y, k),$$

$$(2.22b) \qquad B_l(k) = I_2 + \int_{-\infty}^{\infty} dy\, \mathcal{G}(0, y; k)[Q(y) - Q_f(y)]\Phi(y, k),$$

are each other's inverse. Under the assumptions of Prop. 2.1, in Eqs. (2.21) and (2.22) one needs to take $k \in \mathbb{R}$, where all eigenfunctions (2.10) and (2.11) are simultaneously defined.

Using (2.16) and (2.21) to obtain the asymptotic behavior of the Jost solutions as $x \to \pm\infty$, for $k \in \mathbb{R}$ we can then express each set of Jost solutions as a linear combination of the other set, i.e.,

$$(2.23a) \qquad \begin{pmatrix}\phi(x, k) & \bar{\phi}(x, k)\end{pmatrix} = \begin{pmatrix}\bar{\psi}(x, k) & \psi(x, k)\end{pmatrix} S(k),$$

$$(2.23b) \qquad \begin{pmatrix}\bar{\psi}(x, k) & \psi(x, k)\end{pmatrix} = \begin{pmatrix}\phi(x, k) & \bar{\phi}(x, k)\end{pmatrix} \bar{S}(k),$$

where the *scattering matrices* $S(k)$ and $\bar{S}(k)$ are obviously each other inverses, and they are given by

$$(2.24) \qquad S(k) = W_r^{-1}(k)B_l(k), \qquad \bar{S}(k) = B_r(k)W_r(k).$$

For later convenience we write

$$(2.25) \qquad S(k) = \begin{pmatrix} a(k) & \bar{b}(k) \\ b(k) & \bar{a}(k) \end{pmatrix}, \qquad \bar{S}(k) = \begin{pmatrix} \bar{c}(k) & d(k) \\ \bar{d}(k) & c(k) \end{pmatrix},$$

where the entries of the scattering matrices are usually referred to as *scattering coefficients*, and at this stage they are all in general defined only for $k \in \mathbb{R}$. Moreover, since $\det \Phi(x, k)$ and $\det \Psi(x, k)$ are independent of $x$, from (2.6b) and (2.17) it follows that

$$(2.26) \qquad \det \Phi(x, k) = 1, \qquad \det \Psi(x, k) = \frac{2\lambda_r}{\lambda_r + k},$$

and consequently we obtain for $k \in \mathbb{R}$:

$$(2.27) \qquad \det S(k) = \frac{1}{\det W_r(k)} = \frac{\lambda_r + k}{2\lambda_r}, \qquad \det \bar{S}(k) = \det W_r(k) = \frac{2\lambda_r}{\lambda_r + k}.$$

If we now denote by $\mathrm{Wr}(v_1, v_2) \stackrel{\text{def}}{=} \det \begin{pmatrix} v_1 & v_2 \end{pmatrix}$ the Wronskian of any two vector solutions $v_1$, $v_2$ of the scattering problem (2.1a), then Eqs. (2.23) yield the following "Wronskian" representations for the scattering coefficients in (2.25) for $k \in \mathbb{R}$:

(2.28a)

$$a(k) = \frac{\mathrm{Wr}(\phi, \psi)}{\mathrm{Wr}(\bar{\psi}, \psi)} = \frac{\lambda_r + k}{2\lambda_r}\mathrm{Wr}(\phi, \psi), \qquad \bar{a}(k) = \frac{\mathrm{Wr}(\bar{\psi}, \bar{\phi})}{\mathrm{Wr}(\bar{\psi}, \psi)} = \frac{\lambda_r + k}{2\lambda_r}\mathrm{Wr}(\bar{\psi}, \bar{\phi}),$$

(2.28b)

$$b(k) = \frac{\mathrm{Wr}(\bar{\psi}, \phi)}{\mathrm{Wr}(\bar{\psi}, \psi)} = \frac{\lambda_r + k}{2\lambda_r}\mathrm{Wr}(\bar{\psi}, \phi), \qquad \bar{b}(k) = \frac{\mathrm{Wr}(\bar{\phi}, \psi)}{\mathrm{Wr}(\bar{\psi}, \psi)} = \frac{\lambda_r + k}{2\lambda_r}\mathrm{Wr}(\bar{\phi}, \psi),$$

and

$$(2.28c) \quad c(k) = \frac{\mathrm{Wr}(\phi, \psi)}{\mathrm{Wr}(\phi, \bar{\phi})} = \frac{2\lambda_r}{\lambda_r + k}a(k), \qquad \bar{c}(k) = \frac{\mathrm{Wr}(\bar{\psi}, \bar{\phi})}{\mathrm{Wr}(\phi, \bar{\phi})} = \frac{2\lambda_r}{\lambda_r + k}\bar{a}(k),$$

$$(2.28d) \quad d(k) = \frac{\mathrm{Wr}(\psi, \bar{\phi})}{\mathrm{Wr}(\phi, \bar{\phi})} = -\frac{2\lambda_r}{\lambda_r + k}\bar{b}(k), \qquad \bar{d}(k) = \frac{\mathrm{Wr}(\phi, \bar{\psi})}{\mathrm{Wr}(\phi, \bar{\phi})} = -\frac{2\lambda_r}{\lambda_r + k}b(k),$$

where the arguments $(x, k)$ of the Jost solutions have been omitted for brevity, and the second set of identities in (2.28c) and (2.28d) are obtained from $\bar{S}(k) = S^{-1}(k)$.

These Wronskian representations can then be used to define the values of the scattering coefficients from the right/left edge of the cut $\Sigma_r$, consistently with (2.18). Explicitly, taking into account that $\lambda_r^\pm(k) = \pm\sqrt{A_r^2 - |k|^2}$ on the right/left edge of the cut for $k \in \Sigma_r$ (cf. Fig. 1), one has

$$(2.29\text{a}) \qquad a^\pm(k) = \frac{\lambda_r^\pm(k) + k}{2\lambda_r^\pm(k)}\, \mathrm{Wr}(\phi(x, k), \psi^\pm(x, k)) \qquad k \in [0, iA_r]\,,$$

$$(2.29\text{b}) \qquad \bar{a}^\pm(k) = \frac{\lambda_r^\pm(k) + k}{2\lambda_r^\pm(k)}\, \mathrm{Wr}(\bar{\psi}^\pm(x, k), \bar{\phi}(x, k)) \qquad k \in [-iA_r, 0]\,,$$

$$(2.29\text{c}) \qquad b^\pm(k) = \frac{\lambda_r^\pm(k) + k}{2\lambda_r^\pm(k)}\, \mathrm{Wr}(\bar{\psi}^\pm(x, k), \phi(x, k)) \qquad k \in [0, iA_r]\,,$$

$$(2.29\text{d}) \qquad \bar{b}^\pm(k) = \frac{\lambda_r^\pm(k) + k}{2\lambda_r^\pm(k)}\, \mathrm{Wr}(\bar{\phi}(x, k), \psi^\pm(x, k)) \qquad k \in [-iA_r, 0]\,,$$

and similarly for the scattering coefficients from the left defined by (2.28c)-(2.28d).

Eqs. (2.28) allow one to analytically continue some of the scattering coefficients off the real $k$-axis under the assumption $(\boldsymbol{H}_1)$. In fact, (2.28) and Prop. 2.2 imply:

- $a(k)$ is analytic in $k \in \mathbb{K}_r^+$, and continuous for $k \in \overline{\mathbb{K}_r^+} \setminus \{iA_r\}$ [with values across the cut $a^\pm(k)$ as in (2.29a)]; also

$$(2.30\text{a}) \qquad a(k) \sim \frac{iA_r}{2\lambda_r}\, \mathrm{Wr}(\phi(x, iA_r), \psi(x, iA_r)), \quad k \to iA_r.$$

- $\bar{a}(k)$ is analytic in $k \in \mathbb{K}_r^-$, and continuous in $k \in \overline{\mathbb{K}_r^-} \setminus \{-iA_r\}$ [with values across the cut $\bar{a}^\pm(k)$ as in (2.29b)]; also

$$(2.30\text{b}) \qquad \bar{a}(k) \sim \frac{-iA_r}{2\lambda_r}\, \mathrm{Wr}(\bar{\psi}(x, -iA_r), \bar{\phi}(x, -iA_r)), \quad k \to -iA_r.$$

- $b(k)$ is continuous for $\partial\mathbb{K}_r^+ \setminus \{iA_r\}$ [with values across the cut $b^\pm(k)$ as in (2.29c)], and $\bar{b}(k)$ is continuous for $k \in \partial\mathbb{K}_r^- + \setminus \{-iA_r\}$ [with values across the cut $\bar{b}^\pm(k)$ as in (2.29d)]; at the branch points

$$(2.30\text{c}) \qquad b^\pm(k) \sim \frac{iA_r}{2\lambda_r}\, \mathrm{Wr}(\bar{\psi}(x, iA_r), \phi(x, iA_r)), \qquad k \to iA_r\,,$$

$$(2.30\text{d}) \qquad \bar{b}^\pm(k) \sim \frac{-iA_r}{2\lambda_r}\, \mathrm{Wr}(\bar{\phi}(x, -iA_r), \psi(x, -iA_r)), \quad k \to -iA_r\,.$$

Similar results can be derived for the four scattering coefficients from the left, although the corresponding properties can also be obtained from those above using (2.28c) and (2.28d). Note, in particular, that $a(k)$ [resp., $\bar{a}(k)$] has a branch point singularity at $k = iA_r$ [resp., $k = -iA_r$], where $\lambda_r = 0$, while $c(k)$ [resp., $\bar{c}(k)$] is well-defined there, consistently with (2.28c).

For future convenience we also define the *reflection coefficients* from the right as follows:

$$(2.31\text{a}) \qquad \rho(k) = \frac{b(k)}{a(k)} \qquad \text{for } k \in \mathbb{R}, \qquad \rho^\pm(k) = \frac{b^\pm(k)}{a^\pm(k)} \quad \text{for } k \in [0, iA_r),$$

$$(2.31\text{b}) \qquad \bar{\rho}(k) = \frac{\bar{b}(k)}{\bar{a}(k)} \qquad \text{for } k \in \mathbb{R}, \qquad \bar{\rho}^\pm(k) = \frac{\bar{b}^\pm(k)}{\bar{a}^\pm(k)} \quad \text{for } k \in (-iA_r, 0],$$

and the *reflection coefficients* from the left as:

$$(2.31c) \qquad r(k) = \frac{d(k)}{c(k)} = -\frac{\bar{b}(k)}{a(k)}, \qquad \bar{r}(k) = \frac{\bar{d}(k)}{\bar{c}(k)} = -\frac{b(k)}{\bar{a}(k)}, \quad k \in \mathbb{R},$$

where in the last two expressions we have used that $S(k) = \bar{S}^{-1}(k)$. The reciprocals $1/a(k)$, $1/\bar{a}(k)$, $1/c(k)$, and $1/\bar{c}(k)$ are usually referred to as (right and left) *transmission coefficients*.

Furthermore, as shown in [**19**] for the case of fully asymmetric NZBCs, from (2.24) and (2.22), using (2.14) and (2.16) we also obtain the following integral representation for the scattering matrix:

$$
\begin{aligned}
S(k) = & \int_0^\infty dy\, e^{i\lambda_r y \sigma_3} W_r^{-1}(k)[Q(y) - Q_r]\Phi(y,k) \\
(2.32) \qquad & + W_r^{-1}(k)\left[ I_2 + \int_{-\infty}^0 dy\, e^{iky\sigma_3}[Q(y) - Q_l]\Phi(y,k) \right],
\end{aligned}
$$

which could serve as an alternative to the Wronskian representations to establish the analytic continuation in the appropriate half planes of the scattering coefficients $a(k)$ and $\bar{a}(k)$.

**2.3. Symmetries of eigenfunctions and scattering data.** The scattering problem (2.1a) admits two involutions: $(k, \lambda_r) \to (k^*, \lambda_r^*)$ and $(k, \lambda_r) \to (k, -\lambda_r)$. Correspondingly, eigenfunctions and scattering data satisfy two sets of symmetry relations.

**First symmetry.** Using the asymptotic behaviors (2.6b) and (2.17), the symmetries for the Jost solutions are given by:

(2.33a)
$$\bar{\psi}^*(x,k^*) = i\sigma_2\psi(x,k) \text{ for } k \in \mathbb{K}_r^+ \cup \mathbb{R}, \quad \psi^*(x,k^*) = -i\sigma_2\bar{\psi}(x,k) \text{ for } k \in \mathbb{K}_r^- \cup \mathbb{R},$$
(2.33b)
$$\left(\bar{\psi}^\pm(x,k^*)\right)^* = i\sigma_2\psi^\pm(x,k), \quad \left(\psi^\pm(x,k^*)\right)^* = -i\sigma_2\bar{\psi}^\pm(x,k) \text{ for } k \in [-iA_r, iA_r],$$
(2.33c)
$$\phi^*(x,k^*) = i\sigma_2\bar{\phi}(x,k) \text{ for } k \in \mathbb{C}^- \cup \mathbb{R}, \quad \bar{\phi}^*(x,k^*) = -i\sigma_2\phi(x,k) \text{ for } k \in \mathbb{C}^+ \cup \mathbb{R},$$

where $\sigma_2$ is the second Pauli matrix introduced in (2.2). From (2.23) we then obtain $S^*(k^*) = \sigma_2 S(k)\sigma_2$ on the continuous spectrum $k \in \mathbb{R}$, and wherever all entries in the scattering matrix are simultaneously defined. Under the assumption $(\boldsymbol{H}_0)$ for the potential, the symmetry relations for the scattering coefficients can be written as

(2.34a)   $\bar{a}^*(k^*) = a(k)$ for $k \in \mathbb{K}_r^+ \cup \mathbb{R}$,   $\left(\bar{a}^\pm(k^*)\right)^* = a^\pm(k)$ for $k \in [0, iA_r]$,

(2.34b)   $\bar{b}^*(k) = -b(k)$   for $k \in \mathbb{R}$,   $\left(\bar{b}^\pm(k^*)\right)^* = -b^\pm(k)$   for $k \in [0, iA_r]$.

We note that the above symmetries relate the values of the scattering coefficients in the upper/lower half plane of $k$, and from the same side of the cut. Taking into account (2.31), one can easily establish the symmetry relations satisfied by the reflection coefficients:

(2.35a)   $\bar{\rho}^*(k) = -\rho(k)$   for $k \in \mathbb{R}$,   $\left(\bar{\rho}^\pm(k^*)\right)^* = -\rho^\pm(k)$   for $k \in [0, iA_r]$,

(2.35b)   $\bar{r}^*(k) = -r(k)$   for $k \in \mathbb{R}$.

**Second symmetry.** When using a single sheet for the Riemann surface of the function $\lambda_r^2 = k^2 + A_r^2$, the involution $(k, \lambda_r) \to (k, -\lambda_r)$ can only be considered across the cut. So this second involution relates values of eigenfunctions and scattering coefficients for the same value of $k$ from either side of the cut. On the cut one has

$$(2.36) \qquad \bar\psi^{\mp}(x, k) = \frac{\lambda_r^{\pm}(k) + k}{-iq_r} \psi^{\pm}(x, k), \qquad \text{for } k \in [-iA_r, iA_r],$$

while $\phi^+(x, k) = \phi^-(x, k)$ for $k \in [0, iA_r]$, and $\bar\phi^+(x, k) = \bar\phi^-(x, k)$ for $k \in [-iA_r, 0]$. Using these symmetries in the Wronskian representations for the scattering coefficients (2.28), one obtains:

$$(2.37a) \qquad a^{\pm}(k) = \frac{\lambda_r^{\mp}(k) - k}{iq_r^*} b^{\mp}(k) \qquad \text{for } k \in [0, iA_r],$$

$$(2.37b) \qquad \bar a^{\mp}(k) = \frac{\lambda_r^{\mp}(k) + k}{-iq_r} \bar b^{\pm}(k) \qquad \text{for } k \in [-iA_r, 0].$$

From (2.31a), we then have the following symmetries for the reflection coefficients from the right:

$$(2.38a) \qquad \rho^{\pm}(k) = \frac{iq_r^*}{\lambda_r^{\pm}(k) - k} \frac{a^{\mp}(k)}{a^{\pm}(k)} \qquad \text{for } k \in [0, iA_r],$$

$$(2.38b) \qquad \bar\rho^{\pm}(k) = \frac{-iq_r}{\lambda_r^{\mp}(k) + k} \frac{\bar a^{\mp}(k)}{\bar a^{\pm}(k)} \qquad \text{for } k \in [-iA_r, 0].$$

Note that the above relationships imply that

$$(2.39)$$
$$\rho^+(k)\rho^-(k) = q_r^*/q_r \quad \text{for } k \in [0, iA_r], \quad \bar\rho^+(k)\bar\rho^-(k) = q_r/q_r^* \quad \text{for } k \in [-iA_r, 0].$$

Using (2.36) in (2.28c) and (2.28d), for $k \in \Sigma_r$ the symmetry relations for the scattering coefficients from the left are given by:

$$(2.40a) \qquad c^{\pm}(k) = \frac{-iq_r}{\lambda_r^{\pm}(k) + k} \bar d^{\mp}(k) \qquad \text{for } k \in [0, iA_r],$$

$$(2.40b) \qquad \bar c^{\pm}(k) = \frac{\lambda_r^{\mp}(k) + k}{-iq_r} d^{\mp}(k) \qquad \text{for } k \in [-iA_r, 0].$$

**2.4. Discrete eigenvalues.** A *discrete eigenvalue* is a value of $k \in \mathbb{K}_r^+ \cup \mathbb{K}_r^-$ (corresponding to $\lambda_r \in \mathbb{C} \setminus \mathbb{R}$) for which there exists a nontrivial solution $v$ to (1.2) with entries in $L^2(\mathbb{R})$. These eigenvalues occur for $k \in \mathbb{K}_r^+$ iff the functions $\phi(x, k)$ and $\psi(x, k)$ are linearly dependent (i.e., iff $a(k) = 0$), and for $k \in \mathbb{K}_r^-$ iff the functions $\bar\psi(x, k)$ and $\bar\phi(x, k)$ are linearly dependent (i.e., iff $\bar a(k) = 0$). Equations (2.6b) and (2.17) imply that the corresponding eigenfunctions are exponentially decaying as $x \to \pm\infty$. The conjugation symmetry (2.34a) then ensures that the discrete eigenvalues occur in complex conjugate pairs. The algebraic multiplicity of each discrete eigenvalue coincides with the multiplicity of the corresponding zero of $a(k)$ [for $k \in \mathbb{K}_r^+$], or $\bar a(k)$ [for $k \in \mathbb{K}_r^-$].

In this work we assume that discrete eigenvalues are simple, and finite in number. Also, we assume that there are no spectral singularities, i.e., zeros of the scattering coefficients $a(k)$ and $\bar a(k)$ for $k \in \mathbb{R} \cup \Sigma_r$. Establishing conditions on the asymptotic amplitudes and phases that guarantee absence of spectral singularities is an interesting problem, but is beyond the scope of this paper and will be the subject of future investigation. In any event, we mention that spectral singularities can be incorporated in the inverse problem with slight modifications of the approach presented in Sec. 3.

**2.5. Large $k$ behavior of eigenfunctions and scattering data.** In order to properly pose the inverse scattering problem, one has to determine the asymptotic behavior of eigenfunctions and scattering data as $k \to \infty$. Assuming $\partial_x q \in L^1$, integration by parts on the integral equations (2.19) yields for the asymptotic behaviors of the eigenfunctions as $|k| \to \infty$ in the appropriate half planes on Sheet I:

$$(2.41a) \quad \Psi_d(x,k)e^{i\lambda_r \sigma_3 x} = I_2 + o(1) , \qquad \Psi_o(x,k)e^{i\lambda_r \sigma_3 x} = \frac{iQ(x)\sigma_3}{2k} + o(1/k) ,$$

$$(2.41b) \quad \Phi_d(x,k)e^{ik\sigma_3 x} = I_2 + o(1) , \qquad \Phi_o(x,k)e^{ik\sigma_3 x} = \frac{iQ(x)\sigma_3}{2k} + o(1/k) ,$$

where subscripts $_d$ and $_o$ denote the diagonal and off-diagonal parts, respectively, of the corresponding matrix Jost solutions $\Psi(x,k)$ and $\Phi(x,k)$. From the Wronskian representations (2.28) for the scattering coefficients, and taking into account that $\lambda_r \sim k$ as $k \to \infty$, we then obtain the asymptotic behavior of the scattering coefficients:

$$(2.42a) \quad a(k) = \frac{\lambda_r + k}{2\lambda_r} \mathrm{Wr}\left(\phi(x,k), \psi(x,k)\right) \sim 1 \quad \text{as} \quad |k| \to \infty,\ k \in \mathbb{K}_r^+ \cup \mathbb{R}$$

$$(2.42b) \quad \bar{a}(k) = -\frac{\lambda_r + k}{2\lambda_r} \mathrm{Wr}\left(\bar{\phi}(x,k), \bar{\psi}(x,k)\right) \sim 1 \quad \text{as} \quad |k| \to \infty,\ k \in \mathbb{K}_r^- \cup \mathbb{R}$$

while

$$b(k) = O(1/k^2) , \qquad \bar{b}(k) = O(1/k^2) \quad \text{as} \quad |k| \to \infty,\ k \in \mathbb{R} .$$

Taking into account (2.31), the above also imply that

$$(2.42c) \quad \rho(k) = O(1/k^2) , \qquad \bar{\rho}(k) = O(1/k^2) \quad \text{as} \quad |k| \to \infty,\ k \in \mathbb{R} ,$$

$$(2.42d) \quad r(k) = O(1/k^2) , \qquad \bar{r}(k) = O(1/k^2) \quad \text{as} \quad |k| \to \infty,\ k \in \mathbb{R} .$$

## 3. Inverse scattering problem

**3.1. Triangular representations for the eigenfunctions.** We introduce the following two triangular representations for the fundamental eigenfunctions:

$$(3.1a) \quad \tilde{\Psi}(x,k)e^{-x\Lambda_r(k)} = I_2 + \int_x^\infty ds\, K(x,s)e^{(s-x)\Lambda_r(k)} ,$$

$$(3.1b) \quad \Phi(x,k)e^{ik\sigma_3 x} = I_2 + \int_{-\infty}^x ds\, J(x,s)e^{-ik\sigma_3(s-x)} ,$$

where the kernels $K(x,s) = [K_{ij}(x,s)]_{i,j=1,2}$ and $J(x,s) = [J_{ij}(x,s)]_{i,j=1,2}$ are "triangular" kernels, i.e., such that $K(x,y) \equiv 0$ for $x > y$ and $J(x,y) \equiv 0$ for $x < y$. The above ansatz for the triangular representations is standard for the Jost solutions in the rapidly decaying case (see, for instance [4, 38]). In the NZBC case, the issue of existence of triangular representations such as the above has been addressed in [19]. Here we omit the details for brevity.

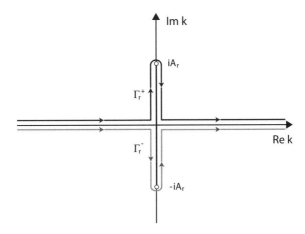

FIGURE 2. The oriented contours $\Gamma_r^{\pm}$.

We note that (3.1) yield the corresponding triangular representations for the Jost solutions (2.6b) and (2.17):

$$(3.2a) \qquad \Psi(x,k) = W_r(k)e^{-i\lambda_r \sigma_3 x} + \int_x^{\infty} ds\, K(x,s)W_r(k)\, e^{-i\lambda_r \sigma_3 s},$$

$$(3.2b) \qquad \Phi(x,k) = e^{-ik\sigma_3 x} + \int_{-\infty}^{x} ds\, J(x,s)\, e^{-ik\sigma_3 s}.$$

Eq. (3.2b) is the standard triangular representation for the Jost solutions in the decaying case, while (3.2a) provides the appropriate generalization to the NZBC case. Inserting the representations (3.2) into the scattering problem (2.1a), and matching terms with the same $k$-dependence, one obtains the reconstruction of the potential $q(x)$ in terms of the entries of the kernels $K(x,y)$ and $J(x,y)$:

$$(3.3) \qquad q(x) = q_r - 2K_{12}(x,x) = 2J_{12}(x,x).$$

In [19] it is shown that in the fully asymmetric case a sufficient condition for the triangular representations and reconstruction formulas to hold is that the potential $q(x)$ satisfies ($\boldsymbol{H}_2$) and $\partial_x q \in L^1(\mathbb{R})$. The analog obviously holds here, although less strict integrability requirements are necessary for $x \in \mathbb{R}^-$.

Note that here we have omitted the time dependence for brevity. If all the above assumptions on the potential hold for all $t \geq 0$, then inserting the time dependence on the Jost and fundamental eigenfunctions (see Sec. 4 for details) yields a parametric $t$-dependence for the Marchenko kernels, and the reconstruction formulas for the potential read

$$(3.4) \qquad q(x,t) = q_r(t) - 2K_{12}(x,x;t) = 2J_{12}(x,x;t).$$

**3.2. Marchenko equations.** In this subsection we formulate the inverse scattering problem in terms of right and left Marchenko integral equations.

3.2.1. *Right Marchenko equations.* Let us write (2.23a) explicitly as:

(3.5a)
$$\frac{\phi(x,k)}{a(k)} = \bar{\psi}(x,k) + \rho(k)\psi(x,k) \qquad k \in \mathbb{R},$$

(3.5b)
$$\frac{\phi^\pm(x,k)}{a^\pm(k)} = \bar{\psi}^\pm(x,k) + \rho^\pm(k)\psi^\pm(x,k) \qquad k \in [0, iA_r),$$

(3.5c)
$$\frac{\bar{\phi}(x,k)}{\bar{a}(k)} = \psi(x,k) + \bar{\rho}(k)\bar{\psi}(x,k) \qquad k \in \mathbb{R},$$

(3.5d)
$$\frac{\bar{\phi}^\pm(x,k)}{\bar{a}^\pm(k)} = \psi^\pm(x,k) + \bar{\rho}^\pm(k)\bar{\psi}^\pm(x,k) \qquad k \in (-iA_r, 0],$$

where $\rho(k)$, $\rho^\pm(k)$ and $\bar{\rho}(k)$, $\bar{\rho}^\pm(k)$ are given by (2.31a) and (2.31b), respectively. Multiplying (3.5a) by $e^{i\lambda_r y}$ for $y > x$, and substituting the triangular representation (3.2a) we obtain

$$\left[\frac{e^{i\lambda_r x}\phi(x,k)}{a(k)} - W_{r,1}(k)\right]e^{i\lambda_r(y-x)} = \int_x^\infty ds\, K(x,s)W_{r,1}(k)\,e^{i\lambda_r(y-s)}$$

(3.6)
$$+ \rho(k)\left[e^{i\lambda_r(x+y)}W_{r,2}(k) + \int_x^\infty ds\, K(x,s)W_{r,2}(k)\,e^{i\lambda_r(s+y)}\right],$$

where $W_{r,j}(k)$ denotes the $j$-th column of the eigenvector matrix $W_r(k)$ in (2.15). We remark that $\lambda_r \sim k$ as $|k| \to \infty$, so that the term in the left-hand side decays as $|k| \to \infty$ in $\mathbb{K}_r^+ \cup \mathbb{R}$. For the purpose of this section, it will be convenient to consider the eigenfunctions as functions of $\lambda_r$, i.e., to use:

$$k = k(\lambda_r) \equiv \sqrt{\lambda_r^2 - A_r^2}\,.$$

Note that $\lambda_r \in \mathbb{R}$ is in one-to-one correspondence with either $k \in \Gamma_r^+$ or $k \in \Gamma_r^-$ (cf. Fig. 2). In the following we will assume $k \in \Gamma_r^+$ for the eigenfunction $\psi(x,k)$ [analytic for $k \in \mathbb{K}_r^+$], and $k \in \Gamma_r^-$ for $\bar{\psi}(x,k)$ [analytic for $k \in \mathbb{K}_r^-$]. We then formally integrate (3.6) with respect to $\lambda_r \in \mathbb{R}$, multiply by $1/2\pi$, exchange the order of integration, and evaluate

$$\frac{1}{2\pi}\int_{-\infty}^\infty d\lambda_r \begin{pmatrix} 1 \\ -iq_r^*/(\lambda_r + k) \end{pmatrix} e^{i\lambda_r(y-s)} = \begin{pmatrix} \delta(y-s) \\ 0 \end{pmatrix}$$

to obtain

$$\mathcal{I} = K(x,y)\begin{pmatrix}1\\0\end{pmatrix} + F(x+y) + \int_x^\infty ds\, K(x,s)F(s+y),$$

where

$$\mathcal{I} \equiv \frac{1}{2\pi}\int_{-\infty}^\infty d\lambda_r \left[\frac{e^{i\lambda_r x}\phi(x,k)}{a(k)} - W_{r,1}(k)\right]e^{i\lambda_r(y-x)},$$

$$F(x) = \frac{1}{2\pi}\int_{-\infty}^\infty d\lambda_r\, \rho(k)W_{r,2}(k)\,e^{i\lambda_r x}\,.$$

As explained above, in the above integrals $k = k(\lambda_r)$ with $k \in \Gamma_r^+$. The procedure and the results are the same shown in [**19**]. We assume discrete eigenvalues $k_1,\dots,k_N$ in $\mathbb{K}_r^+$ are simple and finite in number. Since the eigenfunctions $\phi(x,k_n)$ and $\psi(x,k_n)$ are proportional, i.e., there exists a complex constant $b_n$ such that

$\phi_n(x, k_n) = b_n \psi(x, k_n)$, then denoting by $\tau_n$ the residue of $1/a(k)$ at $\lambda_r = \lambda_r(k_n)$, we can write

$$(3.7) \qquad \lim_{k \to k_n} (\lambda_r(k) - \lambda_r(k_n)) \frac{\phi(x, k)}{a(k)} = C_n \, \psi(x, k_n), \qquad C_n = b_n \tau_n \,,$$

and $C_n$ is referred to as the *norming constant* associated with the discrete eigenvalue $k_n$. The expression of $\mathcal{I}$ can be easily obtained using Residue Theorem and Jordan's Lemma, thus yielding the right Marchenko integral equation as:

$$(3.8) \qquad K(x, y) \begin{pmatrix} 1 \\ 0 \end{pmatrix} + \Omega_r(x + y) + \int_x^\infty ds \, K(x, s) \Omega_r(s + y) = \begin{pmatrix} 0 \\ 0 \end{pmatrix} \,,$$

where

$$(3.9) \qquad \Omega_r(x) := F(x) - F_d(x), \qquad F_d(x) = i \sum_{n=1}^N e^{i\lambda_r(k_n)x} C_n W_{r,2}(k_n) \,.$$

Next, let us multiply (3.5c) by $e^{-i\lambda_r y}$ for $y > x$, substitute (3.2a), and then formally integrate (3.6) with respect to $\lambda_r \in \mathbb{R}$ and multiply by $1/2\pi$. At the (simple) discrete eigenvalues $k_1^*, \ldots, k_N^*$ in $\mathbb{K}_r^-$ [necessarily finite in number, and the complex conjugates of the zeros of $a(k)$ in $\mathbb{K}_r^+$] the eigenfunctions $\bar\phi(x, k_n^*)$ and $\bar\psi(x, k_n^*)$ are proportional to each other, i.e., there exist complex constants $\bar b_n$ such that $\bar\phi(x, k_n^*) = \bar b_n \bar\psi(x, k_n^*)$. Then, denoting by $\bar\tau_n$ the residue of $1/\bar a(k)$ at $\lambda_r = \lambda_r(k_n^*)$, we can write

$$(3.10) \qquad \lim_{k \to k_n^*} (\lambda_r - \lambda_r(k_n^*)) \frac{\bar\phi(x, k)}{\bar a(k)} = \bar C_n \bar\psi(x, k_n^*), \qquad \bar C_n = \bar b_n \bar\tau_n \,,$$

and $\bar C_n$ is referred to as the *norming constant* associated with the discrete eigenvalue $k_n^*$. Proceeding as before, we obtain

$$(3.11) \qquad K(x, y) \begin{pmatrix} 0 \\ 1 \end{pmatrix} + \bar\Omega_r(x + y) + \int_x^\infty ds \, K(x, s) \bar\Omega_r(s + y) = \begin{pmatrix} 0 \\ 0 \end{pmatrix} \,,$$

where

$$(3.12) \qquad \bar\Omega_r(x) := \bar F(x) - \bar F_d(x) \,,$$

with
(3.13)

$$\bar F(x) = \frac{1}{2\pi} \int_{-\infty}^\infty d\lambda_r \, e^{-i\lambda_r x} \bar\rho(k) W_{r,1}(k), \qquad \bar F_d(x) = -i \sum_{n=1}^N e^{-i\lambda_r(k_n^*)x} \bar C_n W_{r,1}(k_n^*) \,.$$

Note that in the integral in (3.13) $\lambda_r \in \mathbb{R}$, and $k = k(\lambda_r) \in \Gamma_r^-$ (see Fig. 2). Using the symmetry relations (2.33) and (2.34), and the definitions (3.7) and (3.10), we get $\bar\tau_n = \tau_n^*$, $\bar b_n = -b_n^*$ and $\bar C_n = -C_n^*$. As a result,

$$(3.14) \qquad F^*(x) = i\sigma_2 \bar F(x), \quad F_d^*(x) = i\sigma_2 \bar F_d(x), \quad \Omega_r^*(x) = i\sigma_2 \bar\Omega_r(x).$$

In conclusion, we can write the Marchenko equations (3.8) and (3.11) and their kernels (3.9) and (3.12) as a single $2 \times 2$ Marchenko equation:

$$(3.15) \qquad K(x, y) + \boldsymbol{\Omega}_r(x + y) + \int_x^\infty ds \, K(x, s) \boldsymbol{\Omega}_r(s + y) = 0_{2\times2} \,,$$

where

$$(3.16) \qquad \boldsymbol{\Omega}_r(x) = \begin{pmatrix} \Omega_r(x) & \bar\Omega_r(x) \end{pmatrix} \,.$$

$\Omega_r$, $\bar{\Omega}_r$ are given by (3.9) and (3.12), and satisfy $\bar{\Omega}_r(x) = i\sigma_2 \Omega_r^*(x)$. Note that the $2 \times 2$ kernel $\Omega_r(x)$ anticommutes with the Pauli matrix $\sigma_3$, and satisfies the conjugation symmetry relation

$$(3.17) \qquad \Omega_r^*(x) = \sigma_2 \Omega_r(x) \sigma_2.$$

3.2.2. *Left Marchenko equations.* In order to derive the left Marchenko equations, let us write (2.23b) explicitly as:

$$(3.18a) \qquad \frac{1}{\bar{c}(k)} \bar{\psi}(x,k) = \phi(x,k) + \bar{r}(k)\bar{\phi}(x,k) \quad k \in \mathbb{R},$$

$$(3.18b) \qquad \frac{1}{c(k)} \psi(x,k) = \bar{\phi}(x,k) + r(k)\phi(x,k) \quad k \in \mathbb{R},$$

where $r(k)$ and $\bar{r}(k)$ are given by (2.31c). Under the same assumptions as in Sec. 3.2.1 regarding the potential and the discrete spectrum, and considering in this case the eigenfunctions as functions of $k$, we multiply (3.18a) by $e^{-iky}$ for $y < x$ and substitute the triangular representations (3.2b). Integrating (3.18a) with respect to $k \in \mathbb{R}$, multiplying by $1/2\pi$, and exchanging the order of integration, we have

$$(3.19) \qquad \tilde{\mathcal{I}} = J(x,y) \begin{pmatrix} 0 \\ 1 \end{pmatrix} + G(x+y) + \int_{-\infty}^{x} ds\, J(x,s)G(s+y),$$

where

$$\tilde{\mathcal{I}} \equiv \frac{1}{2\pi} \int_{-\infty}^{\infty} dk \left[ \frac{e^{-ikx}\psi(x,k)}{c(k)} - \begin{pmatrix} 0 \\ 1 \end{pmatrix} \right] e^{ik(x-y)},$$

$$G(x) = \frac{1}{2\pi} \int_{-\infty}^{\infty} dk\, r(k)\, e^{-ikx} \begin{pmatrix} 1 \\ 0 \end{pmatrix}.$$

In order to compute $\tilde{\mathcal{I}}$ so as to express it in terms of the Marchenko kernel $J(x,y)$, one needs to be able to close the contour at infinity in the upper half-plane of $k$. Unlike what happens for the Marchenko equations from the right, in this case closing the contour at infinity requires including the contribution of the branch cut that corresponds to $k \in [0, iA_r]$, i.e. $\Sigma_r$ in the upper half-plane. To this end, let us consider, for $0 < \varepsilon < R < +\infty$, the closed contour $\Gamma(R,\varepsilon)$ consisting of the following pieces, with the orientation specified in Fig. 3: (i) $[-R, -\varepsilon]$; (ii) $[-\varepsilon + i0, -\varepsilon + iA_r]$; (iii) the semicircle $\{iA_r + \varepsilon e^{i(\pi-\theta)} : 0 \le \theta \le \pi\}$ clockwise; (iv) $[\varepsilon + i0, \varepsilon + iA_r]$; (v) $[\varepsilon, R]$; (vi) $\{Re^{i\theta} : 0 \le \theta \le \pi\}$ counterclockwise. $R$ is assumed large enough and $\varepsilon$ small enough so that all of the finitely many discrete eigenvalues $k_n$ ($n = 1, 2, \ldots, N$) in $\mathbb{K}_r^+$ belong to the interior region of the contour. Since $\psi(x,k)$ and $1/c(k)$ have finite limits as $k \to iA_r$, the integral defining $\tilde{\mathcal{I}}$ with the integration confined to the semicircle around the branch point does not contribute as $\varepsilon \to 0^+$. Because of Jordan's Lemma, the integral defining $\tilde{\mathcal{I}}$ when confined to the large semicircle (vi) does not contribute either as $R \to +\infty$. Then one has $\tilde{\mathcal{I}} = \tilde{\mathcal{I}}_1 + \tilde{\mathcal{I}}_2$, the contribution $\tilde{\mathcal{I}}_1$ pertaining to the residues of the function under the integral sign at the poles $k_n \in \mathbb{K}_r^+$; and the contribution $\tilde{\mathcal{I}}_2$ pertaining to the integral around $k \in [0, iA_r]$ in the upper-half $k$-plane. We shall evaluate the two contributions separately.

Since we assumed that the discrete eigenvalues $k_n$ in $\mathbb{K}_r^+$ are simple poles of $1/c(k)$, and the transmission coefficient is continuous for $k \in \partial\mathbb{K}_r^+$, taking into

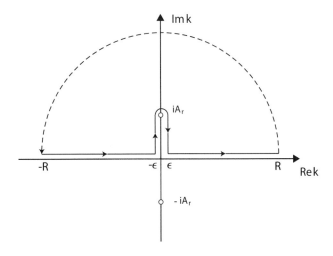

FIGURE 3. The contour $\Gamma(R, \varepsilon)$.

account that $\psi(x, k_n) = \phi(x, k_n)/b_n$, we obtain

$$\tilde{\mathcal{I}}_1 = i \sum_{n=1}^{N} e^{-ik_n y} \tilde{C}_n \phi(x, k_n) , \qquad \tilde{C}_n = \frac{\tilde{\tau}_n}{b_n} ,$$

where $\tilde{\tau}_n$ is the residue of $1/c(k)$ at $k = k_n$, and $\tilde{C}_n$ is the associated norming constant. Therefore, we have

$$(3.20) \qquad \tilde{\mathcal{I}}_1 = G_1(x + y) + \int_{-\infty}^{x} ds\, J(x, s) G_1(s + y) ,$$

where

$$G_1(x) = i \sum_{n=1}^{N} e^{-ik_n x} \tilde{C}_n \begin{pmatrix} 1 \\ 0 \end{pmatrix} .$$

Note that (2.28c) implies the residues $\tilde{\tau}_n$ and $\tau_n$, and hence the norming constants $\tilde{C}_n$ and $C_n$, are related as follows:

$$(3.21) \qquad \tilde{\tau}_n = \frac{\lambda_r(k_n) + k_n}{2\lambda_r(k_n)} \tau_n, \qquad \tilde{C}_n\, C_n = \tau_n^2 \frac{\lambda_r(k_n) + k_n}{2\lambda_r(k_n)} .$$

Let us now look into the second contribution $\tilde{\mathcal{I}}_2$, which arises for $k \in [0, iA_r]$ on either side of the cut and $\lambda_r \in \mathbb{R}$. We have

$$\tilde{\mathcal{I}}_2 = \lim_{\epsilon \to 0} \frac{1}{2\pi} \left( \int_{i0-\epsilon}^{iA_r-\epsilon} - \int_{i0+\epsilon}^{iA_r+\epsilon} \right) dk \left[ \frac{\psi(x, k)}{c(k)} e^{-ikx} - \begin{pmatrix} 0 \\ 1 \end{pmatrix} \right] e^{ik(x-y)}$$

$$(3.22) \qquad = \frac{1}{2\pi} \int_{0}^{iA_r} dk \left[ \frac{\psi^-(x, k)}{c^-(k)} - \frac{\psi^+(x, k)}{c^+(k)} \right] e^{-iky} ,$$

where, as usual, superscripts $^\pm$ denote the limiting values from the left/right edge of the cut, respectively. Using (2.28c) we can write $\tilde{\mathcal{I}}_2$ as

$$\tilde{\mathcal{I}}_2 = \frac{1}{2\pi} \int_{0}^{iA_r} dk \left[ \frac{\lambda_r^+(k) - k}{2\lambda_r^+(k)} \frac{\psi^-(x, k)}{a^-(k)} - \frac{\lambda_r^+(k) + k}{2\lambda_r^+(k)} \frac{\psi^+(x, k)}{a^+(k)} \right] e^{-iky} ,$$

and the symmetry relations (2.36) and (2.37a) allow to express $(\lambda_r^+(k)-k)\psi^-/a^- = -(\lambda_r^+(k)+k)\bar\psi^+/b^+$, so that:

$$(3.23) \qquad \tilde{\mathcal{I}}_2 = -\frac{1}{4\pi}\int_0^{iA_r} dk\frac{\lambda_r^+(k)+k}{\lambda_r^+(k)}\left[\frac{\bar\psi^+(x,k)}{b^+(k)}+\frac{\psi^+(x,k)}{a^+(k)}\right]e^{-iky}.$$

Using first the scattering equation (3.5a) and then again the symmetry relation (2.37a), we finally have (3.24)

$$\tilde{\mathcal{I}}_2 = -\frac{1}{4\pi}\int_0^{iA_r} dk\frac{\lambda_r^+(k)+k}{\lambda_r^+(k)}\frac{\phi^+(x,k)}{a^+(k)\,b^+(k)}e^{-iky} = \frac{iq_r}{4\pi}\int_0^{iA_r}\frac{dk}{\lambda_r^+(k)}\frac{\phi^+(x,k)}{a^-(k)a^+(k)}e^{-iky}.$$

We can now insert into the last expression the triangular representation (3.2b), and obtain

$$(3.25) \qquad \tilde{\mathcal{I}}_2 = G_2(x+y)+\int_{-\infty}^x ds\,J(x,s)G_2(s+y),$$

with

$$G_2(x) = \frac{iq_r}{4\pi}\int_0^{iA_r}\frac{dk}{\lambda_r^+(k)}\frac{e^{-ikx}}{a^+(k)a^-(k)}\begin{pmatrix}1\\0\end{pmatrix}.$$

If we now define

$\Omega_l(x) = G(x) - G_1(x) - G_2(x)$
(3.26)

$$\equiv \left[\frac{1}{2\pi}\int_{-\infty}^\infty dk\,r(k)\,e^{-ikx} - i\sum_{n=1}^N e^{-ik_n x}\tilde{C}_n - \frac{iq_r}{4\pi}\int_0^{iA_r}\frac{dk}{\lambda_r^+(k)}\frac{e^{-ikx}}{a^+(k)a^-(k)}\right]\begin{pmatrix}1\\0\end{pmatrix},$$

use (3.20) and (3.25) to compute $\tilde{\mathcal{I}} = \tilde{\mathcal{I}}_1+\tilde{\mathcal{I}}_2$ and introduce it into (3.19), we finally arrive at the left Marchenko integral equation

$$(3.27) \qquad J(x,y)\begin{pmatrix}0\\1\end{pmatrix}+\Omega_l(x+y)+\int_{-\infty}^x ds\,J(x,s)\Omega_l(s+y) = \begin{pmatrix}0\\0\end{pmatrix}.$$

In a similar way, starting from (3.18b), one can derive the "adjoint" left Marchenko equation

$$(3.28) \qquad J(x,y)\begin{pmatrix}1\\0\end{pmatrix}+\bar\Omega_l(x+y)+\int_{-\infty}^x ds\,J(x,s)\bar\Omega_l(s+y) = \begin{pmatrix}0\\0\end{pmatrix},$$

where

$$\bar\Omega_l(x) = i\sigma_2\Omega_l^*(x).$$

The two Marchenko equations can be written in a compact matrix form as follows:

$$(3.29) \qquad J(x,y)+\boldsymbol{\Omega}_l(x+y)+\int_{-\infty}^x ds\,J(x,s)\boldsymbol{\Omega}_l(s+y) = 0_{2\times2},$$

where

$$(3.30) \qquad \boldsymbol{\Omega}_l(x) = \begin{pmatrix}\bar\Omega_l(x)&\Omega_l(x)\end{pmatrix},\qquad \Omega_l^*(x) = \sigma_2\boldsymbol{\Omega}_l(x)\sigma_2.$$

The asymmetry between left/right Marchenko integral equations is due to the choice of the one-sided NZBC (1.3) with $A_r>0$. Indeed, if one considered boundary conditions such that $q(x,t)\to q_l(t) = A_l\,e^{-2iA_l^2 t+i\theta_l}$ as $x\to-\infty$, and $q(x,t)\to 0$ as $x\to+\infty$, with $A_l>0$, $0\le\theta_l<2\pi$ arbitrary constants, the roles of the two integral equations would be reversed. As in the fully asymmetric case, in the Marchenko integral equations from the left, $\Omega_l(x)$ (cf. Eq. (3.26)) has three separate contributions: one from the discrete spectrum, one from the reflection coefficients from the left, $r(k)$ and $\bar{r}(k)$, integrated over values of $k$ in the continuous spectrum, i.e., $k \in \mathbb{R}$, and a third contribution (sometimes referred to as the dispersive-shock wave, or DSW, contribution) which contains an integral over imaginary values of $k$ where the product of transmission coefficients $1/[a^+(k)a^-(k)]$ appears. On the other hand, $\Omega_r(x)$ in the integral equations from the right (cf. Eq. (3.9)) has only two contributions: one from the discrete spectrum, and one from the reflection coefficients from the right, $\rho(k)$ and $\bar{\rho}(k)$. In (3.9), however, the reflection coefficients are integrated over all $\lambda_r \in \mathbb{R}$, which means that the integral includes, in addition to the continuous spectrum $k \in \mathbb{R}$, also a contribution from $k \in \Sigma_r$. Moreover, the integrand over $\Sigma_r$ can never be set to be identically zero (due to the symmetries (2.38a) and (2.38b)), which implies that when $\Sigma_r \neq \emptyset$ (i.e., whenever one deals with one-sided boundary conditions (1.3) with $A_r \neq 0$), no pure soliton solutions exist, and solitons are always accompanied by a radiative contribution of some sort.

The Marchenko integral equations obtained here provide the necessary setup for the study of the long-time behavior of the solutions by means of matched asymptotics, as was recently done for KdV in [1].

### 3.3. Riemann-Hilbert problem formulation.
The purpose of this section is to formulate the inverse problem as matrix Riemann-Hilbert (RH) problems from the left and from the right for a suitable set of sectionally analytic/meromorphic functions in the cut plane $\mathbb{K}_r$, with assigned jumps across $\mathbb{R} \cup \Sigma_r$.

3.3.1. *Riemann-Hilbert problem from the right.* For the formulation of the Riemann-Hilbert problem in terms of scattering data from the right, we consider the following matrix of eigenfunctions:

(3.31)
$$M(x,k) = \begin{cases} \left[ \dfrac{\phi(x,k)}{a(k)} e^{ikx} \quad \psi(x,k) e^{-i\lambda_r x} \right] & k \in \mathbb{K}_r^+ \\[4mm] \left[ \bar{\psi}(x,k) e^{i\lambda_r x} \quad \dfrac{\bar{\phi}(x,k)}{\bar{a}(k)} e^{-ikx} \right] & k \in \mathbb{K}_r^- \end{cases},$$

such that $M(x,k) \to I_2$ as $k \to \infty$, and formulate the inverse problem as a Riemann-Hilbert problem for the sectionally meromorphic matrix $M(x,k)$ across $\partial\mathbb{K}_r^+ \cup \partial\mathbb{K}_r^-$. Explicitly, we determine three jump matrices as illustrated in Fig. 4: $V_0$ is the jump matrix across the real axis of the complex $k$-plane; $V_1$ across $\Sigma_r^+ = [0, iA_r]$, and $V_2$ across $\Sigma_r^- = [-iA_r, 0]$. All jump matrices depend on $k$ along the appropriate contour in the complex plane, as well as, parametrically, on $(x,t) \in \mathbb{R} \times \mathbb{R}^+$ [the $x$-dependence is explicit, while the time dependence is "hidden" in that of the corresponding reflection coefficients, see Section 4, and will be omitted for brevity].

The RH problem across the real axis can be written in matrix form as: $M^+(x,k) = M^-(x,k)V_0(x,k)$, $k \in \mathbb{R}$, i.e.,
(3.32)
$$\left[ \frac{\phi^+(x,k)}{a^+(k)} e^{ikx} \quad \psi^+(x,k) e^{-i\lambda_r x} \right] = \left[ \bar{\psi}^-(x,k) e^{i\lambda_r x} \quad \frac{\bar{\phi}^-(x,k)}{\bar{a}^-(k)} e^{-ikx} \right] V_0(x,k),$$

where in this case the superscripts $\pm$ denote limiting values from the upper/lower complex plane, respectively. The jump matrix across the real axis can be easily

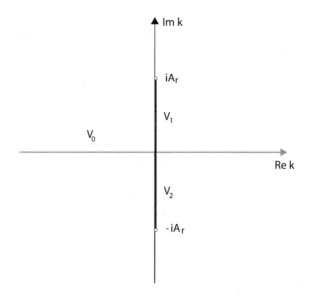

FIGURE 4. The jump matrices $V_j$, $j = 0, 1, 2$ of the RH problem across $\mathbb{R} \cup \Sigma_r^+ \cup \Sigma_r^-$.

computed from (2.23a), and it is given by:

$$(3.33) \qquad V_0(x, k) = \begin{pmatrix} [1 - \rho(k)\bar{\rho}(k)] \, e^{i(k-\lambda_r)x} & -\bar{\rho}(k) \, e^{-2i\lambda_r x} \\ \rho(k) \, e^{2ikx} & e^{i(k-\lambda_r)x} \end{pmatrix}.$$

We then write the RH problem across $\Sigma_r^+$ as: $M^+(x, k) = M^-(x, k)V_1(x, k)$, $k \in \mathbb{C}^+$, where now the superscripts $\pm$ denote limiting values from the right/left edge of the cut across $\Sigma_r^+$ ($\Sigma_r$ in the upper half plane). Taking into account that for $k \in \Sigma_r^+$ one has $\lambda_r^{\pm}(k) = \pm\sqrt{A_r^2 - |k|^2}$ from the right/left edge of the cut, while $k$ and $\phi$ are continuous, we have:

$$(3.34)$$
$$\left[ \frac{\phi^+(x, k)}{a^+(k)} e^{ikx} \quad \psi^+(x, k) e^{-i\lambda_r^+ x} \right] = \left[ \frac{\phi^-(x, k)}{a^-(k)} e^{ikx} \quad \psi^-(x, k) e^{-i\lambda_r^- x} \right] V_1(x, k).$$

In order to compute $V_1(x, k)$, we note that using (2.29) we can write:

$$(3.35) \qquad \phi^{\pm}(x, k) = a^{\pm}(k)\bar{\psi}^{\pm}(x, k) + b^{\pm}(k)\psi^{\pm}(x, k), \quad k \in \Sigma_r^+,$$

and relate $\bar{\psi}^{\pm}(x, k)$ to $\psi^{\mp}(x, k)$ using the symmetry relations (2.36). One then finds

$$(3.36) \qquad \frac{\phi^+(x, k)}{a^+(k)} = -\frac{iq_r^*}{\lambda_r^+(k) + k}\psi^-(x, k) + \rho^+(k)\psi^+(x, k),$$

$$(3.37) \qquad \psi^+(x, k) = -\frac{iq_r}{\lambda_r^+(k) + k}\left[ \frac{\phi^-(x, k)}{a^-(k)} - \rho^-(k)\psi^-(x, k) \right],$$

and inserting (3.37) into (3.36), we obtain:

$$(3.38)$$
$$\frac{\phi^+(x, k)}{a^+(k)} = -\frac{iq_r}{\lambda_r^+(k) + k}\left[ \frac{\phi^-(x, k)}{a^-(k)}\rho^-(k) + \left( \frac{q_r^*}{q_r} - \rho^+(k)\rho^-(k) \right) \psi^-(x, k) \right].$$

Using (2.39) and comparing (3.38) and (3.37) to (3.34), the jump matrix $V_1(x,k)$ is then found to be:

$$(3.39) \qquad V_1(x,k) = -\frac{iq_r}{\lambda_r^+(k)+k}\begin{pmatrix} \rho^+(k) & e^{-i(\lambda_r^+(k)+k)x} \\ 0 & -\rho^-(k)\,e^{-2i\lambda_r^+(k)x} \end{pmatrix}.$$

The RH problem across $\Sigma_r^-$ will be written as $M^+(x,k) = M^-(x,k)V_2(x,k)$, $k \in \mathbb{C}^-$, with superscripts $\pm$ denoting non-tangential limits from the right/left of the cut across $\Sigma_r^-$, i.e., $\Sigma_r$ in the lower half plane. Explicitly, using (2.13) one has:

$$(3.40)$$
$$\left[\bar\psi^+(x,k)\,e^{i\lambda_r^+ x} \quad \frac{\bar\phi^+(x,k)}{\bar a^+(k)}\,e^{-ikx}\right] = \left[\bar\psi^-(x,k)\,e^{i\lambda_r^- x} \quad \frac{\bar\phi^-(x,k)}{\bar a^-(k)}\,e^{-ikx}\right] V_2(x,k).$$

In order to compute $V_2(x,k)$, we again use (2.29) to write:

$$(3.41) \qquad \bar\phi^\pm(x,k) = \psi^\pm(x,k) + \bar\rho^\pm(k)\bar\psi^\pm(x,k), \quad k \in \Sigma_r^-,$$

and relate the eigenfunctions $\psi^\pm(x,k)$ to $\bar\psi^\mp(x,k)$ using the symmetry relations (2.36). We the obtain the following expressions:

$$(3.42) \qquad \frac{\bar\phi^+(x,k)}{\bar a^+(k)} = -\frac{iq_r}{\lambda_r^+(k)+k}\bar\psi^-(x,k) + \bar\rho^+(k)\bar\psi^+(x,k),$$

$$(3.43) \qquad \bar\psi^+(x,k) = -\frac{iq_r^*}{\lambda_r^+(k)+k}\left[\frac{\bar\phi^-(x,k)}{\bar a^-(k)} - \rho^-(k)\bar\psi^-(x,k)\right].$$

Inserting (3.43) into (3.42), one has:

$$(3.44)$$
$$\frac{\bar\phi^+(x,k)}{\bar a^+(k)} = -\frac{iq_r^*}{\lambda_r^+(k)+k}\left[\frac{\bar\phi^-(x,k)}{\bar a^-(k)}\bar\rho^+(k) + \left(\frac{q_r}{q_r^*} - \bar\rho^+(k)\bar\rho^-(k)\right)\bar\psi^-(x,k)\right].$$

Using (2.39) and comparing (3.44) and (3.43) to (3.40), the jump matrix $V_2(x,k)$ is found to be:

$$(3.45) \qquad V_2(x,k) = -\frac{iq_r^*}{\lambda_r^+(k)+k}\begin{pmatrix} -\bar\rho^-(k)\,e^{2i\lambda_r^+(k)x} & 0 \\ e^{i(\lambda_r^+(k)+k)x} & \bar\rho^+(k) \end{pmatrix}.$$

Note that the jump matrices satisfy the following upper/lower half plane symmetry:

$$V_2(x,k) = \sigma_2 V_1^*(x,k^*)\sigma_2.$$

Solving the inverse problem as a RH problem (with poles, corresponding to the zeros of $a(k)$ and $\bar a(k)$ in the upper/lower half planes) then amounts to computing the sectionally meromorphic matrix $M(x,k)$ with the given jumps, and normalized to the identity as $k \to \infty$. Specifically, we can write the problem as $M^+ = M^- + (V - I_2)M^-$, where $V(x,k) = V_j(x,k)$ for $j = 0,1,2$ depending on which piece of the contour is being considered, and superscripts $\pm$ denote non-tangential limits from either side of the contour. Then, subtracting the behavior as $k \to \infty$, and the residues of $M^\pm$ at the poles in $\mathbb{K}_r^\pm$ from both sides we obtain

$$(3.46) \quad M^+ - I_2 - \sum_{n=1}^{N}\frac{1}{k-k_n}\mathrm{Res}_{k_n}M^+ - \sum_{n=1}^{N}\frac{1}{k-k_n^*}\mathrm{Res}_{k_n^*}M^- =$$

$$M^- - I_2 - \sum_{n=1}^{N}\frac{1}{k-k_n}\mathrm{Res}_{k_n}M^+ - \sum_{n=1}^{N}\frac{1}{k-k_n^*}\mathrm{Res}_{k_n^*}M^- + (V - I_2)M^-.$$

The left-hand side of the above equation is now analytic in $\mathbb{K}_r^+$, and it is $O(1/k)$ as $k \to \infty$ there, while the sum of all terms but the last one in the right-hand side is analytic in $\mathbb{K}_r^-$, and is $O(1/k)$ as $k \to \infty$ there. We then introduce projectors $P_\pm$ over $\Gamma_r^\pm \equiv \mathbb{R} \cup \Sigma_r^\pm$:

$$P_\pm[f](z) = \frac{1}{2\pi i} \int_{\Gamma_r^\pm} \frac{f(\xi)}{\xi - k} \, d\xi \,,$$

where $\int_{\Gamma_r^+}$ [resp. $\int_{\Gamma_r^-}$] denotes the integral along the oriented contours in Fig. 2, and when $k \in \Gamma_r^\pm \cap \mathbb{R}$ the limit is taken from the above/below. One can easily prove that if $f^\pm$ are analytic in $\mathbb{K}_r^\pm$ and are $O(1/k)$ as $k \to \infty$, the following holds: $P_\pm f^\pm = \pm f^\pm$ and $P_+ f^- = P_- f^+ = 0$. Then, applying $P_\pm$ to both sides of (3.46), we find for $k \in \mathbb{C}^\pm \setminus \Sigma_r$

$$(3.47) \quad M(k) = I_2 + \sum_{n=1}^{N} \frac{\operatorname{Res}_{k_n} M^+}{k - k_n} + \sum_{n=1}^{N} \frac{\operatorname{Res}_{k_n^*} M^-}{k - k_n^*} + \frac{1}{2\pi i} \int_{\Gamma_r^\pm} \frac{M^-(\xi)}{\xi - k} [V(\xi) - I_2] \, d\xi \,,$$

where the $x$-dependence in eigenfunctions and jump matrices has been omitted for brevity. Taking into account that the second column of $\operatorname{Res}_{k_n} M^+$ is zero for all $n$, while the first column is proportional to the second column of $M^+(x, k_n)$, and vice-versa the first column of $\operatorname{Res}_{k_n^*} M^-$ is zero for all $n$, while the second column is proportional to the second column of $M^-(x, k_n^*)$ according to (3.7), the above integral/algebraic system can be closed by evaluating it at each $k = k_n$ and $k = k_n^*$. The potential is then reconstructed by the large $k$ expansion of the latter, since

$$M_o(x, k) = \frac{i}{2k} Q(x) \sigma_3 + o(1/k) \,,$$

where subscript $_o$ denotes the off-diagonal part of the matrix $M(x, k)$. Note that unlike what happens in the same-amplitude case, the above system cannot be reduced to a purely algebraic one: although the reflection coefficients can be chosen to be identically zero on the continuous spectrum, i.e., for $k \in \mathbb{R}$, the integrals appearing in the right-hand side of (3.47) always exhibit a non-zero contribution from the contours $\Sigma_r^\pm$. In particular, this implies that no pure soliton solutions exist, and solitons are always accompanied by a radiative contribution of some sort. One could nonetheless solve the system iteratively, assuming the reflection coefficients are small for $k \in \Sigma_r^\pm$, and thus obtaining NLS solutions comprising solitons superimposed to small radiation.

3.3.2. *Riemann-Hilbert problem from the left.* The inverse problem can also be formulated as a RH problem from the left, i.e., written in terms of left scattering data, by introducing the sectionally meromorphic matrix of eigenfunctions

$$\tilde{M}(x, k) = \begin{cases} \left[ \phi(x, k) \, e^{ikx} \quad \dfrac{\psi(x, k)}{c(k)} \, e^{-i\lambda_r x} \right], & k \in \mathbb{K}_r^+ \\[3mm] \left[ \dfrac{\bar{\psi}(x, k)}{\bar{c}(k)} \, e^{i\lambda_r x} \quad \bar{\phi}(x, k) \, e^{-ikx} \right], & k \in \mathbb{K}_r^- \end{cases} .$$

The RH problem across the real $k$-axis can be written in matrix form as: $\tilde{M}^+(x, k) = \tilde{M}^-(x, k) \tilde{V}_0(x, k)$, $k \in \mathbb{R}$ i.e.,
(3.48)

$$\left[ \phi^+(x, k) \, e^{ikx} \quad \frac{\psi^+(x, k)}{c^+(k)} \, e^{-i\lambda_r x} \right] = \left[ \frac{\bar{\psi}^-(x, k)}{\bar{c}^-(k)} \, e^{i\lambda_r x} \quad \bar{\phi}^-(x, k) \, e^{-ikx} \right] \tilde{V}_0(x, k) \,,$$

where in this case the superscripts $\pm$ denote limiting values from the upper/lower complex plane, respectively. The jump matrix across the real axis can be easily computed from (2.23b), and it is given by:

$$(3.49) \qquad \tilde{V}_0(x,k) = \begin{pmatrix} e^{i(k-\lambda_r)x} & r(k)\,e^{-2i\lambda_r x} \\ -\bar{r}(k)\,e^{2ikx} & [1-r(k)\bar{r}(k)]\,e^{i(k-\lambda_r)x} \end{pmatrix}.$$

In the RH problem across $\Sigma_r^+$, one has

$$\tilde{M}^+(x,k) = \left[ \phi^+(x,k)\,e^{ikx} \quad \frac{\psi^+(x,k)}{c^+(k)}\,e^{-i\lambda_r^+ x} \right],$$

$$\tilde{M}^-(x,k) = \left[ \phi^-(x,k)\,e^{ikx} \quad \frac{\psi^-(x,k)}{c^-(k)}\,e^{-i\lambda_r^- x} \right].$$

Note, however, that unlike what happens in the RH problem from the right, here one cannot use (2.23b) to determine the jump. The same holds for the RH problem on $\Sigma_r^-$. In fact, in both equations (2.23b), i.e., $\bar{\psi}(x,k) = \bar{c}(k)\phi(x,k) + \bar{d}(k)\bar{\phi}(x,k)$ and $\psi(x,k) = d(k)\phi(x,k) + c(k)\bar{\phi}(x,k)$, the right-hand sides are only simultaneously defined for $k \in \mathbb{R}$, and cannot be extended on either $\Sigma_r^+$ or $\Sigma_r^-$. This is also evident from (2.31c), where it is clear that, unlike $\rho(k)$ and $\bar{\rho}(k)$, which can be respectively continued on $\Sigma_r^+$ and $\Sigma_r^-$, the reflection coefficients from the left $r(k)$ and $\bar{r}(k)$ are only generically defined on the continuous spectrum, i.e., for $k \in \mathbb{R}$.

In order to formulate the RH problem from the left on $\Sigma_r$, one has to consider both pieces of the cut $\Sigma_r^+$ and $\Sigma_r^-$ simultaneously, and take into account that: (i) $\lambda_r$ changes sign across $\Sigma_r$; (ii) $\phi^+(x,k) = \phi^-(x,k)$ for $k \in \Sigma_r^+$, and $\bar{\phi}^+(x,k) = \bar{\phi}^-(x,k)$ for $k \in \Sigma_r^-$; (iii) $\psi^\pm(x,k)/c^\pm(k)$ and $\bar{\psi}^\pm(x,k)/\bar{c}^\pm(x,k)$ are related to each other via the symmetry relations (2.36), (2.40a) and (2.40b), i.e.:

$$(3.50a) \qquad \frac{\psi^\pm(x,k)}{c^\pm(x,k)} = \frac{\bar{\psi}^\mp(x,k)}{\bar{d}^\mp(x,k)} \qquad k \in \Sigma_r^+,$$

$$(3.50b) \qquad \frac{\bar{\psi}^\pm(x,k)}{\bar{c}^\pm(x,k)} = \frac{\psi^\mp(x,k)}{d^\mp(x,k)} \qquad k \in \Sigma_r^-.$$

Solving the RH problem from the left (with poles, corresponding to the zeros of $c(k)$ and $\bar{c}(k)$ in the upper/lower half planes, which, by (2.28c) are the same as the ones from the right) amounts to computing the sectionally meromorphic matrix $\tilde{M}(x,k)$ with the given jumps, and normalized to the identity as $k \to \infty$. The potential is then reconstructed by the large $k$ expansion of the latter, since

$$\tilde{M}_o(x,k) = \frac{i}{2k}Q(x)\sigma_3 + o(1/k),$$

where, as before, subscript $_o$ is used to denote the off-diagonal part of the matrix.

The RH problem formulated in Sections 3.3.1 and 3.3.2 provides the key setup for the investigation of the long-time asymptotic behavior by the Deift-Zhou steepest descent method [13, 14, 17]. The time dependence in the system is simply accounted for by the time dependence of the scattering coefficients, as described in Sec. 4. When one is interested only in capturing the leading order behavior of the solution for large $t$, the jumps across the contours illustrated in Fig. 4 and determined above can be simplified by suitable factorizations and contour deformations, and reduced to certain model problems for which "explicit" solutions (often expressed in terms of Riemann theta functions) can be sought for. This study obviously goes beyond the scope of the present paper, and will be the subject of future investigation.

## 4. Time evolution of the scattering data

The time evolution of the eigenfunctions is to be determined from (2.1b), which, taking into account that $q(x,t) \to 0$ as $x \to -\infty$, yields

$$(4.1) \qquad v_t \sim 2ik^2\sigma_3 v \quad \text{as } x \to -\infty.$$

The time-independent boundary condition (2.6b) for $\Phi = (\phi \ \ \bar\phi)$, however, is not compatible with the above time evolution. Therefore we define time-dependent functions $\varphi(x,k,t) = e^{A_\infty t}\phi(x,k,t)$ and $\bar\varphi(x,k,t) = e^{-A_\infty t}\bar\phi(x,k,t)$, with $A_\infty = 2ik^2$, to be solutions of (2.1b). Then the time evolution equations for the Jost solutions $\Phi = (\phi \ \ \bar\phi)$ are found to be:

$$(4.2) \qquad \partial_t \Phi = \left[i(2k^2 - |q|^2 + Q_x)\sigma_3 - 2kQ\right]\Phi - 2ik^2\,\Phi\,\sigma_3.$$

Similarly, taking into account that $q(x,t) \to A_r\, e^{-2iA_r^2 t + i\theta_r}$ as $x \to +\infty$, one finds the time evolution of the Jost solutions $\Psi = (\bar\psi \ \ \psi)$ to be given by:

$$(4.3) \qquad \partial_t \Psi = \left[i(2k^2 - |q|^2 + Q_x)\sigma_3 - 2kQ\right]\Psi - i(2k\lambda_r - A_r^2)\Psi\,\sigma_3.$$

Differentiating (2.23a) with respect to $t$ and taking into account the time evolution of the Jost solutions (4.2)-(4.3), we obtain for the scattering matrix $S(k,t)$ the following ODE with respect to $t$:

$$(4.4) \qquad \partial_t S = i(2k\lambda_r - A_r^2)\sigma_3 S - 2ik^2 S\,\sigma_3.$$

As a consequence, the scattering coefficients from the right are such that:

$$(4.5a) \quad a(k,t) = a(k,0)\,e^{i[2k(\lambda_r - k) - A_r^2]t}, \qquad \bar a(k,t) = \bar a(k,0)\,e^{i[2k(k-\lambda_r) + A_r^2]t},$$

$$(4.5b) \quad b(k,t) = b(k,0)\,e^{i[-2k(\lambda_r + k) + A_r^2]t}, \qquad \bar b(k,t) = \bar b(k,0)\,e^{i[2k(\lambda_r + k) - A_r^2]t},$$

$$(4.5c) \quad \rho(k,t) = \rho(k,0)\,e^{-2i(2k\lambda_r - A_r^2)t}, \qquad \bar\rho(k,t) = \bar\rho(k,0)\,e^{2i(2k\lambda_r - A_r^2)t}.$$

In particular, Eq. (4.5a) shows that the discrete eigenvalues $k_n$ are time independent, and given by the zeros of $a(k,0)$. Moreover, for the large $k$ behavior of $a(k,t)$, taking into account that

$$\lambda_r - k = \frac{A_r^2}{2k}\left[1 + O(k^{-2})\right],$$

one still finds from (4.5a) that, consistently with (2.42), $a(k,t) \sim 1$ as $|k| \to \infty$ for $k \in \mathbb{K}_r^+ \cup \mathbb{R}$ and for all $t \geq 0$.

Similarly, differentiating (2.23b) with respect to $t$ and taking into account the time evolution of the Jost solutions (4.2) and (4.3), we get for the scattering matrix $\bar S(k,t)$ the following ODE:

$$(4.6) \qquad \partial_t \bar S = -i(2k\lambda_r - A_r^2)\bar S\,\sigma_3 + 2ik^2\sigma_3\,\bar S.$$

Therefore, the time dependence of the scattering coefficients from the left is given by

(4.7a)    $c(k,t) = c(k,0)\, e^{i[2k(\lambda_r - k) - A_r^2]t}$ ,      $\bar{c}(k,t) = \bar{c}(k,0)\, e^{i[2k(k - \lambda_r) + A_r^2]t}$ ,

(4.7b)    $d(k,t) = d(k,0)\, e^{i[2k(\lambda_r + k) - A_r^2]t}$ ,      $\bar{d}(k,t) = \bar{d}(k,0)\, e^{i[-2k(\lambda_r + k) + A_r^2]t}$ ,

(4.7c)    $r(k,t) = r(k,0)\, e^{4ik^2 t}$ ,      $\bar{r}(k,t) = \bar{r}(k,0)\, e^{-4ik^2 t}$ .

Finally, we determine the time dependence of the norming constants. Differentiating $\phi(x, k_n) = b_n \psi(x, k_n)$ with respect to time and evaluating the first column of (4.2) and the second column of (4.3) at $k = k_n$, we get

$$b_n(t) = b_n(0) e^{-i[2k_n(k_n + \lambda_r(k_n)) - A_r^2]t}, \quad n = 1, \ldots, N.$$

Then from the definition of the norming constants in (3.7), we obtain

(4.8)    $$C_n(t) = C_n(0) e^{-2i[2k_n \lambda_r(k_n) - A_r^2]t}, \quad n = 1, \ldots, N.$$

The time dependence of the norming constants $\tilde{C}_n$ appearing in (3.26) can be found in a similar way, or simply taking into account the symmetry relation (3.21).

## 5. Direct and inverse scattering in the uniformization variable $z$

Unlike what happens when dealing with fully asymmetric boundary conditions [19], here we can introduce a uniformization variable $z$ (cf. [12, 20]) defined by the conformal mapping:

(5.1)    $$z = k + \lambda_r(k),$$

with inverse mapping given by:

$$k = \frac{1}{2}\left(z - \frac{A_r^2}{z}\right), \qquad \lambda_r = z - k = \frac{1}{2}\left(z + \frac{A_r^2}{z}\right).$$

The conformal transformation (5.1) maps the two-sheeted Riemann surface defined by $\lambda_r^2 = k^2 + A_r^2$ onto a single complex $z$-plane, as shown in Fig. 5. Specifically, one has:

- Sheet I [resp. Sheet II] of the Riemann surface is mapped onto the exterior [resp. interior] of the circle $C_r$ of radius $A_r$;
- The branch cut $\Sigma_r$ on either sheet is mapped onto $C_r$;
- The real $k$-axis on Sheet I [resp. Sheet II] is mapped onto $(-\infty, -A_r) \cup (A_r, +\infty)$ [resp. $(-A_r, A_r)$];
- $z(\pm i A_r) = \pm i A_r$ from either sheet, $z(0_I^\pm) = \pm A_r$, and $z(0_{II}^\pm) = \mp A_r$;
- $\operatorname{Im} k > 0$ [$\operatorname{Im} k < 0$] on either sheet is mapped into $\operatorname{Im} z > 0$ [$\operatorname{Im} z < 0$];
- The cut half-plane $\mathbb{K}_r^+$ [resp. $\mathbb{K}_r^-$] of Sheet I is mapped into the upper half $z$-plane [resp. lower half $z$-plane] outside the circle $C_r$;
- The cut half-plane $\mathbb{K}_r^+$ [resp. $\mathbb{K}_r^-$] of Sheet II is mapped into the upper half $z$-plane [resp. lower half $z$-plane] inside the circle $C_r$.

We introduce the following regions in the complex $z$-plane:

$$D_+ = \{z \in \mathbb{C} : (|z|^2 - A_r^2)\, \operatorname{Im} z > 0\}, \quad D_- = \{z \in \mathbb{C} : (|z|^2 - A_r^2)\, \operatorname{Im} z < 0\},$$

corresponding to $\operatorname{Im} \lambda_r > 0$ and $\operatorname{Im} \lambda_r < 0$, respectively, on either sheet. The complex $z$-plane is then partitioned into four regions: the upper/lower half $z$-plane

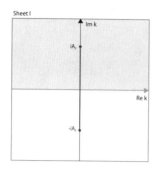
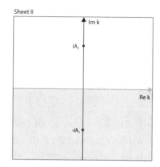
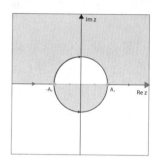

FIGURE 5. Left & Center: the two-sheets of the Riemann surface associated with $\lambda_r^2 = k^2 + A_r^2$. Right: the complex plane for the uniformization variable $z = k + \lambda_r$. The grey regions $(D_+)$ correspond to $\text{Im}\,\lambda_r > 0$, while the white regions $(D_-)$ correspond to $\text{Im}\,\lambda_r < 0$; the circle $C_r$ (red) corresponds to the cut $\Sigma_r$ on either sheet; $(-\infty, -A_r) \cup (A_r, +\infty)$ (blue) and $(-A_r, A_r)$ (green) correspond to the real $k$-axis on sheets I and II, respectively. The oriented contour in the complex $z$-plane for the Riemann-Hilbert problem is also shown.

outside the circle $C_r$, denoted as $D_\pm^{\text{out}}$, respectively, and the upper/lower half $z$-plane inside the circle $C_r$ denoted as $D_\pm^{\text{in}}$, respectively. In the following, we will also denote with $C_r^\pm$ the upper and lower semicircles of radius $A_r$, respectively.

The asymptotic behaviors (2.6b) and (2.17), expressed in terms of $z$, read

(5.2a)
$$\Phi(x,z) \sim I_2\, e^{-ik(z)x\sigma_3}\,, \qquad x \to -\infty\,,$$

(5.2b)
$$\Psi(x,z) \sim \left[ I_2 - \frac{i}{z}\sigma_3 Q_r \right] e^{-i\lambda_r(z)x\sigma_3}\,, \qquad x \to +\infty\,,$$

which then allows one to introduce the Jost solutions $\Psi(x,z) = \left( \bar\psi(x,z)\ \psi(x,z) \right)$ where $\lambda_r(z) \in \mathbb{R}$, i.e., for $z \in \mathbb{R} \cup C_r$, and $\Phi(x,z) = \left( \phi(x,z)\ \bar\phi(x,z) \right)$ where $k(z) \in \mathbb{R}$, i.e., for $z \in \mathbb{R}$. Consistently with Prop. 2.2 and the analogous analyticity properties of the Jost solutions on Sheet II, it then follows that if the potential satisfies $(\boldsymbol{H}_1)$ $\psi(x,z)$ can be analytically continued in $D_+$, $\bar\psi(x,z)$ is analytic in $D_-$, while $\phi(x,z)$ and $\bar\phi(x,z)$ are analytic in $\mathbb{C}^+$ and $\mathbb{C}^-$, respectively.

Unlike what happens with same-amplitude NZBCs [12], here the continuous spectrum, where all four eigenfunctions are simultaneously defined, corresponds to $z \in \mathbb{R}$, while the circle $C_r$ corresponds to the DSW region. In analogy to what discussed in Sec. 2.2, one can express the two sets of Jost solutions $\Phi(x,z)$ and $\Psi(x,z)$ as

(5.3a)
$$\Phi(x,z) = \Psi(x,z)\, S(z)\,, \qquad S(z) = \begin{pmatrix} a(z) & \bar b(z) \\ b(z) & \bar a(z) \end{pmatrix}\,, \qquad z \in \mathbb{R}\,,$$

(5.3b)
$$\Psi(x,z) = \Phi(x,z)\, \bar S(z)\,, \qquad S(z) = \begin{pmatrix} \bar c(z) & d(z) \\ \bar d(z) & c(z) \end{pmatrix}\,, \qquad z \in \mathbb{R}\,.$$

Then, expressing the scattering coefficients as Wronskians of the Jost solutions:

(5.4a) $\quad a(z) = \dfrac{z}{2\lambda_r(z)}\,\mathrm{Wr}(\phi(x,z),\psi(x,z)), \qquad \bar{a}(z) = \dfrac{z}{2\lambda_r(z)}\,\mathrm{Wr}(\bar{\psi}(x,z),\bar{\phi}(x,z)),$

(5.4b) $\quad b(k) = \dfrac{z}{2\lambda_r(z)}\,\mathrm{Wr}(\bar{\psi}(x,z),\phi(x,z)), \qquad \bar{b}(k) = \dfrac{z}{2\lambda_r(z)}\,\mathrm{Wr}(\bar{\phi}(x,z),\psi(x,z)),$

one can establish that:

- $a(z)$ is continuous in $\mathbb{R} \cup C_r^+ \setminus \{iA_r\}$ and analytic in $D_+^{\mathrm{out}}$
- $\bar{a}(z)$ is continuous in $\mathbb{R} \cup C_r^- \setminus \{-iA_r\}$ and analytic in $D_-^{\mathrm{out}}$
- $b(z)$ is continuous in $\mathbb{R} \cup C_r^+ \setminus \{iA_r\}$ and analytic in $D_-^{\mathrm{in}}$
- $\bar{b}(z)$ is continuous in $\mathbb{R} \cup C_r^- \setminus \{-iA_r\}$ and analytic in $D_+^{\mathrm{in}}$

Obviously, the above scattering coefficients are the same as those in (2.28a) and (2.28b), expressed in terms of the uniformization variable. The reflection coefficients are then given by:

(5.5a) $\qquad\qquad \rho(z) = \dfrac{b(z)}{a(z)}, \qquad \bar{\rho}(z) = \dfrac{\bar{b}(z)}{\bar{a}(z)}, \qquad z \in \mathbb{R}\,.$

Note that $\rho(z)$ is also defined on $C_r^+$, and the corresponding values obviously coincide with the limiting values $\rho^\pm(k)$ on either edge of the cut $\Sigma_r^+$. Similarly, $\bar{\rho}(z)$ is defined on $C_r^-$, and the values coincides with the limiting values $\bar{\rho}^\pm(k)$ on either edge of the cut $\Sigma_r^-$.

Similar results can be derived for the scattering coefficients from the left i.e., the entries of $\bar{S}(z)$, using the analog of (2.28c) and (2.28d). In particular, the reflection coefficients from the left are given by:

(5.5b) $\qquad r(z) = \dfrac{d(z)}{c(z)} \equiv -\dfrac{\bar{b}(z)}{a(z)}, \qquad \bar{r}(z) = \dfrac{\bar{d}(z)}{\bar{c}(z)} \equiv -\dfrac{b(z)}{\bar{a}(z)}, \qquad z \in \mathbb{R}\,,$

corresponding to (2.31c) expressed in terms of the uniformization variable. Unlike $\rho(z)$ and $\bar{\rho}(z)$, the reflection coefficients from the left are generically not defined on either $C_r^+$ or $C_r^-$.

### 5.1. Symmetries of eigenfunctions and scattering data.

In terms of the uniformization variable $z$, since $\lambda_{r,\mathrm{II}} = -\lambda_{r,\mathrm{I}}$ when both sheets are considered, the two involutions $(k, \lambda_r) \to (k^*, \lambda_r^*)$ (i.e., upper/lower half $k$-plane), and $(k, \lambda_r) \to (k, -\lambda_r)$ (i.e., opposite sheets) are: $z \to z^*$ (i.e., upper/lower half $z$-plane) and $z \to -A_r^2/z$ (outside/inside the circle $C_r$).

**First symmetry.** The asymptotic behavior (5.2) yields the following symmetries for the Jost solutions:

(5.6a) $\qquad\qquad \bar{\psi}^*(x, z^*) = i\sigma_2 \psi(x, z) \quad \text{for } z \in D_+ \cup C_r \cup \mathbb{R}\,,$

(5.6b) $\qquad\qquad \psi^*(x, z^*) = -\,i\sigma_2 \bar{\psi}(x, z) \quad \text{for } z \in D_- \cup C_r \cup \mathbb{R}\,,$

(5.6c)
$\quad \phi^*(x, z^*) = i\sigma_2 \bar{\phi}(x, z) \quad \text{for } \mathbb{C}^+ \cup \mathbb{R}\,, \qquad \bar{\phi}^*(x, z^*) = -i\sigma_2 \phi(x, z) \quad \text{for } \mathbb{C}^- \cup \mathbb{R}\,.$

Consequently, the symmetry relations for the scattering coefficients can be written as

(5.7)
$\bar{a}^*(z^*) = a(z) \quad \text{for } z \in D_+^{\mathrm{out}} \cup C_r^+ \cup \mathbb{R}\,, \qquad \bar{b}^*(z^*) = -b(z) \quad \text{for } z \in D_-^{\mathrm{in}} \cup C_r^- \cup \mathbb{R}\,.$

Taking into account (5.5a), the reflection coefficients then satisfy the following symmetry relation:

$$(5.8) \qquad \bar{\rho}^*(z^*) = -\rho(z) \quad \text{for } z \in \mathbb{R} \cup C_r^+ .$$

**Second symmetry.** Since $\lambda_r(-A_r^2/z) = -\lambda_r(z)$ and $k(-A_r^2/z) = k(z)$, taking into account the boundary conditions (5.2), one can easily establish the following additional symmetry relations for the Jost solutions:

$$(5.9a) \qquad \bar{\psi}(x,z) = \frac{-iq_r^*}{z} \psi(x, -A_r^2/z) \quad \text{for } z \in D_- \cup C_r ,$$

$$(5.9b) \qquad \psi(x,z) = \frac{-iq_r}{z} \bar{\psi}(x, -A_r^2/z) \quad \text{for } z \in D_+ \cup C_r ,$$

(5.9c)
$$\phi(x,z) = \phi(x, -A_r^2/z) \quad \text{for } z \in \mathbb{R} \cup C_r^+ , \qquad \bar{\phi}(x,z) = \bar{\phi}(x, -A_r^2/z) \quad \text{for } z \in \mathbb{R} \cup C_r^- .$$

Using the above symmetries in the Wronskian representations for the scattering coefficients (5.4) one obtains:

$$(5.10a) \qquad a(z) = \frac{z}{-iq_r^*} b(-A_r^2/z) \quad \text{for } z \in D_+^{\text{out}} \cup \mathbb{R} \cup C_r^+ ,$$

$$(5.10b) \qquad \bar{a}(z) = \frac{z}{-iq_r} \bar{b}(-A_r^2/z) \quad \text{for } z \in D_-^{\text{out}} \cup \mathbb{R} \cup C_r^- .$$

Similarly, the symmetry relations for the scattering coefficients from the left are given by:

$$(5.10c) \qquad c(z) = \frac{-iq_r}{z} \bar{d}(-A_r^2/z) \quad \text{for } z \in D_+^{\text{out}} \cup \mathbb{R} \cup C_r^+ ,$$

$$(5.10d) \qquad \bar{c}(z) = \frac{-iq_r^*}{z} d(-A_r^2/z) \quad \text{for } z \in D_-^{\text{out}} \cup \mathbb{R} \cup C_r^- .$$

### 5.2. Discrete eigenvalues.

A *discrete eigenvalue* is a value of $z \in D_+ \cup D_-$ for which there exists a nontrivial solution $v$ to (1.2) with entries in $L^2(\mathbb{R})$. These eigenvalues occur for $z \in D_+^{\text{out}}$ iff the functions $\phi(x,z)$ and $\psi(x,z)$ are linearly dependent (i.e., iff $a(z) = 0$), for $z \in D_-^{\text{out}}$ iff the functions $\bar{\psi}(x,z)$ and $\bar{\phi}(x,z)$ are linearly dependent (i.e., iff $\bar{a}(z) = 0$), for $z \in D^{\text{in}}$ iff the functions $\phi(x,z)$ and $\bar{\psi}(x,z)$ are linearly dependent (i.e., iff $b(z) = 0$), and finally for $z \in D_+^{\text{in}}$ iff the functions $\psi(x,z)$ and $\bar{\phi}(x,z)$ are linearly dependent (i.e., iff $\bar{b}(z) = 0$). The conjugation symmetry (5.7) and the second symmetry (5.10) then imply that the discrete eigenvalues occur in quartets: $\{z_n, z_n^*, -A_r^2/z_n, -A_r^2/z_n^*\}_{n=1}^N$.

### 5.3. Asymptotic behavior as $z \to \infty$ and $z \to 0$.

The asymptotic properties of the eigenfunctions and the scattering coefficients are needed to properly pose the inverse problem. Note that the limit $|k| \to \infty$ corresponds to $z \to \infty$ in Sheet I, and $z \to 0$ in Sheet II. Standard Wentzel-Kramers-Brillouin (WKB) expansions in the scattering problem (2.1a) rewritten in terms of $z$ yield the following asymptotic behaviors for the eigenfunctions:

(5.11a)

as   $z \to \infty$ :

$$\Psi_d(x, z)\, e^{i\lambda_r(z)\sigma_3 x} = I_2 + o(1)\,, \quad \Psi_o(x, z)\, e^{i\lambda_r(z)\sigma_3 x} = \frac{i}{z} Q(x)\sigma_3 + o(1/z)\,,$$

(5.11b)
as   $z \to 0$ :

$$\Psi_o(x, z)\, e^{i\lambda_r(z)\sigma_3 x} = \frac{i}{z} Q_r \sigma_3 + O(1)\,, \qquad \Psi_d(x, z)\, e^{i\lambda_r(z)\sigma_3 x} = -Q(x)\sigma_3 Q_r^{-1}\sigma_3 + o(1)\,,$$

and

(5.12a)

as   $z \to \infty$ :

$$\Phi_d(x, z)\, e^{ik(z)\sigma_3 x} = I_2 + o(1)\,, \qquad \Phi_o(x, z)\, e^{ik(z)\sigma_3 x} = \frac{i}{z} Q(x)\sigma_3 + o(1/z)\,,$$

(5.12b)
as   $z \to 0$ :

$$\Phi_d(x, z)\, e^{ik(z)\sigma_3 x} = I_2 + O(z)\,, \qquad \Phi_o(x, z)\, e^{ik(z)\sigma_3 x} = -\frac{iz}{A_r^2} Q(x)\sigma_3 + o(z)\,,$$

where, as before, subscripts $d$ and $o$ denote diagonal and off-diagonal part of the matrix. Under the assumption $(\boldsymbol{H}_1)$ for the potential, from the Wronskian representations (5.4) for the scattering coefficients and the above asymptotic behavior of the eigenfunctions, we then obtain the following asymptotic behavior for the scattering coefficients at large $z$:

(5.13a)    $\displaystyle\lim_{z \to \infty} a(z) = 1$   for $z \in D_+^{\mathrm{out}} \cup \mathbb{R}\,,$    $\displaystyle\lim_{z \to \infty} \bar{a}(z) = 1$   for $z \in D_-^{\mathrm{out}} \cup \mathbb{R}\,,$

and

(5.13b)        $\displaystyle\lim_{z \to \infty} z\, b(z) = 0\,,$    $\displaystyle\lim_{z \to \infty} z\, \bar{b}(z) = 0$   for $z \in \mathbb{R}\,.$

Similarly, the asymptotic behavior for the scattering coefficients as $z \to 0$ is as follows:

(5.14a)        $b(z) = \dfrac{i}{q_r} z + O(z^2)$   for $z \in D_-^{\mathrm{in}} \cup (-A_r, A_r)\,,$

(5.14b)        $\bar{b}(z) = \dfrac{i}{q_r^*} z + O(z^2)$   for $z \in D_+^{\mathrm{in}} \cup (-A_r, A_r)\,,$

and

(5.14c)    $\displaystyle\lim_{z \to 0} \frac{a(z)}{z^2} = 0\,,$    $\displaystyle\lim_{z \to 0} \frac{\bar{a}(z)}{z^2} = 0$   for $z \in (-A_r, A_r)\,.$

**5.4. Riemann-Hilbert problem.** In this section we formulate the inverse scattering problem as matrix Riemann-Hilbert problems from the right and from the left for a suitable set of sectionally analytic/meromorphic functions in $D_+ \cup D_-$, with assigned jumps across $\mathbb{R} \cup C_r$, i.e., the oriented contour in the complex $z$-plane indicated in Fig. 5.

5.4.1. *Riemann-Hilbert problem from the right.* For the formulation of the Riemann-Hilbert problem in terms of scattering data from the right, we introduce the following matrix of eigenfunctions:

$$(5.15) \qquad M(x,k) = \begin{cases} \left[ \dfrac{\phi(x,z)}{a(z)}\, e^{ik(z)x} \quad \psi(x,z)\, e^{-i\lambda_r(z)x} \right] & z \in D_+^{\mathrm{out}}, \\[2.5ex] \left[ \dfrac{\phi(x,z)}{b(z)}\, e^{ik(z)x} \quad \bar\psi(x,z)\, e^{i\lambda_r(z)x} \right] & z \in D_-^{\mathrm{in}}, \\[2.5ex] \left[ \psi(x,z)\, e^{-i\lambda_r(z)x} \quad \dfrac{\bar\phi(x,z)}{\bar b(z)}\, e^{-ik(z)x} \right] & z \in D_+^{\mathrm{in}}, \\[2.5ex] \left[ \bar\psi(x,z)\, e^{i\lambda_r(z)x} \quad \dfrac{\bar\phi(x,z)}{\bar a(z)}\, e^{-ik(z)x} \right] & z \in D_-^{\mathrm{out}}. \end{cases}$$

The asymptotic behavior of the eigenfunctions and scattering coefficients (cf. Sec. 5.3) establishes that for $z \in D_\pm^{\mathrm{out}}$

$$(5.16) \qquad M(x,z) = I_2 + O\left(\frac{1}{z}\right), \quad \text{as } z \to \infty,$$

and for $z \in D_\pm^{\mathrm{in}}$

$$(5.17) \qquad M(x,z) = -\frac{i}{z}\sigma_3 Q_r \sigma_1 + O(1), \quad \text{as } z \to 0,$$

where $\sigma_1$ is the first Pauli matrix given by $\sigma_1 = \begin{pmatrix} 0 & 1 \\ 1 & 0 \end{pmatrix}$.

One has to determine four jump matrices: $V_0(x,z)$ across $z \in (-\infty, -A_r) \cup (A_r, +\infty)$; $V_1(x,z)$ across $z \in (-A_r, A_r)$; $V_2(x,z)$ across the semicircle $C_r^+$ in the upper half $z$-plane; and $V_3(x,z)$ across the semicircle $C_r^-$ in the lower half $z$-plane.

The RH problem across $z \in (-\infty, -A_r) \cup (A_r, +\infty)$ can be written in matrix form as $M^+(x,z) = M^-(x,z)\, V_0(x,z)$, i.e.,
(5.18)
$$\left[ \frac{\phi^+(x,z)}{a^+(z)}\, e^{ikx} \quad \psi^+(x,z)\, e^{-i\lambda_r x} \right] = \left[ \bar\psi^-(x,z)\, e^{i\lambda_r x} \quad \frac{\bar\phi^-(x,z)}{\bar a^-(z)}\, e^{-ikx} \right] V_0(x,z),$$

where superscripts $\pm$ denote limiting values from the upper/lower complex $z$-plane, respectively. Using (5.3a), the jump matrix is found to be:

$$(5.19) \qquad V_0(x,z) = \begin{pmatrix} [1 - \rho(z)\bar\rho(z)]\, e^{-i(A_r^2/z)x} & -\bar\rho(z)\, e^{-2i\lambda_r(z)x} \\ \rho(z)\, e^{2ik(z)x} & e^{-i(A_r^2/z)x} \end{pmatrix}.$$

We can write the RH problem across $z \in (-A_r, A_r)$ as $M^+(x,z) = M^-(x,z)\, V_1(x,z)$, where the superscripts $\pm$ again denote limiting values from the upper/lower complex $z$-plane, respectively. Explicitly, from (5.3a) we have
(5.20)
$$\left[ \frac{\phi^+(x,z)}{b^+(z)}\, e^{ik(z)x} \quad \bar\psi^+(x,z)\, e^{i\lambda_r(z)x} \right] = \left[ \psi^-(x,z)\, e^{-i\lambda_r(z)x} \quad \frac{\bar\phi^-(x,z)}{\bar b^-(z)}\, e^{-ik(z)x} \right] V_1(x,z),$$

and (5.3a) yields the following expression for the jump matrix $V_1(x,z)$:

$$(5.21) \qquad V_1(x,z) = \begin{pmatrix} \left[ 1 - \dfrac{1}{\rho(z)\bar\rho(z)} \right] e^{izx} & -\dfrac{1}{\bar\rho(z)}\, e^{2i\lambda_r(z)x} \\ \dfrac{1}{\rho(z)}\, e^{2ik(z)x} & e^{izx} \end{pmatrix}.$$

The RH problem across the semicircle $C_r^+$ in the upper half $z$-plane is written as $M^+(x,z) = M^-(x,z)V_2(x,z)$, i.e.,

(5.22)
$$\left[ \frac{\phi^+(x,z)}{a^+(z)} e^{ik(z)x} \quad \psi^+(x,z) e^{-i\lambda_r^+(z)x} \right] = \left[ \frac{\phi^-(x,z)}{b^-(z)} e^{ik(z)x} \quad \bar{\psi}^-(x,z) e^{i\lambda_r^-(z)x} \right] V_2(x,z),$$

where the superscripts $\pm$ denote limiting values from the exterior/interior of $C_r^+$, respectively. Taking into account that $\phi(x,z)$ is continuous across $C_r^+$, i.e., $\phi^+(x,z) = \phi^-(x,z)$, and that from (5.4) it follows $\phi(x,z) = a(z)\bar{\psi}(x,z) + b(z)\psi(x,z)$ for $z \in C_r^+$, the jump matrix $V_2(x,z)$ is then found to be:

(5.23)
$$V_2(x,z) = \begin{pmatrix} \rho(z) & e^{-izx} \\ 0 & -\dfrac{1}{\rho(z)} e^{-2i\lambda_r(z)x} \end{pmatrix}.$$

Finally, the RH problem across the semicircle $C_r^-$ in the lower half $z$-plane is $M^+(x,z) = M^-(x,z)V_3(x,z)$, and explicitly as

(5.24)
$$\left[ \bar{\psi}^+(x,z) e^{i\lambda_r^+(z)x} \quad \frac{\bar{\phi}^+(x,z)}{\bar{a}^+(z)} e^{-ik(z)x} \right] = \left[ \psi^-(x,z) e^{-i\lambda_r^-(z)x} \quad \frac{\bar{\phi}^-(x,z)}{\bar{b}^-(z)} e^{-ik(z)x} \right] V_3(x,z),$$

where the superscripts $\pm$ denote limiting values from the exterior/interior of $C_r^-$, respectively. Clearly $\bar{\phi}(x,z)$ is continuous across $C_r^-$, i.e., $\bar{\phi}^+(x,z) = \bar{\phi}^-(x,z)$, and since, according to (5.4), the relationship $\bar{\phi}(x,z) = \bar{a}(z)\psi(x,z) + \bar{b}(z) + \bar{\psi}(x,z)$ still holds on $C_r^-$, the jump matrix $V_3(x,z,t)$ is found to be:

(5.25)
$$V_3(x,z) = \begin{pmatrix} -\dfrac{1}{\bar{\rho}(z)} e^{2i\lambda_r(z)x} & 0 \\ e^{izx} & \bar{\rho}(z) \end{pmatrix}.$$

It is worth noting that the RH problems across $z \in (-\infty, -A_r) \cup (A_r, +\infty)$ and $z \in (-A_r, A_r)$ correspond to the RH problems across the real axis in Sheet I and in Sheet II, respectively.

Solving the inverse problem as a RH problem [with poles, corresponding to the zeros of $a(z)$ and $\bar{a}(z)$ in $D_\pm^{\text{out}}$, and to the zeros of $b(z)$ and $\bar{b}(z)$ in $D_\mp^{\text{in}}$] then amounts to computing the sectionally meromorphic matrix $M(x,z)$ with the given jumps, and asymptotic behaviors (5.16) as $z \to \infty$, and (5.17) as $z \to 0$. As in defocusing NLS equation with NZBCs [18], in addition to the behavior at $z = \infty$ and the poles from the discrete spectrum one also needs to subtract the pole at $z = 0$ in order to obtain a regular RH problem.

Once the parametric time dependence of the scattering coefficients is taken into account in the jump matrices, which can be easily obtained from the results of Sec. 4:

(5.26a)  $a(z,t) = a(z,0) e^{-i(A_r^4/z^2)t}$,  $\qquad \bar{a}(z,t) = \bar{a}(z,0) e^{i(A_r^4/z^2)t}$,

(5.26b)  $b(z,t) = b(z,0) e^{i(2A_r^2 - z^2)t}$,  $\qquad \bar{b}(z,t) = \bar{b}(z,0) e^{i(z^2 - 2A_r^2)t}$,

(5.26c)  $\rho(z,t) = \rho(z,0) e^{i(2A_r^2 - z^2 - A_r^4/z^2)t}$,  $\qquad \bar{\rho}(z,t) = \bar{\rho}(z,0) e^{i(z^2 - 2A_r^2 + A_r^4/z^2)t}$,

the potential is then reconstructed by the large-$z$ expansion of the matrix $M(x,z,t)$:

(5.27)
$$M_o(x,z,t) = \frac{i}{z} Q(x,t)\sigma_3 + o(1/z),$$

and by the asymptotic behavior as $z \to 0$ of $M(x, z, t)$:

$$(5.28) \qquad M_o(x, z, t) = -\frac{i}{z}\sigma_3 Q_r(t)\sigma_1 + Q(x, t)Q_r^{-1}(t)\sigma_1 + o(1),$$

with subscript $_o$ denoting the off-diagonal part.

5.4.2. *Riemann-Hilbert from the left.* The inverse problem can also be formulated as a Riemann-Hilbert from the left, considering the following eigenfunctions matrix

$$(5.29) \qquad \tilde{M}(x, z) = \begin{cases} \left[ \phi(x, z)\, e^{ik(z)x} \quad \dfrac{\psi(x, z)}{c(z)}\, e^{-i\lambda_r(z)x} \right] & z \in D_+^{\text{out}}, \\[2ex] \left[ \phi(x, z)\, e^{ik(z)x} \quad \dfrac{\bar{\psi}(x, z)}{\bar{d}(z)}\, e^{i\lambda_r(z)x} \right] & z \in D_-^{\text{in}}, \\[2ex] \left[ \dfrac{\psi(x, z)}{d(z)}\, e^{-i\lambda_r(z)x} \quad \bar{\phi}(x, z)\, e^{-ik(z)x} \right] & z \in D_+^{\text{in}}, \\[2ex] \left[ \dfrac{\bar{\psi}(x, z)}{\bar{c}(z)}\, e^{i\lambda_r(z)x} \quad \bar{\phi}(x, z)\, e^{-ik(z)x} \right] & z \in D_-^{\text{out}}, \end{cases}$$

such that $\tilde{M}(x, z) \to I_2$ both as $z \to \infty$ for $z \in D_\pm^{\text{out}}$, and as $z \to 0$ for $z \in D_\pm^{\text{in}}$.

The RH problem across $z \in (-\infty, -A_r) \cup (A_r, +\infty)$ is written in matrix form as $\tilde{M}^+(x, z) = \tilde{M}^-(x, z)\, \tilde{V}_0(x, z)$, i.e.,

$$(5.30)$$
$$\left[ \phi^+(x, z)\, e^{ik(z)x} \quad \frac{\psi^+(x, z)}{c^+(z)}\, e^{-i\lambda_r(z)x} \right] = \left[ \frac{\bar{\psi}^-(x, z)}{\bar{c}^-(z)}\, e^{i\lambda_r(z)x} \quad \bar{\phi}^-(x, z)\, e^{-ik(z)x} \right] \tilde{V}_0(x, z),$$

where superscripts $^\pm$ denote limiting values from the upper/lower complex $z$-plane, respectively. Using (5.3b), the jump matrix is determined as follows

$$(5.31) \qquad \tilde{V}_0(x, z) = \begin{pmatrix} e^{-i(A_r^2/z)x} & r(z)\, e^{-2i\lambda_r(z)x} \\ -\bar{r}(z)\, e^{2ik(z)x} & [1 - r(z)\bar{r}(z)]\, e^{-i(A_r^2/z)x} \end{pmatrix}.$$

We can write the RH problem across $z \in (-A_r, A_r)$ as

$$\tilde{M}^+(x, z) = \tilde{M}^-(x, z)\, \tilde{V}_1(x, z),$$

where the superscripts $^\pm$ again denote limiting values from the upper/lower complex $z$-plane, respectively. Explicitly, one has

$$(5.32)$$
$$\left[ \phi^+(x, z)\, e^{ik(z)x} \quad \frac{\bar{\psi}^+(x, z)}{\bar{d}^+(z)}\, e^{i\lambda_r(z)x} \right] = \left[ \frac{\psi^-(x, z)}{d^-(z)}\, e^{-i\lambda_r(z)x} \quad \bar{\phi}^-(x, z)\, e^{-ik(z)x} \right] \tilde{V}_1(x, z),$$

and again (5.3b) yields the following expression for the jump matrix $\tilde{V}_1(x, z)$:

$$(5.33) \qquad \tilde{V}_1(x, z) = \begin{pmatrix} e^{izx} & \dfrac{1}{\bar{r}(z)}\, e^{2i\lambda_r(z)x} \\[2ex] -\dfrac{1}{r(z)}\, e^{2ik(z)x} & \left[1 - \dfrac{1}{r(z)\bar{r}(z)}\right] e^{izx} \end{pmatrix}.$$

In formulating the RH problems from the left across the semicircles, we should first of all notice that on $C_r^+$ one has $\tilde{M}^+(x, z) = \left[ \phi^+(x, z)\, e^{ik(z)x} \quad \frac{\psi^+(x,z)}{c^+(z)}\, e^{-i\lambda_r^+(z)x} \right]$ and $\tilde{M}^-(x, z) = \left[ \phi^-(x, z)\, e^{ik(z)x} \quad \frac{\psi^-(x,z)}{d^-(z)}\, e^{i\lambda_r^-(z)x} \right]$, while on $C_r^-$ it is $\tilde{M}^+(x, z) = \left[ \frac{\bar{\psi}^+(x,z)}{\bar{c}^+(z)}\, e^{i\lambda_r^+ x} \quad \bar{\phi}^+(x, z)\, e^{-ikx} \right]$ and $\tilde{M}^-(x, z) = \left[ \frac{\psi^-(x,z)}{d^-(z)}\, e^{-i\lambda_r^- x} \quad \bar{\phi}^-(x, z)\, e^{-ikx} \right]$. Clearly, $\phi(x, z)$ and $\bar{\phi}(x, z)$ are continuous across $C_r^+$ and $C_r^-$, respectively. Moreover, since $k(z) = k(-A_r^2/z)$, then one also has $\phi(x, z) = \phi(x, -A_r^2/z)$ and $\bar{\phi}(x, z) =$

$\bar{\phi}(x, -A_r^2/z)$, consistently with the fact that the eigenfunctions $\phi(x, k)$, $\bar{\phi}(x, k)$ are continuous across $\Sigma_r^+$ and $\Sigma_r^-$, respectively. The jump conditions are then provided by the symmetry relationships (5.9b), (5.10c) and (5.10d), which yield:

(5.34) $\qquad \dfrac{\psi(x, z)}{c(z)} = \dfrac{\bar{\psi}(x, -A_r^2/z)}{\bar{d}(-A_r^2/z)}\,, \qquad \phi(x, z) = \phi(x, -A_r^2/z)\,, \qquad z \in C_r^+\,,$

(5.35) $\qquad \dfrac{\psi(x, z)}{d(z)} = \dfrac{\bar{\psi}(x, -A_r^2/z)}{\bar{c}(-A_r^2/z)}\,, \qquad \bar{\phi}(x, z) = \bar{\phi}(x, -A_r^2/z)\,, \qquad z \in C_r^-\,.$

Note that the RH problem in this case is clearly posed as a nonlocal one, with the jumps relating values of the meromorphic eigenfunctions at symmetric points $z$ and $-A_r^2/z$ on the semicircles $C_r^{\pm}$.

Solving the inverse problem as a RH problem [with poles, corresponding to the zeros of $c(z)$ and $\bar{c}(z)$ in $D_{\pm}^{\mathrm{out}}$, and to the zeros of $d(z)$ and $\bar{d}(z)$ in $D_{\pm}^{\mathrm{in}}$] then amounts to computing the sectionally meromorphic matrix $\tilde{M}(x, z)$ with the given jumps, and normalized to identity as $z \to \infty$ for $z \in D_{\pm}^{\mathrm{out}}$ and as $z \to 0$ for $z \in D_{\pm}^{\mathrm{in}}$. Once the parametric time dependence of the scattering coefficients is taken into account in the jump matrices (cf. Sec. 4):

(5.36a) $\qquad c(z, t) = c(z, 0)\, e^{-i(A_r^4/z^2)t}\,, \qquad\qquad \bar{c}(z, t) = \bar{c}(z, 0)\, e^{i(A_r^4/z^2)t}\,,$

(5.36b) $\qquad d(z, t) = d(z, 0)\, e^{i(z^2 - 2A_r^2)t}\,, \qquad\quad \bar{d}(z, t) = \bar{d}(z, 0)\, e^{i(2A_r^2 - z^2)t}\,,$

(5.36c) $\qquad r(z, t) = r(z, 0)\, e^{4ik^2(z)t}\,, \qquad\qquad\quad \bar{r}(z, t) = \bar{r}(z, 0)\, e^{-4ik^2(z)t}\,,$

the potential is then reconstructed by the large $k$ expansion of the matrix $\tilde{M}(x, z, t)$:

(5.37) $$\tilde{M}_o(x, z, t) = \frac{i}{z}Q(x, t)\sigma_3 + o\left(1/z\right)\,,$$

and by the asymptotic behavior as $z \to 0$ of $\tilde{M}(x, z)$:

(5.38) $$\tilde{M}_o(x, z, t) = -\frac{iz}{A_r^2}Q(x, t)\sigma_3 + o(z)\,.$$

The RH problems from the right and from the left formulated in terms of the uniformization variable $z$ provide an alternative, and possibly more advantageous set-up for the investigation of the long-time asymptotic behavior of NLS solutions with one-sided nontrivial boundary conditions via the nonlinear steepest descent method (see for instance [14]).

## 6. Conclusions

We have developed the IST for the focusing NLS equation with a (one-sided) nonzero boundary condition as $x \to +\infty$. Such kind of boundary conditions are obviously outside the class considered in [12], where the amplitudes of the background field are taken to be the same at both space infinities. One should notice, though, that unlike the case of fully asymmetric boundary conditions, i.e., when the amplitudes of the NLS solutions as $x \to \pm\infty$ are different, and similarly to what happens in the same-amplitude case, here one can still introduce a uniformization variable that allows mapping the multiply sheeted Riemann surface for the scattering parameter into a single complex plane. Nonetheless, important differences with respect to the same-amplitude case still arise both in the direct and in the inverse problem. In particular, in addition to solitons (corresponding to the discrete eigenvalues of the scattering problem), and to radiation (corresponding to the continuous spectrum of the scattering operator, and represented in the inverse problem by the reflection coefficients for $k \in \mathbb{R}$), one also has a nontrivial contribution from the transmission coefficients for $k \in \Sigma_r$, as shown by the last term in (3.26), contributing to the left Marchenko equations. Correspondingly, (2.38a) and (2.38b) show that in the right Marchenko equations one always has a nontrivial contribution from the integral terms in (3.9) and (3.12), since $\rho(k)$ [resp. $\bar{\rho}(k)$] cannot vanish for $k \in \Sigma_r^+$ [resp. $k \in \Sigma_r^-$]. In particular, this implies that no pure soliton solutions exist, and solitons are always accompanied by a radiative contribution of some sort. As a consequence, unlike the symmetric case, here no explicit solution can be obtained by simply reducing the inverse problem to a set of algebraic equations.

The results presented in this paper will pave the way for the investigation of the long-time asymptotic behavior of fairly general NLS solutions with nontrivial boundary conditions via the nonlinear steepest descent method, in analogy to what was done, for instance, in [17] for the modified KdV equation, or in [14] for the focusing NLS with initial condition $q(x,0) = Ae^{i\mu|x|}$, with $A, \mu$ positive constants. Moreover, the Marchenko integral equations obtained here will provide an alternative setup for the study of the long-time behavior of the solutions by means of matched asymptotics, as was recently done for KdV in [1].

The study of the long-time asymptotics, as well as the derivation of solutions describing solitons superimposed to small radiation, will be the subject of future investigation.

**Acknowledgments.** The authors would like to acknowledge Mark Ablowitz, Gino Biondini, Francesco Demontis and Cornelis van der Mee for many valuable discussions. Also, we are grateful to the anonymous referee for a large number of useful comments that have help us improve the manuscript. This research is partially supported by NSF, under grant No. DMS-1311883, by INFN, under INFN IS-CSN4 "Mathematical Methods of Nonlinear Physics", and by MIUR, under 2011 PRIN "Geometric and analytic theory of Hamiltonian systems in finite and infinite dimensions".

# References

[1] M. J. Ablowitz and D. E. Baldwin, *Interactions and asymptotics of dispersive shock waves—Korteweg-de Vries equation*, Phys. Lett. A **377** (2013), no. 7, 555–559, DOI 10.1016/j.physleta.2012.12.040. MR3019830

[2] M. J. Ablowitz, D. J. Kaup, A. C. Newell, and H. Segur, *The inverse scattering transform-Fourier analysis for nonlinear problems*, Studies in Appl. Math. **53** (1974), no. 4, 249–315. MR0450815 (56 #9108)

[3] M. J. Ablowitz, B. Prinari, and A. D. Trubatch, *Discrete and continuous nonlinear Schrödinger systems*, London Mathematical Society Lecture Note Series, vol. 302, Cambridge University Press, Cambridge, 2004. MR2040621 (2005c:37117)

[4] M. J. Ablowitz and H. Segur, *Solitons and the inverse scattering transform*, SIAM Studies in Applied Mathematics, vol. 4, Society for Industrial and Applied Mathematics (SIAM), Philadelphia, Pa., 1981. MR642018 (84a:35251)

[5] M. J. Ablowitz and H. Segur, *Asymptotic solutions of the Korteweg-deVries equation*, Studies in Appl. Math. **57** (1976/77), no. 1, 13–44. MR0481656 (58 #1757)

[6] N.N. Akhmediev, V.M. Eleonskii, and N.E. Kulagin, *Generation of a periodic sequence of picosecond pulses in an optical fiber. Exact solutions*, Sov. Phys. JETP **89** , 1542-1551 (1985) [in Russian].

[7] N.N. Akhmediev, A. Ankiewicz, and M. Taki, *Waves that appear from nowhere and disappear without a trace*, Phys. Lett. A, **373**, 675–678 (2009). DOI:10.1016/j.physleta.2008.12.036

[8] N.N. Akhmediev, A. Ankiewicz, and J.M. Soto-Crespo, *Rogue waves and rational solutions of the nonlinear Schrödinger equation*, Phys. Rev. E, **80**, 026601 (2009). DOI: http://dx.DOI.org/10.1103/PhysRevE.80.026601

[9] N. N. Akhmediev and V. I. Korneev, *Modulation instability and periodic solutions of the nonlinear Schrödinger equation* (Russian, with English summary), Teoret. Mat. Fiz. **69** (1986), no. 2, 189–194. MR884491 (88b:35165)

[10] T.B. Benjamin, *Instability of periodic wavetrains in nonlinear dispersive systems*, Proc. Roy. Soc. A **299** 59–75 (1967). DOI: 10.1098/rspa.1967.0123

[11] T.B. Benjamin and J.E. Feir, *The disintegration of wavetrains in deep water. Part I*, J. Fluid Mech. **27**, 417–430 (1967). DOI: http://dx.doi.org/10.1017/S002211206700045X

[12] G. Biondini and G. Kovačič, *Inverse scattering transform for the focusing nonlinear Schrödinger equation with nonzero boundary conditions*, J. Math. Phys. **55** (2014), no. 3, 031506, 22, DOI 10.1063/1.4868483. MR3221240

[13] A. Boutet de Monvel, V. P. Kotlyarov, and D. Shepelsky, *Focusing NLS equation: long-time dynamics of step-like initial data*, Int. Math. Res. Not. IMRN **7** (2011), 1613–1653. MR2806517 (2012f:35498)

[14] R. Buckingham and S. Venakides, *Long-time asymptotics of the nonlinear Schrödinger equation shock problem*, Comm. Pure Appl. Math. **60** (2007), no. 9, 1349–1414, DOI 10.1002/cpa.20179. MR2337507 (2008j:35167)

[15] F. Calogero and A. Degasperis, *Spectral transform and solitons. Vol. I*, Studies in Mathematics and its Applications, vol. 13, North-Holland Publishing Co., Amsterdam-New York, 1982. Tools to solve and investigate nonlinear evolution equations; Lecture Notes in Computer Science, 144. MR680040 (84f:35121)

[16] M. Chen, M.A. Tsankov, J.M. Nash, and C.E. Patton, *Backward-volume-wave microwave-envelope solitons in yttrium iron garnet films*, Phys. Rev. B **49**, 12773–12790 (1994). DOI: http://dx.doi.org/10.1103/PhysRevB.49.12773

[17] P. Deift and X. Zhou, *A steepest descent method for oscillatory Riemann-Hilbert problems. Asymptotics for the MKdV equation*, Ann. of Math. (2) **137** (1993), no. 2, 295–368, DOI 10.2307/2946540. MR1207209 (94d:35143)

[18] F. Demontis, B. Prinari, C. van der Mee, and F. Vitale, *The inverse scattering transform for the defocusing nonlinear Schrödinger equations with nonzero boundary conditions*, Stud. Appl. Math. **131** (2013), no. 1, 1–40, DOI 10.1111/j.1467-9590.2012.00572.x. MR3081238

[19] F. Demontis, B. Prinari, C. van der Mee, and F. Vitale, *The inverse scattering transform for the defocusing nonlinear Schrödinger equations with nonzero boundary conditions*, Stud. Appl. Math. **131** (2013), no. 1, 1–40, DOI 10.1111/j.1467-9590.2012.00572.x. MR3081238

[20] L. D. Faddeev and L. A. Takhtajan, *Hamiltonian methods in the theory of solitons*, Springer Series in Soviet Mathematics, Springer-Verlag, Berlin, 1987. Translated from the Russian by A. G. Reyman [A. G. Reĭman]. MR905674 (89m:58103)

[21] J. Garnier and K. Kalimeris, *Inverse scattering perturbation theory for the nonlinear Schrödinger equation with non-vanishing background*, J. Phys. A **45** (2012), no. 3, 035202, 13, DOI 10.1088/1751-8113/45/3/035202. MR2871428

[22] A. Hasegawa, and F. Tappert, *Transmission of stationary nonlinear optical pulses in dispersive dielectric fibers. I. Anomalous dispersion* and *II. Normal dispersion*, App. Phys. Lett. **23**(3) 142–144 and 171–172 (1973). http://dx.doi.org/10.1063/1.1654836

[23] A. R. It·s, A. V. Rybin, and M. A. Sall', *On the exact integration of the nonlinear Schrödinger equation* (Russian, with English summary), Teoret. Mat. Fiz. **74** (1988), no. 1, 29–45, DOI 10.1007/BF01018207; English transl., Theoret. and Math. Phys. **74** (1988), no. 1, 20–32. MR940459 (89h:35295)

[24] E.A. Kuznetsov, *Solitons in a parametrically unstable plasma*, Sov. Phys. Dokl. (Engl. Transl.) 22, 507–508 (1977).

[25] Y. C. Ma, *The perturbed plane-wave solutions of the cubic Schrödinger equation*, Stud. Appl. Math. **60** (1979), no. 1, 43–58. MR517602 (80e:81012)

[26] D. Mihalache, F. Lederer and D-M Baboiu, *Two-parameter family of exact solutions of the nonlinear Schrödinger equation describing optical soliton propagation*, Phys. Rev. A **47**, 3285-3290 (1993). http://dx.doi.org/10.1103/PhysRevA.47.3285

[27] S. Novikov, S. V. Manakov, L. P. Pitaevskiĭ, and V. E. Zakharov, *Theory of solitons: The inverse scattering method*, Contemporary Soviet Mathematics, Consultants Bureau [Plenum], New York, 1984. Translated from the Russian. MR779467 (86k:35142)

[28] M. Onorato, A.R. Osborne and M. Serio, *Modulational instability in crossing sea states: A possible mechanism for the formation of freak waves*, Phys. Rev. Lett. **96**, 014503 (2006). http://dx.doi.org/10.1103/PhysRevLett.96.014503

[29] C.J. Pethick, and H. Smith, *Bose-Einstein Condensation in Dilute Gases*, Cambridge University Press, Cambridge, 2002.

[30] D. H. Peregrine, *Water waves, nonlinear Schrödinger equations and their solutions*, J. Austral. Math. Soc. Ser. B **25** (1983), no. 1, 16–43, DOI 10.1017/S0334270000003891. MR702914 (84k:76033)

[31] B. Prinari, M. J. Ablowitz, and G. Biondini, *Inverse scattering transform for the vector nonlinear Schrödinger equation with nonvanishing boundary conditions*, J. Math. Phys. **47** (2006), no. 6, 063508, 33, DOI 10.1063/1.2209169. MR2239983 (2007f:37118)

[32] D.R. Solli, C. Ropers, P. Koonath and B. Jalali, *Optical rogue waves*, Nature **450**, 1054–1057 (2007). DOI:10.1038/nature06402

[33] M. Tajiri and Y. Watanabe, *Breather solutions to the focusing nonlinear Schrödinger equation*, Phys. Rev. E (3) **57** (1998), no. 3, 3510–3519, DOI 10.1103/PhysRevE.57.3510. MR1611665 (98k:35183)

[34] V. E. Zakharov, *Hamilton formalism for hydrodynamic plasma models* (Russian, with English summary), Ž. Èksper. Teoret. Fiz. **60** (1971), 1714–1726; English transl., Soviet Physics JETP **33** (1971), 927–932. MR0312857 (47 #1412)

[35] V.E. Zakharov and A.A. Gelash, *Soliton on unstable condensate*, arXiv:nlin.si 1109.0620 (2011). arXiv:1109.0620 [nlin.SI]

[36] V.E. Zakharov and A.A. Gelash, *Nonlinear stage of modulational instability*, Phys. Rev. Lett. **111**, 054101 (2013). http://dx.doi.org/10.1103/PhysRevLett.111.054101

[37] V. E. Zakharov and L. A. Ostrovsky, *Modulation instability: The beginning*, Phys. D **238** (2009), no. 5, 540–548, DOI 10.1016/j.physd.2008.12.002. MR2591296 (2010h:35037)

[38] V. E. Zakharov and A. B. Shabat, *Exact theory of two-dimensional self-focusing and one-dimensional self-modulation of waves in nonlinear media* (Russian, with English summary), Ž. Èksper. Teoret. Fiz. **61** (1971), no. 1, 118–134; English transl., Soviet Physics JETP **34** (1972), no. 1, 62–69. MR0406174 (53 #9966)

[39] V.E. Zakharov, and A.B. Shabat, *Interaction between solitons in a stable medium*, Sov. Phys. JETP **37**, 823–828 (1973).

[40] A.K. Zvezdin, and A.F. Popkov, *Contribution to the nonlinear theory of magnetostatic spin waves*, Sov. Phys. JETP **57**, 350–355 (1983).

DEPARTMENT OF MATHEMATICS, UNIVERSITY OF COLORADO COLORADO SPRINGS, COLORADO SPRINGS - CO, USA, AND DIPARTIMENTO DI MATEMATICA E FISICA "E. DE GIORGI", UNIVERSITÀ DEL SALENTO AND SEZIONE INFN, LECCE - ITALY
*E-mail address*: bprinari@uccs.edu

DIPARTIMENTO DI MATEMATICA E FISICA "E. DE GIORGI", UNIVERSITÀ DEL SALENTO AND SEZIONE INFN, LECCE - ITALY
*E-mail address*: federica.vitale@le.infn.it

# Selected Published Titles in This Series

For a complete list of titles in this series, visit the
AMS Bookstore at **www.ams.org/bookstore/conmseries/**.